ENVIRONMENTAL SCIENCES AND APPLICATIONS

Series Editors: ASIT K. BISWAS
MARGARET R. BISWAS

Volume 4

THE OZONE LAYER

ENVIRONMENTAL SCIENCES AND APPLICATIONS

Other titles in the series

NOTICE TO READERS

Dear Reader

If your library is not already a standing order customer or subscriber to this series, may we recommend that you place a standing or subscription order to receive immediately upon publication all volumes published in this valuable series. Should you find that these volumes no longer serve your needs your order can be cancelled at any time without notice.

The Editors and the Publisher will be glad to receive suggestions or outlines of suitable titles, reviews or symposia for consideration for rapid publication in this series.

ROBERT MAXWELL
Publisher at Pergamon Press

THE OZONE LAYER

Proceedings of the Meeting of Experts Designated by Governments,
Intergovernmental and Nongovernmental Organizations on the Ozone
Layer Organized by the United Nations Environment Programme in
Washington DC, 1-9 March 1977

Editor

ASIT K. BISWAS

Director, Biswas and Associates, Ottawa, Canada

Published for the
UNITED NATIONS ENVIRONMENT PROGRAMME
by
PERGAMON PRESS
OXFORD · NEW YORK · TORONTO · SYDNEY · PARIS · FRANKFURT

U.K.	Pergamon Press Ltd., Headington Hill Hall, Oxford OX3 0BW, England
U.S.A.	Pergamon Press Inc., Maxwell House, Fairview Park, Elmsford, New York 10523, U.S.A.
CANADA	Pergamon of Canada, Suite 104, 150 Consumers Road, Willowdale, Ontario M2J 1P9, Canada
AUSTRALIA	Pergamon Press (Aust.) Pty. Ltd., P.O. Box 544, Potts Point, N.S.W. 2011, Australia
FRANCE	Pergamon Press SARL, 24 rue des Ecoles, 75240 Paris, Cedex 05, France
FEDERAL REPUBLIC OF GERMANY	Pergamon Press GmbH, 6242 Kronberg-Taunus, Pferdstrasse 1, Federal Republic of Germany

First edition 1979

British Library Cataloguing in Publication Data

UNEP Meeting on the Ozone Layer,
Washington, D.C., 1977
The ozone layer. - (Environmental sciences and applications; vol. 4).
1. Atmospheric ozone - Congresses
2. Stratosphere - Congresses
I. Title II. Biswas, Asit, K III. United
Nations Environment Programme IV. Series
551.5'142   QC879.7   79-42879

ISBN 0 08 022429 6

In order to make this volume available as economically and as rapidly as possible the author's typescript has been reproduced in its original form. This method has its typographical limitations but it is hoped that they in no way distract the reader.

Printed and bound in Great Britain by
William Clowes (Beccles) Limited, Beccles and London.

Contents

Preface

MOSTAFA KAMAL TOLBA

Executive Director, United Nations Environment Programme

The Meeting of Experts on the Ozone Layer, held in Washington in March 1977, was
attended by persons designated by governments, and intergovernmental and non-
governmental organizations. The meeting had been decided upon by the Governing
Council of the United Nations Environment Programme at its fourth session in 1976.
The Council directed the Executive Director of UNEP that the meeting should (a)
review all aspects of the ozone layer; (b) identify relevant ongoing activities and
future plans; and (c) agree on a division of labour and a coordinating mechanism
for research activities, future plans, and the collection of related industrial and
commercial information.

The Governing Council recognized "the potential impact that stratospheric pollution
and a reduction in the ozone layer may have on mankind"; it noted "the plans of the
World Meteorological Organization to execute a project on ozone layer monitoring and
research", and the fact that other international organizations and governments were
also working on the ozone layer and atmospheric pollution.

I think it would be useful to place this important meeting within the setting of
UNEP's interests and responsibilities. The protection of the environment against
risks that could threaten human and other forms of life is no sectoral responsibility
within the United Nations system. It also goes far beyond anything that individual
nations alone can do, for environmental risks and dangers do not stop at border
of any given political or economic system. The long-term purpose of UNEP is to
promote everywhere the management of man's activities in such a way that environmental
degradation is minimized, and its resources are used to provide maximum benefits on
a continuing sustainable basis.

Environmental assessment is essential to intelligent environmental management which
depends completely upon an understanding of the environment and information on trends
and changes. Of particular importance is the assessment of those biospheric conditions which are essential to the maintenance of life. In speaking of potential
dangers in this connection, UNEP uses the term "outer limits". UNEP's activities
dealing with the ozone layer are an element in the assessment of outer limits.

Outer limits is thus a shorthand expression for the notion that the biosphere, of
which man is a part, can safely adapt to only a limited extent to the demands placed
upon it by man's activities. When these limits are transgressed, human welfare,
even life itself, is endangered. No social or economic endeavour can safely ignore
the restrictions imposed by biospheric outer limits. Therefore, threats to the outer

limits must be assessed and timely warning of them given; this responsibility forms a part of the UNEP task of Earthwatch.

The assessment of outer limits is complex, and depends on physical, biological, ecological, social, economic and even political factors. That is why the papers presented in this publication deal not only with stratopheric physics and chemistry, but also biology, ecology, trade and economics.

UNEP's responsibility in this connection clearly derives from its catalytic and co-ordinating role, unique in the United Nations System, and calling for a special sort of relationship between UNEP and other members of the United Nations family. The Governing Council, in formulating environmental policies and strategies, speaks not only to the Executive Director and the Secretariat of UNEP, but to all members of the United Nations family and, indeed, to the world. The Executive Director and the Secretariat of UNEP, in responding to the Governing Council, propose, elaborate and coordinate programmes that are executed by any number of combinations of competent bodies. Thus, in coming to grips with the ozone layer problems, UNEP has and will continue to co-operate closely and fully with other organizations, both within and outside the United Nations System.

Following the presentation of the papers reproduced within this publication, the Conference adopted a World Plan of Action on the Ozone Layer. This is divided into three sections:

I.   The natural ozone layer and its modification by man's activities.

II.   The impact of changes in the ozone layer on man, the biosphere and climate.

III.   Socio-economic aspects.

Under these sections, the following 21 actions were agreed upon:

I.   The natural ozone layer and its modification by man's activities;

      1.   Ozone monitoring.

      2.   Solar radiation monitoring.

      3.   Simultaneous species measurements.

      4.   Chemical reactions.

      5.   Development of computational modelling.

      6.   Large-scale atmospheric transport.

      7.   Global constituent budgets.

II.   The impact of changes in the ozone layer on man and the biosphere:

      8.   UV-B radiation monitoring.

      9.   Development of UV-B instrumentation.

      10. UV-B research promotion.

      11. Statistics on skin cancer.

      12. Research on induction mechanisms.

      13. Other health aspects.

      14. Responses to UV-B.

      15. Terrestrial ecosystems.

      16. Aquatic ecosystems.

17. Other agricultural effects.

18. Development of computational modelling.

19. Regional climate.

III. Social-economic aspects:

20. Production and emission data.

21. Methodology for comprehensive assessment.

With regard to institutional arrangements, it was agreed at the meeting that UNEP should exercise a broad coordinating and catalytic role aimed at the integration of research efforts by arranging for:

(a) Collection and dissemination of information on ongoing and planned research activities.

(b) Presentation and review of the results of research.

(c) Identification of further research needs.

(d) Appropriate encouragement of such research.

To this end, it was recommended that in order for UNEP to meet this responsibility, a Co-ordinating Committee on the Ozone Layer should be set up, composed of representatives of the agencies and non-governmental organizations actively participating in the Plan of Action, as well as representatives of countries which have major scientific programmes contributing to it.

At the first meeting of the Committee, held at Geneva in November 1977, it was agreed that UNEP should issue a half-yearly bulletin (January and July) giving information on ongoing and planned research activities on the ozone layer relevant to the World Plan of Action. This bulletin, which has achieved a wide circulation, together with the proceedings of the annual meetings of the Co-ordinating Committee on the Ozone Layer, now rank as the prime co-ordinating mechanisms in the worldwide research efforts in this most important field.

List of Participants

United Nations Bodies

UN/ESA

Mr. Michael Betts
Consultant
Office of Science and Technology

Mr. Bertrand Chatel
Chief Science Application
Office for Science and Technology

UNITAR

Mr. Joseph Therattil
Fellow and Deputy Bursar

UN Specialized Agencies

FAO

Prof. Martyn M. Caldwell
Food and Agriculture Organization
Utah State University

ICAO

Mr. L.F. Mortimer
Technical Officer

UNESCO

Ms. Josephine Doherty
Program Coordinator, MAB Program

Mr. Vernon Gilbert
Coordinator - MAB Project No. 8

Mr. Vernon J. Laurie
Program Coordinator, MAB Program

WHO

Dr. Gerald P. Hanson
Scientist
Regional Office for Americas/Pan American Health
Organization

Dr. Frederick Urbach
Professor and Chairman
Temple University Health Science Center

WMO

Mr. Oliver Ashford (Head of Delegation)
Director, Program Planning and UN Affairs

Dr. Rumen Bojkov
Chief of Atmospheric Sciences Division

Inter-Governmental Organization

EEC

Mr. Stanley P. Johnson (Head of Delegation)
Prevention of Pollution and Nuisances Division

Dr. Roberto Fantechi

Non-Governmental Organizations

ICC

Dr. J. Richard Soulen
Penwalt Corporation

ICSU/SCOPE

Dr. Lester Machta
Director, Air Resources Laboratory NOAA

Mr. Ted Munn
Executive Officer

Mr. W. Robertson II
Executive Secretary
National Academy of Science

Ms. D. Vince-Prue
Assistant to the Executive Secretary
National Academy of Science

AUSTRALIA

    Dr. James H. Whittem
    Scientific Attaché
    Embassy of Australia

BELGIUM

    Dr. Louis Groven (Head of Delegation)
    Scientific Counselor
    Embassy of Belgium

    Prof. Marcel Nicolet
    Institut Belge Aéronomie
    Gouvernement Belge

BOLIVIA

    Mr. Juan-Carlos Gumucio
    Counselor
    Embassy of Bolivia

CANADA

    Dr. W.F.J. Evans (Head of Delegation)
    Atmospheric Processes Research Branch
    Atmospheric Environment Services

    Dr. James E. Brydon
    Director, Contaminants Control Branch
    Canadian Department of Environment

    Dr. Ernest G. Letourneau
    Deputy Director, Radiations Protection Bureau
    Radiation Medicine Division
    Environment Health Directorate
    Canadian Department of National Health and Welfare

CENTRAL AFRICAN EMPIRE

    Mr. Francois Farra-Frond
    Directeur Général
    Ministére de l'urbanisme et de l'Aménagement
      du Territoire

    Mr. Catchy Ngakoudou
    Ingénieur
    Ministere des Eaux et Forets

CHINA

    Mr. Peng Chun-Chiao (Head of Delegation)
    Second Secretary
    Chinese Mission to UN

    Mr. Wu Tsien-Min
    Second Secretary
    Chinese Mission to the UN

DEMOCRATIC PEOPLE'S REPUBLIC OF KOREA

Mr. Kim Hyong Ik (Head of Delegation)
Minister Plenipotentiary
Office of the Democratic Republic of Korea to the UN

Mr. Kim Chol U
First Secretary
Mission to UN

Mr. Kim Chung Gol
First Secretary
Mission to UN

DENMARK

Mr. Allan Astrup Jensen
Environmental Chemist
Agency of Environmental Protection
Ministry of Environment

EGYPT

Prof. Dr. Ahmed M. Azzam (Head of Delegation)
Cultural Counselor & Director
Egyptian Education Bureau
Embassy of the Arab Republic of Egypt

Dr. Ezzeldin A. Moustafa
Cultural Counselor
Embassy of the Arab Republic of Egypt

FEDERAL REPUBLIC OF GERMANY

Dr. Wolfgang Ulrici (Head of Delegation)
Federal Ministry of Science and Technology

Dr. Winfried K. Muttelsee
Regierungs Director
Ministry of Economics

Dr. Rolf Sartorius
Doctor in Chemistry
Federal Environment Agency

Prof. Dr. Hartmut Uehleke
Professor Doctor of Medicine
Federal Health Administration

Prof. Dr. Dieter Ehhalt
Professor
Nuclear Research Center Julich

Dr. Jurgen Russow
Doctor of Chemistry
Hoechst AG

Mr. Hans Bobrowski
First Secretary
Embassy of the Federal Republic of Germany

Mr. Reinhard Marks
Counselor
Embassy of the Federal Republic of Germany

FINLAND

Mr. Hannu Laine
Assistant to Scientific Attaché
Embassy of Finland

FRANCE

Mr. Louis de Lamare (Head of Delegation)
Conseiller Scientifique à la Dgrst.
Ministère de l'Industrie et de la Recherche

Mr. Gérald J. Emptoz
Scientific Attaché
Embassy of France

GUATEMALA

H.E. Abundio Maldonado (Head of Delegation)
Ambassador of Guatemala

Dr. Hernan Hurtado-Prem
Minister Counselor
Embassy of Guatemala

INDIA

Dr. M. Anandakrishnan
Counselor (Science)
Embassy of India

ITALY

Mr. Dante Cadorin
Doctor in Physics
Foreign Affairs Ministry

IVORY COAST

Mr. Mathieu Capet
First Secretary
Embassy of Ivory Coast

LIBYAN ARAB REPUBLIC

Mr. Muftah Elarbash
Engineer
Environmental Protection Department
Ministry of Municipalities

NETHERLANDS

>   Mr. D.J. de Geer (Head of Delegation)
>   Director
>   Ministry of Public Health & Environment
>
>   Mr. Jan Jonkman
>   Second Secretary (Scientific)
>   Embassy of the Netherlands

NICARAGUA

>   Mr. Raul Chavez S.
>   Minister Counselor
>   Embassy of Nicaragua

NORWAY

>   Prof. Eigil Hesstvedt
>   Professor, Institute of Geophysics
>   University of Oslo

PHILIPPINES

>   Mr. Jaime J. Yambao
>   Third Secretary
>   Embassy of the Philippines

REPUBLIC OF KOREA

>   Mr. Kyung-Mok Cho (Head of Delegation)
>   Scientific Attaché
>   Embassy of Korea
>
>   Mr. Doo Byong Shin
>   First Secretary
>   Embassy of Korea

SWEDEN

>   Mr. Ulf Ericsson (Head of Delegation)
>   Scientific Adviser
>   Ministry for Foreign Affairs
>
>   Mr. Nils Daag
>   Second Secretary
>   Royal Swedish Embassy
>
>   Ms. Eva C. Permert
>   Assistant to the Scientific Counselor
>   Royal Swedish Embassy

SWITZERLAND

>   Dr. Christian Favre
>   Scientific Counselor
>   Embassy of Switzerland

THAILAND

    Dr. Pakit Kiravanich
    Head of Environmental Quality Standard Division
    National Environment Board
    Office of the Prime Minister

TOGO

    Mr. Etsri Kpotogbey
    Counselor
    Embassy of the Republic of Togo

UGANDA

    Mr. Abraham C. Nabeta
    Second Secretary
    Embassy of the Republic of Uganda

UNION OF SOVIET SOCIALIST REPUBLICS

    Mr. Sedunov Yuri (Head of Delegation)
    First Deputy Chief
    Hydrometeorogical Service

    Mr. Ryabkov Gennaoly
    Second Secretary
    Embassy of the Union of Soviet Socialist Republics

    Mr. Yuri Yezhov
    Second Secretary
    Embassy of the Union of Soviet Socialist Republics

UNITED KINGDOM

    Dr. Leslie Reed (Head of Delegation)
    Senior Principal Scientific Officer
    Unit on Environmental Pollution
    Department of Environment

    Dr. R. Murgatroyd
    Deputy Chief Scientific Officer
    United Kingdom Meteorological Office

    Dr. Gordon Diprose
    Manager Environmental Section
    Imperial Chemical Industries

    Mr. Alan Smith
    Counselor (Science & Technology)
    British Embassy

    Dr. T. Moynehan
    Attaché (Science)
    British Embassy

UNITED STATES

Mr. Edward Epstein (Head of Delegation)
Associate Administrator for Environmental Monitoring and Prediction
Department of Commerce

Mr. Anthony J. Broderick
Chief, High Altitude Pollution Program
Federal Aviation Agency
Department of Transportation

Ms. Margaret Kripke
Scientist, Litton Bionetics
Frederick Cancer Research

Mr. Herbert L. Wiser
Principal Physical Science Adviser
Office of Research and Development
Environmental Protection Agency

Mr. Paul J. Glasoe
Office of Environmental Affairs
Bureau of Oceans and International Environmental and
  Scientific Affairs
Department of State

Mr. L.R. Greenwood
Director
Upper Atmospheric Research Office
National Aeronautics and Space Administration

Dr. Alan J. Grobecker
Division Director
Office of Atmospheric Sciences
National Science Foundation

URUGUAY

Mr. Alfredo Giro
Counselor
Embassy of Uruguay

VENEZUELA

Dr. Otto Paz
Scientific Adviser, (Counselor)
Embassy of Venezuela

PAPER 1

*Environmental Aspects of Stratospheric Ozone Depletion**

Presented by the United Nations Environment Programme

<u>INTRODUCTION</u>

The Governing Council of UNEP at its meeting in April 1976

"Recognizing the potential impact that stratospheric pollution and a reduction of the ozone layer may have on mankind

Requests the Executive Director to convene a meeting of appropriate international governmental and non-governmental organizations:

- to review all aspects of the Ozone Layer,

- identify ongoing activities and future plans, and

- agree on a division of labour and a coordinating mechanism for: inter alia,

- the compilation of research activities and future plans; and the collection of related and commercial information."

This paper, for presentation at the UNEP Ozone Meeting in Washington, D.C. March 1-9 1977, represents an overview of the natural stratosphere and its impacts on man, the biosphere and climate. It has been prepared as a general review of the subject to provide a background for the more detailed and substantive material to be presented by governments and agencies. It does not deal specifically with research gaps, recommendations and organizational matters. These will emerge from the meeting and be included in the final report. It is based on existing state of the art documentation provided by participating organizations and countries. Major parts of the paper are taken from the material supplied by ICSU-SCOPE. Significant material was obtained from documentation by WMO, FAO, ICAO, OECD and UNESA. Other information was gleaned from supporting documents supplied to UNEP or available in national and international publications.

---

*Prepared by Dr. B.W. Boville, UNEP Consultant.

THE PROBLEM

The existence of the earth's ozone layer and its vital importance to life on this
planet have been known for many years. It is also known that man evolved and
adapted over millions of years to a particular level of ozone in the atmosphere.
But only recently has it been realized, that some of man's activities could lead to
stratospheric pollution and to a significant reduction in ozone amounts, in a few
decades.

The absorption of solar ultraviolet radiation (UV) by ozone, provides the heat
source which maintains the stable stratosphere (the cap to the turbulent weather
systems of the troposphere) and is thus one of the ultimate controls on global
climate. More importantly, this absorption process shields the earth's surface from
most biologically damaging UV (i.e. UV-B)* and increased penetration of this radiation is expected to have detrimental effects on many forms of life.

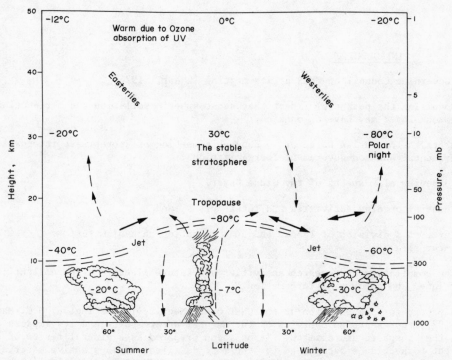

Fig. 1.1. The natural stratosphere. This is warmed from above by the
absorption of ultraviolet radiation by ozone. Hence the temperature increases with height, a temperature inversion, so the stratosphere is
strongly stratified and vertical mixing is very slow (order of years).
Much of the lower-level water vapour is apparently filtered out by the
-80°C cold trap at the tropical tropopause, hence the stratosphere is
virtually cloudless. In contrast the troposphere is heated from below,
temperatures decrease with height and vertical overturning is fast
(order of days), except in low-level inversions. Most pollutants introduced into the troposphere are subject to rain out and only those that
are very stable, such as N_2O, CFMs, etc. reach the stratosphere in
significant amounts.

*UV-B = 290-320 nm.

There is thus a great need to understand the physical and chemical processes involved in the ozone layer, the effects of global ozone changes on climate and the
effects of increased UV-B on life systems, and to monitor the atmospheric parameters,
including radiation, and the active trace gases (primarily nitrogen oxides and
chlorofluoromethanes, CFMs) - to provide a sound basis for any necessary action to
safeguard this fundamental part of the global environment.

THE NATURAL STRATOSPHERE AND OZONE LAYER

The Stratosphere

The vertical structure of the atmosphere is largely determined by the way in which
incoming solar radiation, interacts with atmospheric constituents (nitrogen 78%,
oxygen 21% .....). Above 100 km is found the very high temperature and complex
thermosphere, where all constituents are eventually broken down, and where most of
the very short wave radiation is absorbed. Between 50 km and 30 km, short wavelength UV breaks down very small amounts of molecular oxygen (O_2) into atomic
oxygen (O) and these recombine to form ozone (triatomic oxygen O_3). The resulting
concentration of ozone is only about one millionth of that of oxygen. The ozone in
turn absorbs most of the solar sunburning UV-B, thereby creating a warm region
where temperatures may approach surface values ($\approx 0^{o}C$). The bulk of the solar
energy, which is in the visible part of the spectrum, passes through the transparent atmosphere, to be absorbed at the surface of the earth where it creates the
warmth of the land and oceans.

The region of the atmosphere closest to the surface of the earth is called the
troposphere. Away from the solar heating at the surface temperatures normally
decrease with height up to the tropopause, the base of the stratosphere.

The tropopause is found near 17 km (temperature/$-80^{o}C$) over the warm equatorial
regions, but only 12-8 km (temperature/$-50^{o}C$ to $-60^{o}C$) above the temperate and
polar regions. Above the tropopause there is normally a temperature inversion, and
temperatures increase with height up to the major heating region near 50 km, the
top of the stratosphere.

The troposphere is relatively turbulent and well mixed, compared to the stratosphere
where the temperature inversion prevails. Moreover the active weather systems and
precipitation, in the form of rain and snow, are largely confined to the troposphere.
For these reasons most pollutants injected into the troposphere will be removed
from the atmosphere through physical processes such as rain-out or contact with the
surface within a week. In contrast, a pollutant introduced into the stratosphere
can remain there for several years before being transported vertically into
another layer. It should be mentioned that the broad horizontal wind currents in
the stratosphere tend to be as strong as they are in the troposphere, hence injected stratopheric pollutants may be widely distributed over the globe in a
relatively short time even though their vertical exchange is very slow.

Because the upper stratosphere is heated by the sun's radiation, it must show large
seasonal variations following the sun. Hence, temperatures change little near the
equator (50 km/$0^{o}C$, 30 km/$-30^{o}C$) but the summer pole is very warm (50 km/$+12^{o}C$,
30 km/$-20^{o}C$) and the winter pole, without sunshine, is very cold (50 km/$-20^{o}C$,
30 km/$-80^{o}C$). The summer hemisphere is then dominated by a warm stable anticyclone,
centred on the pole, with winds from the east (easterlies) covering the entire
hemisphere. The tops of the tropospheric weather systems are quickly damped by
this flow and can scarcely be seen in the wind patterns at 20 km. On the other
hand, the winter polar stratospheres feature a cold cyclonic vortex (low pressure
area) and west winds extend outwards towards a sub-tropical high pressure (anti-
cyclonic) belt. The largest scale weather patterns - planetary waves - which are

generated in the tropospheric westerlies can propagate to great heights in the stratospheric polar westerlies. Interactions between these two systems, produce large oscillations in the stratospheric polar vortex (which has much less mass than its tropospheric counterpart), at times leading to its breakdown in what are called sudden stratospheric warmings. It is not yet known, to what extent these inter-actions modify the tropospheric patterns and their long range prediction. As the tropospheric systems vary considerably from the northern to the southern hemisphere, so do the details of the warmings in the two winter stratospheres. In between these great monsoonal changes at higher latitudes, the equatorial regions tend to show a superimposed two-year change, from east to west to east winds (quasi-biennial oscillation).

The Ozone Layer

The small amount of ozone present in the stratosphere, is the result of a balance between processes which form it and others which destroy it. The production processes involve rather fast interactions, between short wavelength UV (250 nm) from the sun and normal oxygen in the upper stratosphere. Since neither the strength of the sun's radiation nor the amount of normal oxygen in the atmosphere[*] can be changed[*] by human intervention, nothing man does will likely alter the production rate.

The destruction processes of ozone in the natural stratosphere were thought to be the recombination of ozone and atomic oxygen and some ultraviolet breakdown. How-ever, it is now a well established fact that nitrogen oxides (NO derived from N_2O) participate in the natural destruction and may account for 70% of it, even though the concentrations of such trace constituents are generally only a few parts per billion (i.e. about 1000 times less than ozone itself). The reason why a small amount of a substance such as nitric oxide can destroy a much larger amount of ozone is that it reacts catalytically, i.e. one molecule of nitric oxide decomposes about 1000 molecules of ozone, before it itself is removed. (It is the destruction phase that can be perturbed by man's activities.)

Nevertheless the main production and destruction processes in the stratosphere take place on a very short time scale, and in the presence of short wavelength UV. They are very effective, in sunlit regions above 25 km and determine the more or less steady amounts of ozone there. Any ozone transported out of this region is quickly replaced. Below 25 km, the effective UV radiation and its by-products rapidly peter out and ozone becomes a quasi-conservative constituent of the atmosphere, until it reaches the ground where it again decomposes. Ozone may thus be said to be produced in the sunlit stratosphere above 25 km, transported to and stored in the middle and lower stratosphere below 25 km, and destroyed at the ground. Average concentrations of ozone increase downwards to about 10 parts per million (10 ppm) at 25 km; they may remain relatively high through the lower stratosphere, then drop very rapidly across the tropopause and remain fairly con-stant, at about 0.1 ppm, through the mixed troposphere to the sink at the ground.

The total amount of ozone overhead[**] is an important quantity, for it determines the amount of UV-B penetration to the earth's surface. Since the bulk of the ozone (85-90%) lies in the stratosphere, between the tropopause and 30 km, it is the depth of this layer which, first of all, defines the horizontal variations of total ozone, modified to some extent by the vigour of stratospheric transports. Hence the ozone layer is thin, about 250 units[**] over the equator where the tropopause is

---

[*]Reference is to changes significant for ozone layer processes.
[**]The average total ozone overhead is equivalent to a column at STP of about 0.3 cm or 300 Dobson units.

at 17 km, and its thickness increases rapidly to about 300 units at mid-latitudes, as the tropopause drops to about 10-12 km. At the middle and high latitudes travelling weather systems also affect the lower stratosphere and the height of the tropopause, so that ozone variations of 100 units over a few days are often observed, being perhaps 400 units behind surface low pressure areas and 300 units behind moving anticyclones; also, the vigorous oscillations in the stratospheric winter vortex result in large poleward ozone transports, and accumulations may bring total ozone amounts up to 500 units, in the late winter and early spring, in the north polar regions. In the south polar regions, spring values are somewhat less due to the reduced frequency of the stratospheric warmings. During the quiescent summer period, in the stratospheres of both hemispheres, total ozone amounts gradually subside generally being slightly less than 300 units, at middle and high latitudes in autumn.

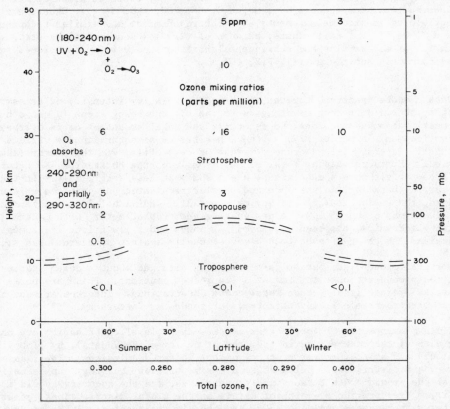

Fig. 1.2. The ozone layer. Ozone is created by the action of short wavelength ultra-violet radiation (UV), on a short time scale, at levels above 25 km where it is found in fairly steady concentrations. At lower stratospheric levels it is almost conservative (neither created nor destroyed) and moves about with the air, so that amounts are quite variable. For motions, the conservative quantity is the mixing ratio which does not change as the air moves up and down. The ozone mixing ratio tends to have a maximum of about 16 ppm above 25 km. However, the contributions to the total ozone column have to be mass-weighted so that the layer from the tropopause to 30 km largely determines the total ozone and its variations. Thus, the total ozone amount is small in the equatorial regions (0.250 cm) and greater at middle and high latitudes (0.300 to 0.500 cm) and varies in space and time in relation to the height of the tropopause.

Using a global mean of 300 units, total ozone varies: latitudinally by 35%;
seasonally by less than 10% at the equator, about 25% at high southern latitudes
and 50% at high northern latitudes; over short periods (days) up to 35%; and over
longer periods in the 10-20 years of records (after removal of the annual oscillation) variations from the mean at single stations are generally within $\pm$5%, and
averaged over the globe within $\pm$2%. It would thus appear that a shift in the
global mean, of 2% in 20 years, might be difficult to interpret, but that a 5%
shift in 20 years would stand out clearly from natural variations.

OZONE AND RELATED MEASUREMENTS

Most atmospheric constituents can be sensed, either through their absorption at
solar wavelengths (e.g. O_3, NO_2, CH_4, N_2O, C1O) or by their emission at terrestrial
wavelengths (e.g. temperature/CO_2, O_3, H_2O, HNO_3 OH, N_2O, N_2O_5). Others are
susceptible to in situ measurements, by electro-chemical and similar techniques
(e.g. H_2O, O_3, NO). Still others, because of very low concentrations etc., require
the taking of air samples and subsequent laboratory analysis by techniques such
as gas chromotography (e.g. FII, F12, CCI_4).

The Dobson ozone spectrophotometer measures the relative intensity of transmitted
sunlight, at two different wavelengths in the UV, one being strongly and other
only weakly absorbed by ozone, to determine the total amount of ozone overhead.
Observations on the zenith sky, as the sun rises or sets through the various
atmospheric layers, yield a crude vertical ozone profile, via the umkehr (inversion)
technique. Mearsurements of total ozone, at most of the 80 station WMO network, are
with Dobsons, although some stations use a somewhat less stable filter instrument.
With careful handling, maintenance and regular recalibration against a reference
standard, the Dobson has an accuracy of about 2%. Individual inaccuracies across
the network may be up to 10%. A new grating spectrophotometer, with modern
digital electronics, has been developed in Canada and a new filter instrument in
New Zealand. Both need to be tested operationally against standard Dobsons.

The vertical ozone distribution is measured in situ, at about a dozen stations
using radiosonde attachments, and at a few rocket stations. These are normally
calibrated against local Dobson total ozone observations. Intercomparisons of the
various electro-chemical and other types of sondes are necessary.

Satellite measurements of total ozone and its upper level distribution are made
using ultraviolet back-scattered from the ozone layer (B-UV data), by limb-scanning
techniques and by radio-meter scanning of the infrared emissions. Their absolute
accuracy and distributions below 25 km are somewhat low, but they can be calibrated
against the ground-truth Dobsons and ozone sondes, and the accuracy should improve
steadily. They provide a wealth of detail on the global distributions, relationship to weather system and geography, and are almost essential to the study of
special events. Future trends will be much easier to identify with satellite data.

Atmospheric measurements of trace constitutents are very difficult to obtain as
often amounts are only in parts per billion. However, considerable efforts have
been made, in a number of countries, using aircraft and skyhook type balloon platforms. These will be enhanced by planned satellite and space shuttle experiments.
Many of the active trace species in the hydrogen - nitrogen - chlorine systems have
been measured, and new data are appearing regularly. Much further work needs to be
carried out, as this is the only practical way to verify the laboratory and model
schemes for predicting ozone depletion on a reasonable time scale.

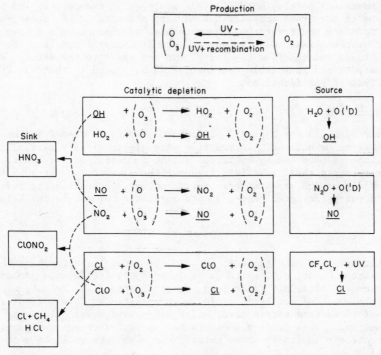

Fig. 1.3. Simplified ozone photochemistry. The stratospheric production and destruction of ozone occur mainly above 25 km where short
wavelength ultraviolet radiation is available for photolysis. On the
production side the photolysis of molecular oxygen, O_2, creates atomic
oxygen, O which recombines with O_2 to form ozone, O_3. In the absence
of catalytic substances, the ozone production would be balanced partially
by photolysis and partially by recombination. The three main catalytic
cycles thought to be active in the atmosphere are shown below the
production cycle. The most important one in the natural stratosphere
is the nitrogen system, the source being nitrous oxide and the sink
nitric acid. It is now thought to account for about 70% of the high-
level ozone destruction. The chlorine cycle may become significant
through the man-made CFM sources.

Note that all three systems interact and that this is a highly
simplified version of a complex process.

PHOTOCHEMISTRY AND MINOR CONSTITUENTS

The Role of Minor Constituents

The distribution of ozone in the atmosphere, as has been shown, is governed by
photochemistry and atmospheric transport in the stratosphere. The basic photochemistry is set by the photolysis and recombination of oxygen species, which is
modulated by the catalytic action of trace substances, primarily found in the
hydrogen, nitrogen and chlorine systems. To be effective, these trace substances
must be products of compounds that diffuse upwards from the surface of the earth
and that are stable enough to survive breakdown and rainout in the troposphere;
or must be injected directly into the stratosphere; or must be created there. In
order to understand and predict these processes, the full chemical reactions
schemes must be determined by laboratory procedures; the processes must be simulated

by complex numerical models, which include atmospheric transports; and finally the distributions of the trace constituents must be measured in the atmosphere, to verify and modulate the first two. Fortunately, advanced work has been carried out in a number of countries. Complex one-and two-dimensional models, with oxygen-hydrogen-nitrogen and chlorine chemistry, can now simulate the natural distributions quite accurately, and reasonable confidence can be placed in their ability to pre-dict the effect of modifications.

The Hydrogen System

The catalytic agent is the hydroxyl radical, OH, which is formed by the fraction-ation of water or methane CH_4, in reaction with electronically excited oxygen, $O(1_D)$. Methane is produced by bacterial action in marshland and lakes. In the natural stratosphere these reactions increase the destruction rates by less than 10%. A doubling of the stratospheric water content, in atmospheric models, yields less than 1% change in total ozone, so its modulation by man is unlikely to be significant.

The Nitrogen System

The catalytic agent is nitric oxide, NO. In the natural stratosphere the source is primarily nitrous oxide, N_2O, which is produced by denitrifaction processes in the soils and oceans. After diffusion into the stratosphere, N_2O reacts with excited oxygen to form NO. The catalytic cycle is interrupted by interaction with a hydroxyl radical to form nitric acid, which diffuses downwards into the tropsphere and is rained out. This process accounts for up to 70% of the destruction side of the ozone process - in its absence, natural ozone amounts would be much higher (1.5 to 2 times).

At high temperatures, molecular nitrogen and oxygen, the major constituents of the atmosphere, combine to form nitric oxide, which is thus produced in combustion processes and nuclear explosions. Other NO sources are lightning, galactic cosmic rays and solar proton events. Only the NO produced in the stratosphere is effec-tive in destroying ozone there.

Fertilizers. The application of nitrogen fertilizer over the globe is currently about 50 megatons per year and may increase to 100-200 MT/yr by the year 2000. In all other processes the nitrogen fixation is about 200 MT/yr. Fertilizer use would thus be expected to have some effect on the denitrification process, which produces molecular nitrogen, N_2 and a small amount (~7%) of nitrous oxide, N_2O. It is the latter which eventually reaches the stratosphere and accounts for a large part of the ozone destruction. The process of denitrification, in the soil and the sea is not well understood. In particular, there may be a very long time lag between fertilizer application and denitrification. Denitrification may also be affected by the acidity of the soil, another pollution problem.

Estimates of the fertilizer effect on ozone destruction, range from pratically nil to as high as 15% by the end of next century. Further research on the nitrogen cycle is needed before any confidence can be placed in these figures.

Aircraft. The advent of the supersonic aircraft, whose high temperature combustion engines inject nitric oxide directly into the stratosphere, provided the spur for major research programmes on the perturbed stratosphere. These definitive studies show that the ozone reduction is related to the injection rate, and to the injection altitude (closeness to ozone production zone). Hence subsonic aircraft injections in the troposphere are not expected to have a noticeable effect, nor are small numbers of concorde-type aircraft operating at 17 km or lower. On the other hand, models indicate that a fleet of advanced SSTs, operating at 21 km with an injection rate of 1.8 MT/yr, could result in a 12% steady state ozone depletion. Continuing studies will be required to monitor current emissions and to determine whether emission controls would be required if such fleets come off the drawing board.

Nuclear explosions. Nitric oxide produced by the high temperatures in a nuclear explosion should result in ozone depletion. Some ozone may be created locally by UV radiation from the firball, but after a few days, this effect should die out and depletion from nitric oxode grow in importance. The magnitude of the depletion effect again depends critically on the injection level, and on subsequent atmospheric diffusion and transport of the nitric oxide. So far intensive studies using total ozone, ozone sondes and, more recently, satellite measurements have failed to reveal any noticeable effect subsequent to nuclear explosions. On track measurements of ozone and nitrogen oxides may help to resolve the apparent discrepancy between theory and observation.

Galactic cosmic rays and solar proton events. Cosmic rays cause ionization and the production of nitric oxide, mainly in the polar lower stratosphere. Suggestions that periodic variations in this production rate might explain some ozone data variations have not been supported by model simulations. Variations of NO production, equivalent to that suggested for maximum to minimum, produce a small percentage change in polar stratospheric mixing ratios, but no noticeable model variations in total ozone.

Energetic solar particles, entering the atmosphere at high geomagnetic latitudes, produce nitric oxide, mainly at levels above 30 km. The strong sloar proton event of August 1972, produced a significant perturbation in stratospheric chemistry, and further studies of major events should cast light on high level processes.

Large amounts of nitric oxide are also produced, above 100 km, by UV action in auroral activity. It is unlikely that this affects the stratospheric ozone budget.

The Chlorine System

The catalytic agent is atomic chlorine, Cl, which is released when stable chlorine compounds are disassociated by short wavelength UV, in the middle and upper stratosphere above 25 km. At lower levels, these compounds are shielded from that UV by the ozone. There is minor ozone destruction in the natural stratosphere due to methyl chloride, CH_3Cl. However the major new source is from the man-made chloroflueremethanes (CFMs), primarily, F11, $CFCl_3$ and F12, CF_2Cl. When free chlorine is released by the action of UV on a CFM molecule at a median altitude of 30 km, it goes through many catalytic cycles in which ozone is destroyed and moved back to the molecular oxygen reservoir. The cycling is interrupted, to some extent, by interaction with the nitrogen system to form chlorine nitrate, $ClNO_3$, but mostly by interaction with the hydrogen system ($Cl+CH_4$) to give hydrogen chloride, HCl. These compounds diffuse downwards into the troposphere and are eventually rained out.

The time scale for a CFM molecule to reach the 30 km level may be as long as 15 years, and its eventual return to the surface, as HCl in rain, about 50-100 years.

The CFMs are manufactured from carbon tetrachloride, hydrofluoric acid and chloroform. The annual global production of F11 and F12 rose from 0.1 million tons in the late fifties to close to 1 million tons in 1974. Since then the production has declined by 10 to 15%, possibly due to economic conditions. About one-half of the CFMs are produced in the United States and most of the rest in Europe.

The major uses of CFMs are as propellants in aerosol spray cans, as cooling agents in refrigeration and air conditioning, and as blowing agents in foams. The aerosal sprays account for about 75% of the emissions into the atmosphere, and about 75% of their use is for personal care products - hair sprays, antiperspirants/deordorants.

A large research effort has gone into verifying the chemical reaction schemes in the laboratory, running these in numerical models of the atmosphere, and measuring the trace constituents and the reactive species in the real atmosphere. The latter measurements are particularly difficult, because of the very low concentrations in the atmosphere and the need to build new precision instruments. Nevertheless, even though there are still significant gaps, much progress has been made. The vertical distribution of the CFMs has been measured a number of times by sampling techniques. They show a constant mixing ratio in the troposphere, and then a decay with height after the sampling balloon has entered the stratosphere - just as the models predict. Measurements of trace constituents, in particular HCl and more recently ClO, have been reported but their distributions are still rather uncertain. The search for alternative chemical reactions, sources and sinks, the many modelling efforts and the atmospheric measurements, albeit still very preliminary, have all failed to reveal any major flaw in the theory. Since it would require at least 15 to 20 years of additional ozone data to sort out an expected atmospheric depletion trend, one must follow the path of chemical and atmospheric prediction systems verified by an atmospheric measurement effort, especially simultaneous measurements of important trace species.

At the present time all reported atmospheric models, including ones by government, universities and industry, yield ozone depletion predictions which fall within the 2 to 20% range at steady state, with a median value of about 8%. These have the following approximate modelling time scales for ozone depletion.* Taking the upper limit case of final depletion at 20% for emissions constant at 1973 level, the models indicate 2% depletion would be reached in 3 years, 5% in 15 years, 10% in 30-40 years and 20% in 100-200 years; a drop in emissions by one-half over the decade changes these to 2% in 3 years, 5% in 30 years, and 11% at steady state. For the median case of final ozone depletion at 8%, the constant 1973 emission case figures are 2% in 17 years, 5% in 70 years and 10% never; a drop to one-half over the next decade, changes these to 2% in 32 years and 5% never. Whether one assumes the drop to take place rapidly, or over a decade, makes a difference of no more than 1% or 5 years in the overall figures. For planning purposes one may work with these figures, updated as required by new research results, and turn to the important question of assessing the environmental impact of such ozone reductions.

IMPACT OF CLIMATE AND WEATHER

Man's sensitivity to the vagaries of climate and weather is ever present in the tragedies of cold regimes, drought and flood. Even small changes is global temperature values may have large consequences - a change of only one degree made the difference between the Little Ice Age and the current climate which has nurtured advanced technological and agricultural developments. Even though the models predict sizeable temperature changes in the stratosphere associated with ozone reductions, the predicted temperature changes at the earth's surface are scarcely separable from the model noise.

However, there are still two aspects of the climate problem that require further consideration. It is well known that very small shifts, in the atmospheric planetary waves and large circulation patterns, can result in dramatic changes in temperature and precipitation regimes. These planetary waves are neither well understood, nor well predicted in atmospheric models. It is suspected that their vertical structure may be important; hence until the problem is resolved, one cannot rule out the possibility of changes, in the ozone layer and stratospheric temperature, having an impact on weather patterns.

---

*See also Fig. 1.4

Of more importance, in this context, may be the fact that the CFMs absorb atmospheric radiation strongly in wavelengths where other absorptions are weak (in the atmospheric radiation window). Hence radiation, that would normally escape to space, is trapped by the CFMs in the atmosphere, just as other wavelengths are trapped by carbon dioxide and water vapour – keeping the earth warm by the greenhouse effect. Continued CFM release at the 1973 rates yields a contribution to the global temperature trend of about plus 0.5^oC at steady state – about one-sixth the predicted effect in increasing carbon dioxide concentrations. These are part of the general climate problem and, since they are as large as observed natural variations on the time scale of a hundred years, they need serious research consideration. (In the words of WMO – it is not yet possible to predict even naturally induced climate changes, for the next year, decade or century, with any useful degree of confidence.)

IMPACT OF INCREASED ULTRAVIOLET RADIATION

The UV Spectrum

Ultraviolet light is invisible and each of its photons is of shorter wavelength and more energetic than visible light which has a range from 400 to 750 nm.

UV shorter than 180 nm does not reach the stratosphere; from 180-240 nm it is absorbed primarily by O_2 to make O_3 and does not reach the lower stratosphere; from 240-290 nm it is absorbed by O_3 and does not reach the earth's surface; from 290-320 nm (UV-B) the radiation is partially absorbed by ozone and the portion which does reach the earth's surface is largely responsible for sunburn, skin cancer and other biological effects, and from 320-400 nm (UV-A) the radiation is mostly transmitted to the ground and may be important in the formation of photo-chemical smog.

Even though the shorter the UV wavelength the greater the biological effect, the transmitted UV-B is effectively terminated at 290 nm since shorter wavelengths do not reach the ground. The relative effectiveness (or the action spectrum) of UV-B in altering DNA, the material carrying the genetic information of all living cells, increases by a factor of about 15,000 as wavelengths drop from 320 to 290 nm.* It follows that wavelengths in the middle of the band (~305 nm) contribute most to biological effects. Under clear skies the transmission at the middle of the band is about 30% for 250 units of total ozone (tropical values) and less than 10% for an extreme value of 500 units of ozone. The biological effect integrated over the spectrum is found to vary about twice as fast as the total ozone variation – i.e. a decrease in total ozone of 10% would result in an increase of about 20% in the weighted sum of the UV-B (henceforth simply called SUV-B).

If one sets the SUV-B at 10 arbitrary units at the equator, then the annual average is about 4 units in middle latitudes. In winter it varies from 10 to 1, and in summer from 10 to 6 units, from equator to mid-latitudes. These values are in general agreement with SUV-B measurements at the ground. It should be noted that, although long-term exposure values are much lower at middle and high latitudes than at tropical latitudes, the main summer growing and exposure seasons at higher latitudes receive only fractionally lower amounts than the tropics.

One latitudinally dependent model gives a somewhat greater ozone depletion in the summer hemisphere for nitrogen (NO) perturbations, and in the winter hemisphere for chlorine (CFM) perturbations. These are probably within error limits; hence a given percentage ozone depletion may as well be considered the same at all latitudes.

---

*This variation defines the action spectrum of UV-B for DNA, deoxyribonucleic acid.

Thus the major errors will not lie in estimations of the change in transmitted UV-B for a given change in total ozone, but rather in the assessment of the damaging effect on living organisms, or their action spectrum. Whilst the action spectrum for unrepaired damage to DNA gives a useful approximation, the real effects can only be assessed from a knowledge of the wavelength to damage response, in each case. In general these action spectra are complicated by screening and photorepair mechanisms and by photosensitization effects. They are rather poorly known, if at all, and require much further research and documentation.

The Photobiology of Damage by UV

There is a wide range of deleterious effects from solar UV, including the well-known bactericidal and erythemal (skinburn) actions. Unfortunately, most of the basic photobiological research has used lowpressure mercury lamps with most of the emission at 254 nm, which is close to the absorbance peak for DNA and RNA,* but is not present in the solar spectrum reaching the ground. Extrapolations from results at 254 nm to the 290 - 320 nm UV-B are not straightforward. The research has concentrated on effects of UV absorbed by proteins and nucleic acids and has led to a good understanding of the photochemistry and photobiology of DNA damage.

The most common cause of UV damage in DNA is the formation of thymidine dimers, which prevents the separation of the DNA double strands, resulting in cell death in microorganisms, including phytoplankton. Lesions in cells may be repaired by dark recovery or by photoreactivation, involving an enzymatic process. During the important process of photoreactivation, the thymidine dimers formed by UV radiation below 300 nm are separated by the action of 300-475 nm UV, and thymidine monomers are regenerated. The synthesis of a new component of DNA, from information on the complementary strand, can then be effected.

In single cells, which have unusually effective DNA repair mechanisms, reactions of proteins become of relatively greater importance. UV irradiation of proteins causes a poorly characterized assortment of reactions in their component amino - acids. The repair mechanisms are also not well known.

Higher organisms show the same principal kinds of photochemical damage to nucleic acids and proteins as microorganisms, with similar repair processes. However, the prediction of UV-B effects is complicated by the fact that the net unrepaired damage, unlike the initial damage, may not be proportional to UV exposure.

To access and predict possible damage one needs to know not only the relative effectiveness of different wavelengths but also dose-response relationships, (i.e. whether low intensities for a long time give similar results to high intensities for a short time, and whether interactions in higher organisms can differ under natural conditions). Simple systems can be studied experimentally by the use of solar simulators, but the effects on man and large ecosystems must be derived from theory or from correlation studies, which are often difficult to interpret.

Impact on Man

The effects of UV-B exposure on man include sunburn and snow-blindness, as well as eye damage and ageing and wrinkling on the skin. These can be avoided or lessened by suitable protection. The production of vitamin D_3 is generally thought to be beneficial, but excessive intake may possibly cause intoxication.

---

*DNA - deoxyribonucleic acid; RNA - ribonucleic acid.

There is a substantial evidence to link UV-B with skin cancer. Sun-induced skin
cancers are generally diseases of less pigmented peoples and are probably caused
by the action of UV-B on DNA. Work with laboratory animals is consistent with this
conclusion. The most common skin cancers (non-melanoma) are only locally invasive
and respond to treatment. They occur, most frequently, on habitually exposed areas
of the body and show a strong correlation of incidence with latitude. These observations, together with differences linked with occupational exposure, point clearly
to sunlight as prime cause of cancer induction, and to increased incidence as a
consequence of increase in UV-B. Generally disseminated skin cancer (melanoma) is
much rarer but has a very high mortality rate and, therefore, represents a significant health hazard. The evidence linking ultraviolet radiation with melanoma is
not so strong as for non-melanoma. The latitude dependence is well established,
but melanoma is not concentrated on the habitually exposed areas of the body. However, it is rarely seen in the regions which are least exposed (e.g. those usually
covered by a bathing suit), and most cases are found in those areas which are
lightly covered or occasionally uncovered. Differences in patterns of locations
between sexes correlate with differences in exposure to sun, with more on the chest
in males and on the legs in females. There is, therefore, still a strong likelihood
that solar UV contributes to the induction and/or to the development of melanoma,
and that an increase in UV-B reaching the ground would result in an increase in
melanoma incidence and mortality. Where UV contributes to the induction of cancers,
the dose-response relationships are complex and unknown and quantitative predictions
can only be speculative. A decrease in the uncertainty, associated with these
predictions, requires not only a better basic understanding of the origin and growth
of skin cancers, but also better epidemiological data.

Impact on Plants

Most of the early experimental work with plants was carried out at 254 nm; only
relatively recently have possible effects of UV-B received attention. Degradation
of chlorophyll, depressed rates of photosynthesis, and cell death, which are among
the observed responses to 254 nm UV light, are assumed to be caused by damage to
nucleic acids and proteins. Longer wavelengths of UV may also destroy some plant
hormones and induce the release of a related substance, ethylene. Experiments with
lamp-filter combinations simulating the increase in UV-B, which would be associated
with a 35 to 50% decrease in ozone, have been carried out in glasshouses, controlled
environment rooms and to a limited extent in the field. These studies have revealed
a great range of sensitivity to damage. Many plants are adapted to grow in full
sunlight and mechanisms to cope with potentially damaging UV-B have developed.
These include elaboration of effective screening pigment systems in the outer layers
which protect the photosynthetic apparatus, temporal separation of UV-sensitive
reactions such as cell division so that they occur mainly during darkness, repair
systems, and possibly some degree of avoidance by positional movements. It was
found that some plants tolerated an increase in UV-B, such as might accompany a
50% depletion of ozone, while others showed signs of stress with regimes that are
considered normal fluxes of UV-B. In sensitive species, the increased UV inhibited
growth and development, depressed photosynthesis and biomass, and enhanced somatic
mutation rate. Pollen germination was also found to be inhibited. Depressions in
growth rate and biomass have been documented in the field for some species in Utah,
and in glasshouses with the roof removed in Florida. However, there is a great
heterogeneity of response, depending on species and environmental conditions; and a
greater degree of damage is observed, more commonly and to an enhanced degree, under
growth room conditions than in the field. This is possibly because of the greater
amount of light available for photorepair under the latter conditions. The
importance of photorepair for plants growing in the open has been demonstrated for
some alpine plants, where one day of exposure to normal levels of solar UV without
visible light resulted in photochemical damage. It is possible that some higher
plants may be able to adapt to a gradual increase in UV-B e.g. by increased

pigmentation, alteration of surface wax layers, etc.

In most experiments solar UV has been supplemented to correspond roughly to a 35 to 50% reduction in ozone at 60° solar altitude; the effects of exposing sensitive plants to lesser degrees of ozone depletion have not yet been well explored. The dose-response relationships have not been well investigated but certain types of damage, such as the depression of photosynthesis, appeared to be cumulative. Other types of damage, for example, in inhibition of leaf expansion, seemed to be more affected by peak dose. In general, exposing plants to UV-B causes damage similar to that found in the shorter wavelength region. Photosynthesis, in particular, appears to be sensitive to UV irradiation, which results in ultra-structural changes in the chloroplasts and an overall loss of photosynthetic capacity.

Impact on Terrestrial Animals

The effects of UV-B radiation on animals other than man have generally received little attention, as it is assumed that they are protected by their outer coverings. A few experiments have shown that large doses of UV-B given during the development stage could lead to abnormalities.

In latitudes where long periods of high natural solar UV are prevalent, three effects attributable to UV have been noted in farm animals. These are cancer eye, a form of ocular carcinoma of cattle; pink-eye, a bacterial infection of cattle which is aggravated by UV; and photosensitization, a complex reaction which results in hypersensitivity following ingestion of certain sensitizing agents (often found in plants) together with UV exposure. Carcinomas are also found in goats, sheep and horses, and in most cases they are found in parts of the body lacking protective melanin, such as eyelids and genitals. Most animals will not expose themselves to full sunlight unless they have a way of preventing excessive light from reaching sensitive areas (e.g. by the presence of UV-absorbing pigments).

Although UV-B is within the visible spectrum of many insects and might affect their behaviour, it is not yet possible to assess the relative importance of the UV-B component compared with other factors. Preliminary experiments, with simulated increases in UV-B, have so far revealed no adverse effects on pollination behaviour, in which UV-dependent perception (mostly UV-A) of flower markings (honey guides) is often important. Little information is available on growth and development responses to UV; to date only the larvae of a few insect species have been found to be sensitive to prolonged irradiation with large UV doses.

Impact on Ecosystems

A broader view of the possible biological impact of increased UV-B includes the non-agricultural aquatic and terrestrial ecosystems, which are important to man for timber, recreation, pest and predator balance, and as a food resource. Experiments with simulated solar sources have revealed considerable difference in sensitivity to UV between species; additional UV might therefore influence the ability of some species to compete in a particular natural ecosystem. Changes in plant metabolites could affect the animals that feed on them.

Aquatic ecosystems might also be affected because solar UV has been shown to penetrate to considerable depths in clear waters, and because some plants crucial to the food chain (phytoplankton) are found close to or on the surface of water, as well as the eggs and larvae of some aquatic animal species. The attenuation of UV-B largely depends on the amount of organic matter present: 10% of the UV penetrated to a depth of 23 m in clear waters of the west equatorial Pacific but only to 20 cm in Douglas Lake, Michigan.

Calculations indicated that about 2% of the euphotic zone (the zone with insufficient light for photosynthesis) might be affected by UV-B with a stratospheric ozone reduction of 25%, resulting in 3 to 10% of the phytoplankton of the world's oceans receiving more UV-B than is now received at the surface. Preliminary experiments indicate that phytoplankton may be quite sensitive to UV and consequences to the aquatic food chain might, therefore, be far-reaching.

The prediction of effects in large and complicated ecosystems requires knowledge of the penetration of UV into the canopies, the sensitivity and behaviour patterns of the component species and the species interactions. At present, both the basic data and the models, which would permit reasonably confident predictions for complex natural ecosystems, are lacking.

The Reduction of Harmful Effects

Some of the radiation effects could be mitigated through educational programmes on the risks involved in excessive exposure to sunlight, and through a conscious effort to breed crops with increased UV resistance. Because predicted solar UV increases take place gradually over decades, selective screening for the increased UV stress would be a natural part of crop development programmes. Long-lived crops, such as trees, would be at a greater risk from the cumulative effects of UV-B exposure.

In natural ecosystems the impact of increased UV-B can only be minimized internally. Some organisms have spare capacity through repair, screening and avoidance mechanisms; others are already near their tolerance levels. Since the behavioural cue for UV-B avoidance is the intensity of visible light, which does not change appreciably with ozone changes, the effectiveness of this cue may be very low.

The UV-B sensitive species of green plants are likely to be the most vulnerable organisms in terrestrial ecosystems because they cannot readily avoid exposure. Long-lived species of forest canopies and grassland plants are likely to be the most subject to cumulative exposure effects.

In aquatic environments, particularly in the open oceans, the avoidance by an increase in depth may not be taken, due to the lack of the visible cue. In some aquatic plants such as phytoplankton, the movement to greater depth may also be associated with greater attenuation of visible light and reduced photosynthesis.

Summary

Research to date indicates that likely consequences of increased UV-B are: (1) an increase in the incidence of skin cancers with a probable increase in mortality; (2) an increase in carcinoma and perhaps in other UV-B related conditions in cattle, leading to some economic loss; (3) a reduction in crop and timber yields, which could have social and economic consequences, and (4) a decrease in plankton production with consequences for the aquatic food chain. There might also be changes in the competitive ability among organisms which could alter entire ecosystems, although at present these are impossible to predict. A possible benefit of increased UV-B is the increased production of vitamin D, although one author concludes that "any hope that an increase in UV might be beneficial should be tempered with caution."

At present, biologists and epidemiologists are being asked to make quantitative predictions from insufficient data. The only predictions that have so far been made are those for the increase in incidence of certain types of skin cancer. The nature of the inherent problems has been stressed: the main targets for UV damage are the important biological macromolecules, proteins and nucleic acids, although in plants destruction of other molecules (e.g. the plant hormones) may add to the damaging effect and be a cause of depressed growth rates. However, extrapolation from laboratory data, obtained with single-celled organisms or cell cultures, to multi-cellular organisms under natural conditions is immensely complicated by screening effects of the outer layers, by behavioural characteristics, by antagonistic and synergistic effects between UV-B light and other wavelengths, and by the presence of repair mechanisms of varying degrees of effectiveness. Experiments with UV-B have now begun in may places, but a realistic evaluation of the possible damage to man and the biosphere that might arise as a consequence of ozone depletion to any particular level requires the input of new information, a refining of the models for predicting damage to individuals and to ecosystems, and a programme of long-term research.

SUMMARY OF FINDINGS ON OZONE DEPLETION AND ITS IMPACT ON MAN AND THE BIOSPHERE

The total amount of ozone in the natural stratosphere is strongly modulated by nitrogen oxides (emanating from the surface as N_2O), being 1.5 to 2 times less than it might be in a pure oxygen/nitrogen atmosphere.

The perturbation of the ozone layer by aircraft emissions of nitrogen oxides is probably negligible at the present time, but could be a matter of concern for future fleets of higher flying, advanced (mach 3) Super Sonic Transports.

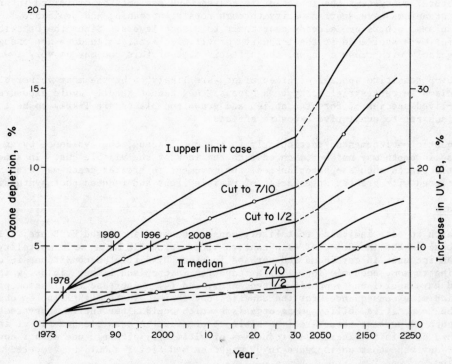

Fig. 1.4. Model ozone depletion. The time scales for ozone depletions predicted by atmospheric photochemical models under various scenarios are depicted schematically above. These will vary with the type of model and the assumptions with regard to chemical reactions and diffusion. The diagram is meant to be illustrative rather than definitive. The input in this case is equivalent to constant CFM emissions at the 1973 level. Curve I illustrates the upper limit case of 20% final ozone depletion and curve II the median or 8% final ozone depletion. These are modulated by hypothetical reductions in emission levels in 1978 to 70% and then to 50%.

The perturbation of the ozone layer by CFM emissions is a matter of concern - best estimates now for long-term (100-200 years) ozone depletions based on recent emission levels are: upper limit 20%, median 8%, lower limit 2%, with a half life of 40 to 50 years (time to reach one half the percentage depletion).

The ozone perturbations due to nitrogen fertilizers, other halogens and as yet unknown sources cannot be estimated with any accuracy at this time.

The impact of ozone depletions would be:

- an effect on climate, although the magnitude is uncertain;

- an increase in skin cancer, eye-damage, sunburning and skin ageing in man (and some animals);

- mutagenic and other effects on plants and animals which have limited protective and repair mechanisms; and

- possible changes in ecosystems (effects on phytoplankton etc.), although the magnitude and the dependencies are largely unknown.

On the ozone depletion side recent developments have tended to narrow the uncertainities and increase the general level of confidence in the predictions - conversely, one tends to be more concerned with what is not known than what is known on the impact side.

SOCIO-ECONOMIC FACTORS

Introduction

The ozone layer is a global environmental resource that is vital to life on this planet. It appears to have existed in its current natural state throughout the process of evolution. There have been suggestions that variations due to extreme natural events may have caused the disappearance of some ancient species but these are highly speculative. In light of present knowledge, it is reasonable to assume that the average global ozone amount has been relatively constant but future values may change as a result of man's present activities. That is why the problem falls so naturally within UNEP's 'Outer Limits' programme.

When a potential environmental hazard is identified, it is usual to carry out an environmental assessment prior to formulating and control strategy. Such an assessment should be as quantitative as possible and should involve evaluations of the benefits accruing to society from elimination or reduction of the risk and of the costs inherent in the various ways of achieving such reductions. Quantitative evaluation of the benefits that would result from reducing man-induced risks to the ozone layer is extremely difficult. Not only are the effects indirect and incompletey understood, but also in some cases they are expected to occur several generations after the pollutant emission. Thus the effects are, within the economists' time frame, virtually irreversible in that recovery of the ozone layer to its natural state after cessation of damaging emissions may take as long as a century.

Nonetheless, it is not sufficient simply to identify a threat to a vital component of the environment; there is a further responsibility to press ahead with investigations to elucidate and quantify the effects and thereby facilitate some estimation of the benefits that would flow from a reduction of the threat.

In contrast to the benefits, the costs involved in dislocating an industry and in developing alternative technologies and processing and delivery methods are susceptible to fairly precise calculations.

While the benefits that might result from specific actions to reduce the risk to the

ozone layer are not now easily quantifiable and it is not yet possible to justify
them on the basis of definitive benefit/cost analyses, the existence of a number of
risks to human health and welfare have been identified. One can conclude that the
first practical international step is to recommend realistic environmental objectives aimed at setting generally acceptable limits to emissions, which could form
the basis for specific control actions as appropriate. In the achievement of such
objectives, agencies, nations and industries can utilize fairly precise cost/bene-
fit analyses in the assessment of ways, means and alternatives. It is within these
limitations that the following very general qualitative assessments are made.

The Nitrogen System

The photochemistry and behaviour of the stratospheric nitrogen system have been
assessed very thoroughly by the US Climatic Impact Assessment Programme, CIAP, and
related programmes in other countries. These programmes demonstrated that the
current SSTs do not cause significant (detectable) harm to the ozone layer.

Should the production of advanced SSTs be contemplated it will be necessary to
evaluate their potential effects on the ozone layer. Engine emission reductions
might prove to be very costly relative to the lower nitrogen oxide emissions of
current subsonic fan jets. It should be noted that the SST effects may be large
because injection is directly into the ozone layer – for the same reason recovery
after any necessary control action would be relatively fast. In any case, thorough
cost/benefit studies will have to be carried out by the agencies, nations and industry prior to launching a new endeavour.

Man's increasing demands for energy, food and transportation are responsible for more
and more interventions in the natural nitrogen cycle. It is now realized that the
whole fluid envelope of the planet, as it relates to man and the biosphere, is
tightly bound to this nitrogen cycle, which is itself not well understood. The
perturbation of the ozone layer, through the input of additional nitrogen oxides,
is only one basis for concern about nitrogen pollution. A cost/benefit analysis of
the nitrogen fertilizer impact on the ozone layer probably cannot be made at this
time and in any case, it should not be made independently of other nitrogen problems.
What can be said is that the results of intensive research programmes to clarify
the physio-chemical, biological and socio-economic aspects of human modifications
of the nitrogen cycle could be very useful indeed.

The Chlorine System

Unquestionably, it is concern over the impact of CFMs on the ozone layer that has
stimulated current national, international and industrial action in this area. It
is thus imperative to have careful analyses of the socio-economic factors to provide a basis for any decision aimed at environmental protection. Such studies have
been and are being sponsored by a number of agencies, including UNEP, UNESA, OECD
and EPA/USA, and the manufacturing industries. These all stress the lack of
quantitative knowledge on the impact side, and hence the impossibility of evaluating benefits.

However, two quantitative approaches are available. The first involves modelling
the impacts of alternative scenarios on the ozone layer and the second requires
evaluation of the various societal and industrial options.

The model aspects, which have been reviewed earlier, are calculations that can be
easily changed to fit different criteria and easily amended as new research data
become available. For instance, if one arbitrarily sets a critical level at 5%
ozone depletion (10% increase in UV-B) – then for the upper limit (20% final ozone
depletion) case, the critical level for continuing 1973 production is reached in

1990, a drop to 1970 production (to 70%) postpones this date to 1978 and a further drop to 1967 production figures (to 50%) moves this to the year 2008. For the medium (8% final depletion) case, the critical points are respectively 2050, 2150 and never. The significance of such delays lies in the time allowed for more definitive research, which would narrow the uncertainties and clarify the need for either lesser or greater reductions. What they illustrate is that even modest restraints on production or use (and subsequent release) would have a significant impact on future ozone depletions and, presumably, on long-term socio-economic benefits.

Considering the consumer and industrial options, one can make preliminary relative evaluations from data already published. These show that there are three main sectors using CFMs: (1) aerosol propellants, (2) refrigeration and air conditioning, and (3) foaming agents in plastics. The approximate ratio of their shares of CFM release is 15: 4: 1 and of the respective total economic values 2: 25: 1. The data also indicate that 75% of the aerosol propellants are used in personal care products (hair sprays, antiperspirants, deodorants). Significant reductions in CFM release, with minimum economic loss, can clearly be achieved in the aerosol sector. In addition, the societal factor is largely a matter of personal convenience and there are already some substitutes and alternative delivery systems on the market. Reductions in the refrigeration/air conditioning sector would have a serious impact on that industry and the food industry, and from a socio-economic viewpoint would be highly undesirable, if not unacceptable. Fortunately, a large fraction of their CFM emissions could be eliminated by better servicing and recovery systems. In the case of foam manufacture from polyurethane and polystyrene, the substitutes for CFMs probably would lead to a substantial loss in efficiency, but better manufacturing control procedures could cut emissions. It follows from these cost/ benefit considerations, and on the prudent assumption of the validity of current prediction models, that to continue to reap the great societal benefits from CFM use in medico-pharmaceutical aerosols, refrigeration/air conditioning and some foaming products, society may have to limit its use of those non-essential aerosol sprays containing CFM propellants.* This course of action, along with better industrial control procedures and the wider use of alternatives, may even permit modest growth in essential areas. Once again it has to be stressed that the assessments are still quite imprecise and that major benefits are to be expected from the acceleration of monitoring and research programmes on all fronts.

Future Needs:

In order to elucidate the socio-economic factors including both the benefit and cost sides of this problem there is a need to:

- accelerate monitoring and research on all physio-chemical aspects of the strato-spheric ozone layer;

- monitor world-wide production and emission of substances identified as potential modifiers of stratospheric ozone;

- encourage research on climatic impact;

- promote research and investigative studies on the impacts of increased UV-B on man and the biosphere;

- support broader studies of the socio-economic aspects; and

- move forward on the organizational aspects of coordinating these activities and on the general problem of global environmental protection.

---

*More specifically those CFMs such as F11 and F12, lacking hydrogen.

PAPER 2

Effects of Changing Levels of Ultraviolet Radiation on Phytoplankton*

Presented by the Food and Agriculture Organization

<u>INTRODUCTION</u>

Several books and articles on terrestrial and underwater radiation exist. "Solar radiation" edited by N. Robinson in 1966 and "Daylight and its Spectrum" by S.T. Henderson (1970) give extensive information on the whole spectrum of solar radiation reaching the earth's surface. Works by Jerlov (1968), Tyler and Smith (1970), and the book "Optical Aspects of Oceanography" edited by Jerlov and Steemann Nielsen in 1974 deal with underwater spectral radiance and the general optical aspects of underwater light. Ecological problems are treated in "Light as an Ecological Factor" edited by Bainbridge and Evans in 1966. A new edition of this book appeared in 1976. The effect of ultraviolet action on algae has been treated by Halldal (1967) and by Halldal and Taube (1972). In the series "Photophysiology", started in 1964 with A.C. Giese as editor, numerous examples are given of UV effects on different organisms, and in various biological and biochemical systems.

The effect of UV on phytoplankton is the subject of this study. Many UV responses in phytoplankton have much in common with reactions observed in other organisms. It seems fairly certain that responses to UV irradiation are identical in different biological and biochemical systems. The study also of non-photosynthetic forms and photobiological model systems have yielded valuable information on the effect of UV irradiation on photosynthetic organisms. Such information will therefore be used here to enlighten and predict phytoplankton responses to UV where such analyses have not been performed for this group of organisms.

<u>SOLAR RADIATION</u>

Spectral Energy Distribution in Air

The electromagnetic spectrum from ultraviolet at 220 nm and that from the infrared at around 1500 nm is often divided in the following way: _Far UV_ between _220 and 290 nm;_ _Near UV_ between _290 and 380 nm;_ _Visible_ between _380 and 760 nm;_ and _Near IR_ between _760 and 1500 nm._ This division has both physical and biological implications. The boundary towards shorter wavelengths at 220 nm is set by the absorption of radiation in air. The boundary 290 nm demarcates the shorter limit to terrestrial radiation. 380 to 760 nm is the region of human vision, and 1500 nm is around the

---

*Prepared by Per Halldal, University of Oslo.

longer wavelength limit of solar radiation. These divisions will be followed in this article.

The solar radiation which penetrates the atmosphere extends from around 290 nm to about 1500 nm as illustrated in Fig. 2.1.

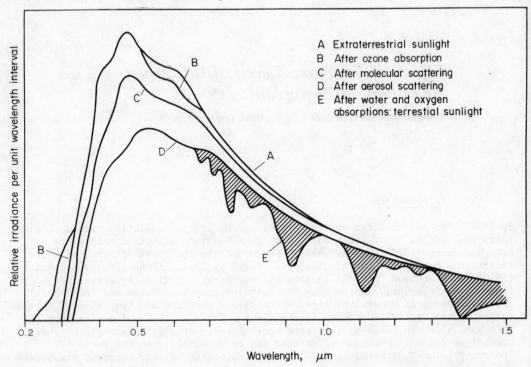

Fig. 2.1. Successive processes affecting sunlight during penetration of the atmosphere. (From S.T. Henderson, 1970)

Terrestrial sunlight does not contain radiation below 290 nm since shorter wavelengths are stopped by the ozone layer formed by far UV radiation at a distance of 10–40 km above the earth (the "ozonesphere"). Maximum concentration is 10 parts per million which is reached at about 25 km (Henderson, 1970).

Terrestrial UV solar radiation varies with latitude and sun direction. Figure 2.2 illustrates terrestrial radiation as part of extra-atmospheric solar radiation.

Figure 2.2 shows that the shorter wavelength limit of terrestrial near UV radiation varies with a special span of 14 nm between 290 and 304 nm.

Spectral Energy Distribution Underwater

As sunlight penetrates natural water drastic spectral changes take place. Great variations are also observed in sea water masses of different types and the effect on lakes and rivers may be considerably greater than is normally observed in the sea.

All types of water show a high attenuation of red and infrared radiation. Only minor radiation can be detected at wavelengths longer than 650 nm at depths greater than 10 m (see Fig. 2.3).

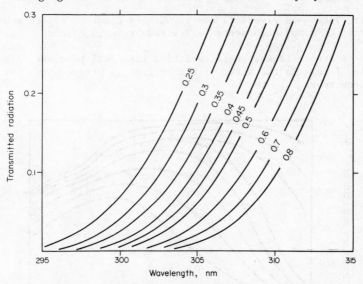

Fig. 2.2. UV solar flux at the earth's surface expressed as extra-
terrestrial flux. Total ozone 0.25 to 0.8 cm NTP, overhead
sun. (Hesstvedt, 1976)

In the near UV spectral region variations in attenuation differ greatly with water
type. Curves for per cent transmittance of surface waters of different types
are given in Fig. 2.4.

In clear oceanic waters outside the continental shelf the transmittance per meter
of downward irradiance of different types lies between 50 and 80 per cent at 300
nm. See Fig. 2.4 I, II, and III. For the coastal water types 1, 3, 5, 7, and 9
the transmittance per meter at 300 nm lies between zero and 7 per cent. For water
type 9 the transmittance is zero at 340 nm.

Most inland lakes and rivers have transmittance characteristics resembling that
illustrated in Fig. 2.4, curves 1 to 9. The high attenuation rate of radiation in

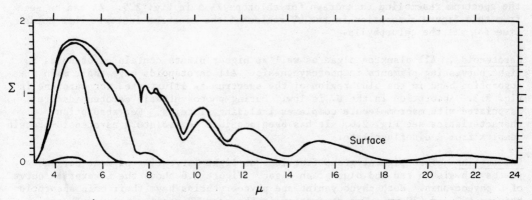

Fig. 2.3. The complete spectrum of downward irradiance in the sea.
Wavelength scale in u (1 u = 1000 nm),
(Jarlov, 1968)

coastal waters, lakes and rivers is due to what is denoted "yellow substance" and large quantities of humus components yellow and brown in colour.

A few lakes have transmittance characteristics identical to clear oceanic waters. A good example of this is the water of Crater lake in Oregon which has been studied by Tyler and coworkers.

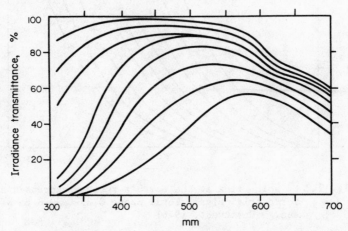

Fig. 2.4 Transmittance per meter of downward irradiance in the surface layer for optical water types. Oceanic types *I, II, III* and coastal types *1, 3, 5, 7, 9*. (Jerlov, 1968)

PHOTOSYNTHESIS BY PHYTOPLANKTON

Pigment Composition of Plankton Algae

Chlorophylls. All plankton algae contain chlorophyll *a*. In addition to this pigment green algae have chlorophyll *b* as light harvesting pigment during photosynthesis. The additional chlorophyll of diatoms, dinoflagellates and cocclithophorids is chlorophyll *c*. Blue-green and red algae have only chlorophyll *a*. All these chlorophylls have two main absorption bands in the visible region of the spectrum resembling that drawn for chlorophyll *a* in Fig. 2.5. As can be seen from this figure absorption in the UV region of the spectrum is low. This holds true for all the chlorphylls.

Carotenoids. All plankton algae as well as higher plants contain carotenoids as light harvesting pigments in photosynthesis. All carotenoids have their main absorption band in the blue region of the spectrum as illustrated for carotene in Fig. 2.3. Absorption in the UV is low. During photosynthesis carotenoids are associated with macromolecule complexes including proteins. For absorption characteristics see Fig. 2.6. It has been possible to isolate a carotenoid-protein complex from a dinoflagellate.

Phycobilins. This pigment group functions as light harvesting pigment during photosynthesis in red and blue-green algae. Figure 2.6 shows the absorption curve of C-phycocyanin. Both phycocyanine and phycoerythrins have their main absorption between 480 and 630 nm. The absorption in the near UV region is low. These pigments are tightly bound to proteins and the absorption spectrum below 300 nm will reflect the absorption of protein as illustrated in Fig. 2.7.

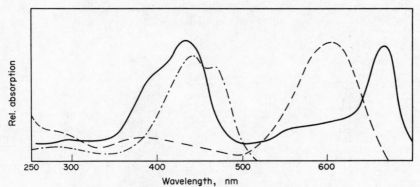

Fig. 2.5. The absorption spectrum of chlorophyll *a in vivo* based upon
calculation in the visible region (Halldal, 1958) and on
assumption from absorption characteristics in the UV of
extracted chlorophyll *a*.   (Halldal, 1966)   The absorption
spectrum of 8-carontene in hexane (Harrer and Junker, 1950).
The absorption curve of C-phycocyanin.   (Ehocha, 1962)

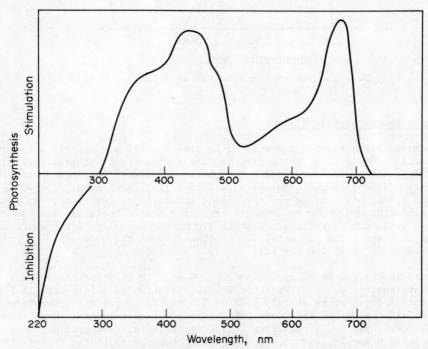

Fig. 2.6. Action spectrum of photosynthetic stimulation and inhibition
in the green alga Ulva lactuca. (From Halldal, 1964)

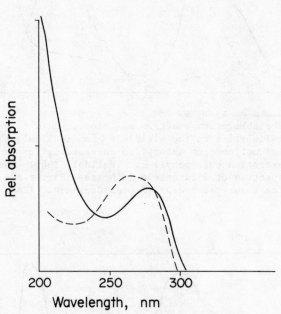

Fig. 2.7. The absorption spectrum of protein and of deoxyribonucleic
         acid (DNA).

Function of Photosyntehtic Pigments

Photosynthetic pigments when functioning in living cells are closely associated with
macromolecular complexes of which proteins are an important integral part both as
structural proteins and as enzymes. The photostability of the chromophore groups
of pigments involved in photosynthesis (Chlorophylls, carotenoids, and phycobilins) is fairly high at the moderate light levels found in nature. In plants a
photostablilizing system for chlorophyll involving a carotenoid is also found. It
is assumed that near UV radiation down to 320 nm has normally little destructive
effect of photosynthetic pigments. As mentioned above the absorption of radiation
at these wavelengths is low for all.

The picture is changed below approximately 300 nm. In this spectral region towards
shorter wavelengths, proteins begin to absorb radiation as illustrated in Fig. 2.7.
The absorption curve drops sharply towards 300 nm, but a protein absorption "tail"
may extend all the way up to 320 nm. Proteins at all wavelengths where they absorb
are easily destroyed by UV radiation. Since proteins are integral parts of all
photosynthetic pigments, and as these protein-chromophores in the photosynthetic
apparatus are arranged in biological membranes, and as enzymes are involved in
numerous photochemical and biochemical reactions during photosynthesis damage to
proteins may be destructive to photosynthesis in many ways.

The photosynthetic apparatus also contains its separate genetic apparatus different
from that in the cell nucleus. The genetic characteristics of an organism or an
organelle are determined by the construction of a nucleic acid denoted

deoxyribonucleic acid (DNA). The action of UV radiation on both proteins and nucleic acids is treated later in this article.

Photosynthetic Action spectra of Phytoplankton

An action spectrum provides information on the pigments involved in photochemical and photobiological processes. The photosynthesis action spectrum of an alga may be determined by irradiating the alga with an equal number of quanta over a fixed period at different spectral wavelengths and measuring the amount of oxygen evolved. Consumption of oxygen, reduction of oxygen produced when certain wavelengths of light are added to white light is proof of photosynthetic inhibition.

Several such action spectra have been determined for different algal groups over the whole visible and UV spectrum. Reviews are given by Halldal (1967), and by Halldal and Taube (1972).

Figure 2.6 gives the action spectrum of photosynthesis (oxygen evolution) of the green alga Ulva. This action spectrum is one of the most detailed analyses performed on a green alga and, with small adjustments, may be representative for several algal groups (diatoms, dinoflagellates, unicellular green alga). What is important for treatment in this article is the region 320 to 290 nm. So far the response in this wavelength region has been demonstrated as very similar in all algal groups investigated (green, blue-greens, reds, dinoflagellates).

In Fig. 2.6 it is shown that the action spectrum has two main peaks, one in red reflecting chlorophyll a and b, and one in the blue reflecting chlorophyll a and b plus carotenoids. From 400 nm towards shorter wavelengths the curve drops sharply towards 300 nm. Below this wavelength photosynthetic inhibition was observed at all wavelengths with a sharp increase towards 220 nm. This far UV response most probably means destruction of proteins. The correlation of this curve with the absorption spectrum of proteins is not particularly good (see Fig. 2.7) but the deviations can be explained on grounds of quanta efficiency of protein destruction at different wavelengths (Halldal, 1967).

The sharp drop from 400 nm towards 300 nm may indicate photosynthetic inhibition. If, however, the effect of different radiation wavelengths in this spectral region is tested on algae with a constant photosynthetic rate in white light, photosynthetic stimulation is always observed in all algal groups so far tested. It should be noted that moderate levels of monochromatic radiation were used in all these experiments.

Below 300 nm photosynthetic inhibition was observed in experiments of the same type as those performed in the spectral region 400 to 300 nm. The alga was kept at a constant photosynthetic rate in white light and radiation between 300 and 220 nm was added to the system. The action spectrum curve for 300 to 220 nm was then calculated.

It should be noted here that the action of UV on an alga in the spectral region 300 to 220 nm has two stages. (1) Photosynthesis is stimulated, demonstrated by enhanced oxygen$_2$ evolution. At moderate radiation intensities in the order of 1000 to 2000 ergs/cm^2 sec this stimulation lasts for from two to five minutes (Halldal, 1966). (2) Inhibition and reduced oxygen evolution then follows. The action spectrum presented in Fig. 2.6 is based upon the smallest UV dose to cause reduction in oxygen evolution.

ACTION OF UV ON DEOXYRIBONUCLEIC ACID (DNA)

Destruction of DNA by Far UV Radiation

The long wavelength absorption of DNA stops around 300 nm. An absorption "tail"
extends up to 320 nm. The absorption curve is given in Fig. 2.7.

When UV radiation is absorbed by the DNA molecule it may be damaged in different
ways. The main products of damage to DNA are pyrimidine hydrates and pyrimidine
dimers. The most common cause of far UV-damage is the formation of thymidine dimers.
Such damage may undergo dark repair and photoreactivation (see Fig. 2.7).

When thymidine dimers are formed separation of the DNA double strands is prevented
during cell division resulting in cell death in microorganisms including phytoplankton. UV action on DNA may also result in mutations.

Photoreactivation of Far UV-DNA-Damage

Lesions in cells after far UV irradiation may be repaired by dark recovery in the
absence of light and by photoreactivation involving an enzymatic process with a
yellow pigment as the radiation absorber. Action spectra for this process have
been determined for several microorganisms including photosynthesis, growth, and
motility in phytoplankton. The action spectra of photoreactivation show two different types of curves which, in many respects, are rather similar, as illustrated
in Fig. 2.8.

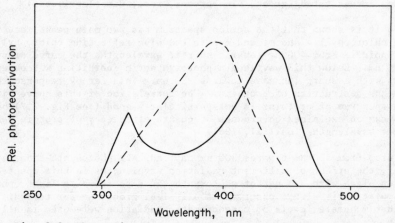

Fig. 2.8. The action spectrum of photoreactivation for survival of the
          blue-green alga Agmenellum (von Baalen, 1968) and for far
          UV-killing in the bacterium Escherichia coli. (Jagger, et
          al., 1970)

Photoreactivation spans the spectral region 300 to 460-480 nm. During the process
of photoreactivation the thymine dimers formed during UV radiation below 300 nm are
separated and thymine monomers regenerated. The synthesis of a new component of
DNA from information on the complementary strand can be effected.

The process of photoreactivation in nature is undoubtedly of great importance. In
natural sunlight radiation around 300 nm will be absorbed by the DNA molecule and
thymine dimers will be formed. At the same time, however, the enzyme process
involved in photoreactivation will come into action as the organisms are irratiated
over the spectral region 300 to 475 nm where the photoreactivating enzyme absorbs.

A direct example of how the UV in natural sunlight inactivated DNA, and how this
inactivation is photoreactivated by the photoreactivating enzyme was given by

Rupert in a model system using transforming DNA, before and after the addition of the photoreactivating enzyme from yeast, as illustrated in Fig. 2.9.

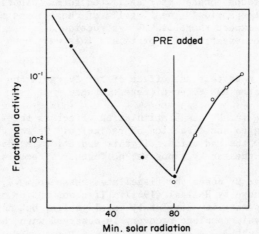

Fig. 2.9. Solar inactivation of transforming DNA (streptomycin
          resistance) in a quartz flask, and its repair in sunlight by
          yeast photoreactivating enzyme (PRE). DNA alone
          -------O------, DNA mixed with PRE-------O-------. (After
          Rupert, 1964)

This model experiment by Rupert clearly demonstrates that the UV contained in solar radiation has a directly damaging effect on DNA, and also that biological systems have a mechanism controlled by solar radiation to adjust for this damage. It should be stated here that the act of photoreactivation is the general rule in biological systems. Microorganisms which do not have this photoprocess are exceptions.

ACTION OF UV RADIATION ON PROTEINS

Proteins are also affected by far UV radiation, though the effects are less well analyzed than for DNA. The effect of UV irradiation on proteins may be most readily assessed with enzymes, but the mechanism of protein inactivation by UV radiation is not clear.

It is known that hydrogen bonds and cystine -SS- bonds are disrupted by UV irradiation, the latter altering the tertiary structure which links polypeptide chains in a protein molecule. In addition, -CS- bonds are also disrupted.

The repair mechanisms after UV damage to proteins are little known. It can be stated that protein destruction after gentle UV irradiation is repaired in the dark or on removal from exposure in the far UV region. High doses of far UV irradiation on proteins result in irreversible damage and inevitably in cell death of microorganisms including phytoplankton.

As the UV in natural sunlight undoubtedly contains quanta which will be absorbed by protein molecules, cells exposed to direct sunlight evidently have protective mechanisms of different kinds to prevent UV damage. How strong these protective mechanisms are is unknown.

The absorption spectrum of protein is shown in Fig. 2.7. It shows that the absorption extends to around 300 nm.

PHYTOPLANKTON MOTILITY

Important plankton groups consist of flagellated forms (dinoflagellates, motile green algae, coccolithophorids). The flagella of these forms mainly consist of proteins. One would expect therefore UV destruction of the motor apparatus to reflect the absorption spectrum of a protein. This was shown to be the case (Halldal, 1961).

It is interesting to note that the effect of far UV radiation on the motor apparatus in flagellates has three different stages.

- In the first stage far UV has a stimulating effect as the flagellates direct swimming according to the direction of radiation. We say that the alga responds phototactically to the radiation. If this radiation is gentle and the flagellate removed from the far UV exposure, no further effect is noticed.

- With increasing far UV doses the flagellates cease to move, evidently by protein denaturation as shown by Halldal (1961). If a population of Platymonas is irradiated to stop motility completely in all cells, but not with doses markedly exceeding this level, complete recovery is observed when the cells are transferred to white light.

- The third stage occurs when the cells approach cell division which takes place after two to four days. During these days cells seem to have recovered from the far UV radiation received, but now a sign of serious damage appears. The cells cease to move and death occurs. The action spectrum for this damage reflects the absorption spectrum of DNA and demonstrates that the damage was due to thymine dimer formation. This damage can be repaired by after irradiation in the spectral region 300 to 450 nm. Light at wavelengths above 500 nm have no repair effect demonstrating photoreactivation.

CONCLUSIONS

UV Action on Surface Waters

The shorter wavelength limit of terrestrial solar radiation is around 300 nm. Depending on latitude and sun direction the spectral span for this lower limit is 14 nm, between 290 and 304 nm. The UV in solar radiation below 320 nm is of utmost importance for life on earth, particularly is this the case for microorganisms including phytoplankton. Far UV radiation (220 to 290 nm) has serious damaging effects on both proteins and nucleic acids. The damage of deoxyribonucleic acid (DNA) has particularly been subjected to analyses and we know that several different types of lesions take place. The most important of these is the formation of the photoproduct thymine dimers. When thymine dimers are formed the two DNA strands fail to separate during cell division which ultimately results in cell death. The mechanism of the UV damage to the protein molecule is less well known, but UV damage to this macromolecule is easily observed even after short UV radiation around 290 nm.

Both DNA and proteins have a sharp fall in their absorption curve from shorter towards longer wavelengths ending around 300 nm, however, absorption tails extend up to 320 nm and UV radiation from the sun will consequently act on both DNA and proteins.

In an experiment performed by Rupert (1964) it was directly demonstrated that the UV content of solar radiation destroyed the DNA molecule drastically.

Both denaturation of proteins and DNA photoinactivation by UV radiation have repair mechanisms for the restoration of these macromolecules.

If proteins have been damaged by gentle doses of UV radiation and thereafter are removed from the UV source either by placing the material in the dark or in moderate intensity of white light proteins may regain their original structure and function. The mechanism for this repair is still obscure.

In the case of the DNA molecule several repair mechanisms are known. Dark repair may take place, but the most important process is that of photoreactivation. Photoreactivation involves a light activated enzymatic reaction in the wavelength region 300 to around 475 nm. By this photoprocess the thymine dimer bonds formed by UV radiation below 300 nm are released and thymine monomers generated. That both DNA photoinactivation and photoreactivation takes place simultaneously in natural sunlight was demonstrated by Rupert, and is illustrated in Fig. 2.9. In the absence of the photoreactivating enzyme DNA was rapidly destroyed, but the addition of photoreactivating enzyme and exposure to the same solar radiation which produced photoinactivation of DNA now showed increased activity in the test system used.

Irradiation at wavelengths above 320 nm in moderate doses at the level which normally occurs in nature rarely inactivates biological systems in their normal environment. The damaging effect of solar radiation is found between 320 and 290 nm. The threshold 320 is somewhat flexible. This may be because of the purity of the monochromatic irradiation source used during the experiments. Usually a spectral half band width of around 15 to 20 nm was used. The correct limit of damaging effect is therefore probably closer to 310-305 than 320.

The lower limit of terrestrial solar radiation is determined by the depth of the ozone layer 10 to 40 km above the surface of the earth, and by latitude and sun direction. It is important to note that both proteins and DNA begin absorption significantly at the same wavelength and that it is at this wavelength that UV cell damage starts, with increasing effect towards shorter wavelengths.

Biological systems evidently have protective mechanisms to prevent damage from UV radiation in natural sunlight as DNA and protein absorption "tails" extend to 320 nm, this protective system must be given some attention.

One of these processes, photoreactivation, has been extensively studied, and we know that it functions under natural conditions and I refer once more to Rupert's model experiment (1964). We do not, however, know how strong this repair mechanism is. If the ozone layer is in some way reduced or damaged, UV doses which are absorbed by both proteins and DNA between 320 and 290 nm will increase. More important than that is undoubtedly that the boundary of solar radiation will move towards shorter wavelengths and simultaneously in the direction where DNA and proteins start to absorb significantly.

It is reasonable to assume that the balance between UV destruction of DNA and proteins, and the repair mechanisms, is a result of evolution and adaptation over millions of years, and as both the destruction and repair mechanisms are so similar in quite different biological systems, some general principles must be involved.

We do not know the strength of these repair systems. Are they strong enough to take care of increased UV damage in cells? I believe no one can give the correct answer. At any rate, when we manipulate with UV doses in the spectral region 320 to 290 nm we are touching a very sensitive biological system indeed and utmost care should be taken when doing so. Increasing UV doses may upset the positive balance which now evidently exists between UV damage and repair mechanisms in natural sunlight. The whole boplogical world, so dependent upon microorganisms, may, if doses increases, be in serious trouble.

Underwater UV Radiation

Under Conclusions (above) solar radiation in air or on organisms living on surface waters was considered. Many such examples of this can be cited. Blue-green algae frequently float on surface films in fresh water and in tropical oceanic waters, the blue-green alga Trichodesmium often gives the water a red colour. Most phytoplankton forms are also at times brought to the surface by mixing of water masses.

When penetrating water, solar radiation undergoes great changes both in intensity and spectral energy distribution. One should, however, note that clear oceanic waters are highly transparent to ultraviolet radiation, at 300 nm as much as 80% of the radiation is transmitted per meter. This means that in the open sea UV radiation penetrates to considerable depths and acts on both proteins and DNA, though the doses will be considerably lower than those in air. The effects of UV radiation should not, however, be neglected in these waters.

In coastal waters the situation is different due to "yellow substance" and brown and yellow humic components. This situation is also true for most inland lakes and rivers. In these water bodies the organisms living there have a much higher protection against UV radiation and an increase in the UV dose around 300 nm in air will have little effect here even in shallow waters. One must, however, keep in mind that even in these water masses mixing occurs and most phytoplankton will at times be in the surface layer of the water. This short exposure to UV radiation is probably of little importance for the cells, and the UV damaging effect is unlikely to occur. This deduction is based upon results from laboratory experiments where it can be shown that a minimum UV dose must be given before any damage is noticeable.

SUMMARY

Terrestrial solar radiation extends towards shorter wavelengths down to between 290 and 304 nm depending upon latitude and sun direction. This lower wavelength limit is set by the ozone layer 10 to 40 km above the earth's surface.

Oceanic water types have high transmittance for UV radiation down to 300 nm. In clear oceanic water the transmittance is as high as 80% per meter. UV radiation below 350 nm is effectively absorbed in coastal waters and in lakes and rivers.

UV radiation in the spectral region between 320 and 380 nm has photobiological effects similar to those of visible light. Photosynthesis is stimulated as well as growth, nucleic acids and proteins are not destroyed by radiation in this spectral region since no absorption takes place there.

Between 290 and 320 nm UV damage to biological systems takes place. Proteins are denaturated and DNA photoinactivated. In phytoplankton these damages are seen as cell death, reduced or inhibited photosynthesis, destruction of the motor apparatus in flagellated forms.

According to results obtained in the laboratory these damages must take place in the air and on water surfaces. In clear oceanic water masses too where the transmittance at 300 nm may be as high as 80% per meter such damage may take place down to a considerable depth.

The protein denaturation may be of a different nature, one of the most important is the altering of the tertiary structure.

The most important result of UV action on DNA is the formation of thymine dimers.

When these are formed the two DNA strands cannot separate when the cell approaches cell division, resulting in cell death. Other DNA photoproducts are also known.

Repair mechanisms exist for UV damage in both proteins and DNA. If gentle-UV damaged cells are removed from environments containing damaging radiation (UV) dark repair mechanisms are activated at least partly restoring both protein structure and DNA function.

DNA damaged by UV radiation under the formation of thymine dimers can be repaired by the process of photoreactivation. In this process an enzyme with a yellow chromophore group releases thymine dimer bonds and restores thymine monomers. It has been shown in model systems that the process of DNA photoinactivation takes place simultaneously with the process of photoreactivation. In microorganisms including phytoplankton, under natural conditions during sunshine, UV radiation continuously destroys DNA by forming thymine dimers while the process of photo-reactivation simultaneously splits these thymine dimers and forms monomers.

The destructive effect of solar UV radiation or biological material is evidently dependent upon different types of repair mechanisms for survival. The boundary between harmful and non-harmful or stimulating quanta in the spectral region where terrestrial solar radiation has its short wavelength limit around 290 nm is very sharp. The boundary is common for most organisms so far tested and the span between different materials is perhaps not more than 10 to 15 nm, between 295 and 310 nm.

Different biological systems evidently have repair mechanisms to adjust for UV damage. These systems for repair seem to be rather similar in nature. It is therefore reasonable to assume that these systems have been destroyed during evolution over millions of years to meet the requirement of protection from the damaging effects of UV radiation from the sun. We do not know the strength of these systems. Can they stand a higher load than they have been developed to take of protein and DNA repair caused by UV damage today?

It is reasonable to assume that evolution has developed a suitable balance between damage and repair in the cells to meet the requirements on earth as they are today. If the ozone layer is reduced this will not only affect the total dose of UV radiation on earth; at the same time the shorter wavelength limit of solar radiation will move towards shorter wavelengths, while simultaneously increased absorption by nucleic acids and proteins will take place as the absorption of these two macromolecules increases sharply towards shorter wavelengths. A reduction of the ozone layer will thus have a doubly negative effect on microorganisms including phytoplankton.

The organisms will be exposed to increased doses of damaging UV radiation. In addition these doses will be more effectively absorbed as the wavelength shift of the shorter wavelength limit of solar radiation is in an unfavourable direction causing more effective absorption of damaging radiation by proteins and nucleic acids.

REFERENCES

Bainbridge, R., Evans, G.C. and Rackham, O. (eds.) 1966. Light as an Ecological Factor. Blackwell, Oxford.

Giese, A.C. (ed.) 1964-1973. Photophysiology Vol. I-VIII. Academic Press.

Halldal, P. 1958. Pigment formation and growth in blue-green algae in crossed gradients of light intensity and temperature. Physiol. Plant., 11: 401-420.

Halldal, P. 1961. Photoinactivation and their reversals in growth and motility
    of the green alga Platymonas. Physiol. Plant. 14: 558-575.

Halldal, P. 1964. Ultraviolet action spectra of photosynthesis and photosynthetic
    inhibition in a green and a red alga. Physiol. Plant., 17: 414-421.

Halldal, P. 1966. Induction phenomena and action spectra analyses of photosynthesis
    in ultraviolet and visible light studied in green and blue-green algae, and in
    isolated chloroplasts. Z. Pflanzenphysiol. 54: 28-48.

Halldal, P. 1967. Ultraviolet action spectra in algology. Photochem. Photobiol.,
    6: 445-460.

Halldal, P., and Taube, C. 1972. Ultraviolet action and photoreactivation in
    algae. In Photophysiology Vol. VII: 163-188. Ed. A.C. Giese.

Henderson, S.T. 1970. Daylight and its Spectrum. London, 1970.

Hesstvedt, E. 1976. On the filtering effect of atmospheric ozone. Report No. 15
    from Institute of Physics, Univ. of Oslo, Norway.

PAPER 3

Effects of Changing Levels of Ultraviolet Radiation on Plant and Timber Production

Presented by the Food and Agriculture Organization

INTRODUCTION

The ozone layer is the principal shield preventing harmful short-wave ultraviolet
radiation from reaching the surface of the earth. Suggestions that man's activities
may be leading to a moderate reduction of the stratospheric ozone layer is of justifiable concern. Several threats such as exhaust emissions from high altitude aircraft and halomethanes diffusing up from the troposphere appear to have a sound
theoretical support to serve as a basis of concern. Several other threats such as a
contribution of NO_2 to the stratosphere from the widespread use of nitrogen fertilizers are more questionable though of sufficient magnitude to warrant further
investigation. The several potential causes of stratospheric ozone reduction would
have more or less additive effects. Ozone reduction would take place over long
periods of time and would not be quickly reversible.

The treat to higher plants used in agriculture and timber production is of particular concern since plants have evolved to be efficient collectors of solar radiation
which is of necessity in effective photosynthetic energy capture. This also maxi-
mizes exposure of plant leaves and shoots to solar UV radiation. Any threat to
plant productivity, no matter how subtle, is of concern in view of the existing
global demands for food and fiber production.

Unfortunately, it is presently impossible to predict the magnitude of the effect on
plant life due to any particular stratospheric ozone reduction. Most of the effects
on plants are of a detrimental nature. However, the increased UV radiation that
would occur with moderate reduction in stratospheric ozone concentrations would
probably result in rather subtle effects on plants. Therefore, it is very difficult to quantitatively assess the overall magnitude of the potential threat - particularly when contemplating global crop production or the productivity from forests.
Even though many effects would be subtle, this does not lessen the cause for
concern, particularly when a perturbation of global proportions and long time constants is being considered. This paper will present a problem statement as to the
nature of the effects presently anticipated and attempt to bring this threat into
context relative to the natural fluctuations to ozone that are already known to
occur.

---

*Paper compiled from the work of Prof. M.M. Caldwell and Dr. E. Wellmann.

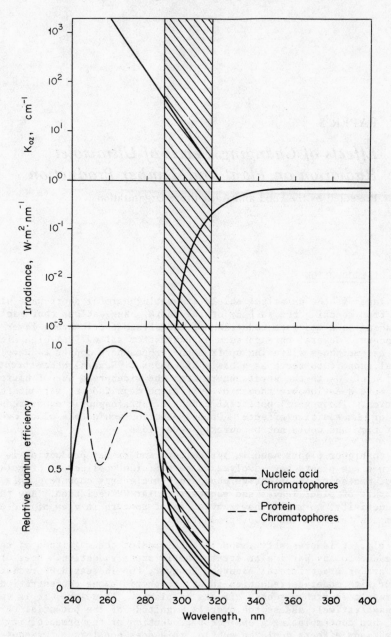

Fig. 3.1. Ozone extinction coefficient adapted from Green et al., [1];
the solar ultraviolet flux (direct sun plus diffuse sky radiation) at
ground level for conditions of 0.300 atm. cm ozone thickness and a solar
angle of 30° from the zenith taken from calculations using the model of
Green et al., [1] and measurements of Bener [3]; generalised action
spectra for biological responses to UV radiation involving protein
(dashed line) and nucleic acid (solid line) chromophores. The hatched
area designates that portion of the spectrum which is involved in the
stratospheric ozone depletion question. (Adapted from Giese [47]).

BASIC BIOLOGICAL IMPLICATIONS OF OZONE REDUCTION

The waveband of concern in the ozone reduction question is confined to a rather small part of the solar UV spectrum. Of particular potential detriment to biological systems are wavelengths shorter than 315 mm since nucleic acids and proteins have an appreciable absorption cross section in this waveband. The absorption cross section of ozone also becomes appreciable only at wavelengths shorter than 315 mm. With decreasing wavelengths, the ozone absorption coefficient increases exponentially (Fig. 3.1). Even though the average total atmospheric ozone column is only on the order of 3 mm if condensed to standard temperature and pressure (commonly denoted as 0.300 atm. cm), it is sufficient to effectively attenuate the solar UV flux to less than $10^{-3} W.m^{-2}.nm^{-1}$ at wavelengths below 295 nm. If the atmospheric ozone column were reduced to only 10% of this thickness, calculations using the model of Green et al. [1], suggest that UV irradiance at ground level, with the sun directly overhead, would still not include wavelengths shorter than 280 nm at intensities greater than $10^{-3} W.m^{-2}.nm^{-1}$. Thus, any consideration of biological effects of increased solar UV flux resulting from reduced atmospheric ozone must be confined to the waveband between 280 and 315 nm (commonly designated as UV-B) and for moderate ozone reduction only the 290 to 315 nm waveband would be of concern.

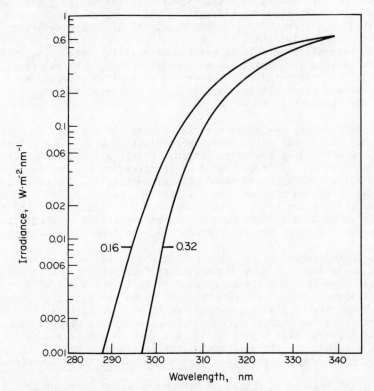

Fig. 3.2. Global (direct sun plus sky) spectral irradiance calculated for a solar angle of 30° from the zenith and atmospheric ozone concentrations of 0.32 and 0.16 atm. cm from the model of Green et al.

With changes in optical thickness of the atmospheric ozone column there is a predictable change in ground-level UV-B irradiance as illustrated in Fig. 3.2. As would be expected from the ozone absorption coefficient, a change in ozone thickness is reflected much more in spectral irradiance at shorter wavelengths. The greater

representation of shorter wavelengths with reduced ozone should be of biological
importance if the biological effectiveness of UV-B radiation corresponds to the
action spectra for effects as mediated by nucleic acid and protein chromophores as
shown in Fig. 3.1. To accommodate this relationship, a useful expression of biologically effective UV-B irradiance, UV-B$_{BE}$, can be defined as

$$UV-B_{BE} = \int_{280}^{315} I_\lambda \cdot E_\lambda \cdot d\lambda \qquad (1)$$

where I_λ is the spectral irradiance at wavelength λ, and E_λ is the relative biological effectiveness at wavelength λ as defined by Caldwell [2]. This effectiveness term closely approximates the action spectra for biological effects involving
nucleic acid and proteins as shown in Fig. 3.1 in the 280 to 315 nm waveband. This
function was, however, derived from a range of action spectra for both higher and
lower plants involving several physiological responses as shown in Fig. 3.3. Except
in the case of 3I, the chromophores were considered to be nucleic acid or protein
compounds. Despite the fact that action spectra can be modified by wavelength-
dependent filtration through plant tissues covering physiological targets, the
correspondence of this sundry of action spectra is reasonably good. In the case of
3I, which deviates most from the generalized spectrum, the chromophore was considered to be a plant hormone, indoleacetic acid. Although the number of available
action spectra for physiological phenomena of higher plants in the UV-B is quite
limited, some generality of this E_λ relationship still seems reasonable. When the
effects of decreasing atmospheric ozone concentrations on UV spectral irradiance is
calculated [1] and integrated with respect to biological effectiveness according to
equation 1, the net effect is an approximate 2% increase in UV-B$_{BE}$ for each 1%
decrease of the atmospheric ozone column.

Although UV photobiology has progressed far during the last few decades, this research has only occasionally involved higher plants and has been primarily concerned with the effects of 254 nm radiation. Nearly all monochromatic radiation of
this wavelength can be easily generated by low-pressure mercury lamps and this
conveniently falls within the waveband of maximum inactivation cross section of
nucleic acids. Although much of what has been learned (particularly the elucidation of various radiation repair mechanisms) may also likely apply to the response
of higher plants to UV-B radiation, a simple extrapolation or prediction of ecological consequences of ozone depletion cannot be made from this information.

Not only are there appreciable differences in the quantitative responses to UV-B
radiation as opposed to 254 nm radiation as is indicated by the action spectra for
physiological responses, but there may also be qualitative differences in the
patterns of response as has been demonstrated by Bachtwey [5] for phytoplankton.
Furthermore, most higher plants have apparently evolved defense mechanisms to minimize the UV-B radiation insult. These include absorption of UV-B radiation in outer
tissue layers and molecular repair mechanisms such as photoreactivation and dark
repair systems. Thus, of ultimate biological interest is the net unrepaired UV-B-
induced damage that would result in nature if the ozone layer were diminished.

SIGNIFICANCE FOR PLANTS OF THE UV-B PORTION OF SUNLIGHT

The life of plants depends on different metabolic (cf. photosynthesis) or developmental (cf. photomorphogenesis) processes which are known to be regulated by certain spectral ranges in the visible radiation. Compared to total sunlight the UV-B
(280 - 320 nm) portion is too small to be a factor in these basic photoreactions.
Furthermore one must consider that UV-B radiation is strongly absorbed at the plants'
surface and will not penetrate either to the inner cells of green leaves or of other
plant organs which show reactions to light. So far there is no evidence for UV
being essential for normal development of any higher plant species. Damaging effects

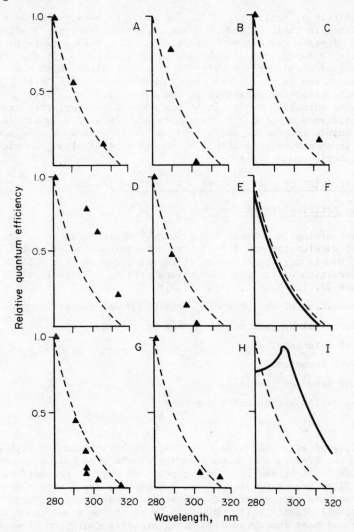

Fig. 3.3. A generalized plant UV-B action spectrum relative to 280 nm
(dashed line) as it corresponds to action spectra data for: A, mutation
of liverwort spores, *Sphaerocarpus donnellii* [6]; B, mutation in the
fungus *Trichophyton mentagrophytes* [7]; C, inhibition of photosynthesis
in *Chlorella pyrenoidosa* [8]; D, cessation of cytoplasmic streaming in
epidermal cells of *Allium cepa* [9]; E, frequency of "endosperm deficiencies" in maize due to chromosomal deletion or mutation [10]; F, germi-
dical action spectrum [11]; G, epidermal cell damage in *Oxyria digyna*
leaves [12]; H, induction of chromosomal aberrations in *Tradescantia
paludosa* pollen [13]; I, base curvature of *Avena coleoptiles* [14]. All
of the action spectra have been expressed in terms of relative quantum
efficiency which denotes either relative quantum effectiveness (spectra
A, B, E, F, H and I) or relative quantum sensitivity (spectra C, D and
G). (From Caldwell [2]).

which can definitely be explained by the influence of sunlight-UV have not been
observed on plants in their natural habitat. The few harmless UV-effects known in
a few specific plants could either be related to UV stress effects or to protection
against possible UV damage and will be discussed in detail. It will be shown that
plants have evolved highly effective protective mechanisms which allow them to
tolerate permanent and in many cases even unreduced sunlight. Plants should therefore prove essentially less sensitive to increasing levels of sunlight UV than man
and possibly some animals. It should be emphasized, however, that present knowledge on UV resistance of plants is very limited. As will be specified in this
review we are unable to predict definitely the consequences of even a two- or three-
fold increase in shortwave sunlight UV which could result from man made disturbances
of the stratospheric ozone layer.

EVALUATION OF UV-B RADIATION EFFECTS SO FAR OBSERVED IN PLANTS

UV-Effects from Artifical Light Sources

Most of the photobiological work dealing with UV-actions has been carried out using
UV radiation of wavelengths and intensities not representative of the natural solar
UV radiation. Quartz-mercury lamps emitting monchromatic UV at 254 nm have been
used for the induction of various deleterious effects. Typical stress phenomena
observed include [2, 12, 15, 16, 17, 18, 19]:

- strong accumulation of phenolic or flavonoid like pigments;

- loss of chlorophyll by indirect enzymatic destruction;

- occurrence of mutations;

- loss of growth hormones;

- inhibition of seed germination;

- disturbances in ion uptake and transport;

- reduced and irregular growth.

These and similar effects will also occur with unphysiologically high doses of
radiation from the UV-B range, because as a rule the same cell compounds (proteins
or nucleic acids) will be the primary receptors of UV. No conclusions concerning
the limits of UV tolerance of plants can, however, be drawn from these experiments.
UV radiation has been demonstrated to elicit detrimental effects resulting in reduced photosynthetic capacity [20, 21, 22]. Most of the research on UV-effects on
photosynthesis has been conducted with high intensity radiation from germicidal
lamps at 254 nm. UV primarily affects not physiological processes such as photosynthesis per se but specific cellular constituents. Studies with algae and isolated chloroplasts yielded fundamental information on mechanisms of UV induced injury did not, however, reveal what will happen to intact plants in their natural
environment, as plants are able to avoid UV injury to internal parts by physical or
chemical shielding. Considerable additional research is needed in order to better
define the effects of UV on photosynthesis and the potential changes in agricultural productivity.

UV-Effects from Natural Solar Radiation

There is so far no evidence that solar UV is a limiting factor for any plant species
in its natural environment. For a long time the stout habit and intense pigmentation of alpine plants were believed to be the result of intense UV irradiation;
however, other environmental factors of this extreme habitat (e.g. low temperatures,
high total solar irradiance) and also genetic reasons have now been found to explain
these phenomena [2, 12]. Solar UV-B radiation is known to produce specific effects
on the synthesis of phenolic and flavonoid pigments like anthocyanin. This

UV-induced pigment formation will be discussed as a protective mechanism. The UV part of sunlight is thought to cause reduced or, in some cases, abnormal growth of several wild and cultivated plants, under extreme growth conditions. UV is in particular active in interaction with additional stress factors like low temperatures, distinct UV effects being observed in high intermountain arid regions [19, 23, 24]. These experimental results cannot be interpreted in terms of potential effects of certain UV levels, because quantitative interrelationships between UV loads and effects are unknown. Furthermore, the experimental conditions (deletion experiments) were of questionable validity.

UV-EFFECTS TO BE EXPECTED WITH USEFUL PLANTS IN CONSEQUENCE OF REDUCED LEVELS OF STRATOSPHERIC OZONE

Experimental Limitations

Experiments to answer the question of whether an increase in UV-B radiation flux can pose a problem to plant production are difficult to perform for the following main reasons.

- Large scale supplementation experiments (selectively enhancing UV-B loads) have so far not been carried out in the field because of technical difficulties and financial limitations (energy requirements).

- Deletion experiments in the field have been performed but are of questionable validity for evaluating UV-tolerance. Concomitant reduction of radiation from the visible range will result in side effects of photosynthesis, morphogenesis or photoreactivation (UV-protection).

- Most investigations of the effectiveness of UV have been carried out with single plants or plant organs under laboratory conditions. In order to answer the posed question, however, climate has to be controlled simulating natural growth conditions. Furthermore, this type of experiment only describes the UV action on the most exposed parts of a vegetation system. Plants as a rule grow in communities (wild plants in ecological niches) partially sheltered from continuous direct beam solar radiation and in part from skylight as well. Plant species adapted to low intensity light, the UV part of which had been filtered out by shading leaves of other species, might be more susceptible to changes in UV radiation than plants thriving in high radiation regimes.

- If irradiations are carried out using solely UV in the absence of radiation of longer wavelenghts, plants would be deprived of photoreactivation and other photoprotective mechanisms. Therefore, damage might become apparent which would not have otherwise. On the other hand supplemental irradiations with light sources emitting besides UV large amounts of longer wavelength radiation might enhance photoreactivation to an extent not to be achieved under natural sunlight conditions. Irradiation techniques are badly in need of improvement [25].

Potentially UV-Sensitive Target Areas of Plants in Different Developmental States

In order to estimate potential UV damage on plants, UV sensitivity throughout onto-genesis has to be taken into account. This has not been considered in former investigations. During seed germination, the root until it has penetrated the ground, should be as UV-sensitive as other meristems or reproductive tissues. In many plants vegetative buds or floral primordia are highly exposed to high radiation loads. Pollen typically not distributed by wind may be susceptible. Special interest should be cast upon potential UV-effects on stomatal movement. Since the stomata of many plants including agricultural crops are also located on the upper leaf surfaces, the guard cells surrounding the stomata being the only photosynthetically active cells directly exposed to UV. The guard cells would be the most vulnerable of all chlorophyll containing cells. The light reactions of photosynthesis in these cells may be required to initiate stomatal opening. Even if UV does not directly affect photosynthesis, the overall efficiency of photosynthesis could be reduced as a result of CO_2 uptake limited by UV promoted stomatal closing.

In order to approach the problem of whether plant productivity will suffer from increased UV loads, susceptible targets have to be identified and analysed as to how far they would intercept radiation in the vegetative system.

Differential UV-B Sensitivity of Higher Plant Species

Another complicating factor in the assessment of the potential impact of reduced stratospheric ozone is the apparent differential sensitivity of higher plants to UV-B radiation.

Following the suggestions during the 1972-73 period that high altitude aircraft exhaust emissions might catalyze a partial destruction of the stratospheric ozone layer, the U.S. Department of Transportation initiated the Climatic Impact Assessment Program. Although the program largely revolved around atmospheric chemistry and dynamics, a later phase involved biological studies to investigate the potential consequences of partial atmospheric ozone reduction. The duration of the actual experimentation was rather brief — on the order of 18 months. Although necessarily of very limited scope, these preliminary studies provided some valuable information. It was demonstrated that UV-B radiation can elicit a number of detrimental plant responses such as reduced photosynthesis and plant growth, enhanced stomatic mutation rates, decreased pollen germination and some plant growth pattern abnormalities. Most of these effects were the result of rather intense UV-B irradiation corresponding to a much more severe ozone reduction than is presently contemplated. It would be very imprudent to attempt to extrapolate from these results to a consideration of the potential consequences of a moderate reduction of the ozone layer.

These studies also indicated that there is a substantial variation in the basic sensitivity of different higher plant species to UV-B radiation. Earlier studies by Cline and Salisbury [16] which employed very intense 254 nm radiation, also indicated substantial differences in species sensitivity to the radiation.

From the various experiments it is difficult to generalize about the basis of the differential species sensitivity to UV-B radiation. Although many of the sensitive species tended to be broadleaf species such as some of the leguminous crops like pea and bean, there is no general categorization that could be applied. Many graminoids were often less sensitive. However, several important graminoides of agricultural interest such as maize and rice displayed sensitivity in some of these experiments. The basic reasons for the differences in sensitivity between species are not at all clear. There were no striking correlations between sensitivity and parameters such as taxonomic affinity, growth form, or other overt plant characteristics. The differential sensitivity of species must, however, relate to their differences in capacity for coping with UV-B radiation.

The degree to which variations in optical properties of plants, such as leaf epidermal UV-B transmittance, and the efficacy of repair systems contribute to the apparent differential sensitivity of higher plants has not yet been elucidated. Nor have extensive studies been undertaken to evaluate the acclimatization potential of individual plants to accommodate increased intensities of UV-B irradiance. This acclimation might be elicited by prior exposure to UV-B radiation or perhaps other environmental factors which might lead to a heightened tolerance of UV-B radiation.

Because of the tremendous diversity of plant life — particularly in non-agricultural ecosystems, some predictors of UV-B radiation sensitivity must be developed. It is inconceivable that even in an ambitious screening program that but a very small sample of the Earth's flora could be empirically tested for UV-B sensitivity. An assessment of the impacts of atmospheric ozone reduction on plant life should obviously be centered around those species which would be most sensitive to change in UV-B radiation. Thus, elucidation of the basis of differential sensitivities of species should be a very high priority.

Although the recent preliminary experiments undertaken during the Department of Transportation program did not include any forest species, a limited number of coniferous species investigated in the study of Cline and Salisbury [16] indicated many of these trees to be highly resistant to short term exposures to the very actinic 254 nm radiation. However, this should not be taken as an indication that forest productivity would not be affected in the event of stratospheric ozone decrease. Chronic exposure to even moderately increased UV-B radiation may be of consequence for coniferous species. The fact that most conifer species do maintain photosynthetically active needles for long periods of time would provide the opportunity for the accumulation of a substantial total UV radiation dose as was discussed in the earlier section.

Interactions of UV with Other Stress Factors

UV radiation has to be regarded as one of several stress factors acting on plants in the environment. If a plant has to cope with another factor such as temperature stress or infection, an additional UV load might result in a detrimental response, while the same UV load was tolerated by the plant under optimal growth conditions.

In this context one positive prospect of an increased UV irradiation should be mentioned. It could be that some plant pathogens are damaged by UV to a greater extent than hosts themselves, or UV might induce alterations within the host (e.g. formation of antibiotic substances or morphological changes decreasing the susceptibility to pathogens. There are some minor experimental hints supporting even such a conception of (some) potentially beneficial consequences of an increase in solar UV radiation [26, 27, 28, 29, 30]). Susceptibility of bean plants to virus [21] was found to be reduced and tobacco mosaic virus-RNA to be inactivated by UV [31, 32]. In any case interactions of this sort must be considered as possible complications for the interpretation of UV effects.

Experimental Evidence for Potential UV Effects

In a series of recent laboratory studies, sponsored by the U.S. Department of Transportation, supplemental UV-B irradiations were used to simulate the effect of a 50% reduction of ozone [33, 34]. From these investigations it can be concluded that, in response to the increased UV load, some agricultural plants (in certain developmental states) will be reduced in growth and in their capacity to photosynthesis. Seedlings were found to be more sensitive to UV than are mature plants. Great differences in UV-tolerance were observed among different species. The experiments, however, did not conclusively enough predict far-reaching consequences to increased levels of UV radiation on plant production. Nevertheless, these results support the view that an increased UV flux could pose a problem to agriculture.

It must be taken into consideration that unreduced plant productivity will require UV resistance not only of the mature plants. In many cases high germination percentage and vigorous seedling growth are essential for economical plant production. These young developmental stages might also suffer from UV stress. The influence of UV-B radiation from FS40 fluorescent sun lamps on germination and seedling development has recently been studied by Krizek [35]. The UV effect was analysed on tomato, radish, cucumber, lettuce and bean and on field crops, such as wheat, cotton, soybean and millet. Germination was found to be not significantly influenced by UV. Abnormal seedling growth, however, was observed in response to continuous exposure to UV alone with all species but wheat. Seedlings were not

grown under field conditions. Photoreactivation or photoprotection mechanisms
could not become effective during UV treatments. UV-loads applied to the plants
were much greater than those projected for a 50% decrease in stratospheric ozone.
From these results one can only conclude that wheat seedlings will be extremely UV
resistant and that seed germination of all tested species will not be effected by
increased UV irradiation.

Cotton seedlings turned out to be very sensitive to UV stress [20]. Laboratory
studies of early seedling growth and of translocation of radiozinc from cotyledons
showed that slight damage of the seedlings might occur under projected decreases in
the ozone layer. These results also must be reproduced under field conditions using
better simulated UV-B solar irradiation.

Two effects which are of particular ecological import are reductions in photosynthetic capacity and curtailment of leaf once expansion, both of which would contribute to an impaired carbon economy of the plant. Studies recently completed by
W.B. Sisson (M.M. Caldwell, personal communication) with *Rumex patientia* exposed to
various levels of polychromatic UV-B irradiance corresponding to solar UV-B
irradiance at various concentrations of atmospheric ozone, provide some indication
of the pattern of response to UV irradiance. For impairment of photosynthesis,
response to UV-B_{BE} was quite nonlinear and there did not appear to be a particular
threshold of UV-B_{BE} to which the plant must be exposed before photosynthetic
damage ensured. Instead, reduction of photosynthetic capacity appeared to be only
a function of the total accumulated exposure to UV-B_{BE} and even UV-B irradiance
corresponding to that under present-day ozone conditions resulted in some impairment of photosynthesis relative to control plants exposed to no UV-B radiation.
Reciprocity in the response to UV-B_{BE} applied over a considerable period of time
(5 to 45 days) for individual leaves exposed to 3 levels of UV-B_{BE} irradiance.
Thus, damage to the photosynthetic system appears to be accumulative over a
considerable portion of the ontogeny of individual leaves and is apparently only a
function of the total UV-B_{BE} to which the leaf is exposed. In terms of leaf
enlargement, reciprocity did not apply over long periods of time. Instead, UV-B_{BE}
was most effective in curtailment of leaf expansion during only a period of two or
three days during early developmental stages. The response was, however, a function
of the level of UV-B_{BE} irradiance during this short period of time. The reduction
of leaf expansion rates appears to be due to effects of UV radiation apart from the
simple limitation of photosynthate supply [36].

This study suggests solar UV-B irradiance, even under present-day conditions, may
be a significant ecological factor for *Rumex patientia*. This species occurs in
open habitats exposed to full insolation and there is a minimal degree of self-
shading among leaves of the same plant. Individual leaves of this species are
active for periods of time sufficient to accumulate appreciable exposure to solar
UV-B radiation. The degree to which other higher plant species would correspond to
this pattern of reduction in growth and photosynthesis as a function of UV-B
irradiation is difficult to predict. *Rumex patientia* was selected in this study because it is both sensitive to UV-B radiation and exposed to full sunlight in nature.
Many other higher plant species are, however, much less sensitive to UV-B radiation
[37]. Nevertheless, if the accumulative mode of damage to the photosynthetic
apparatus is generally applicable to higher plants, even moderately UV-B resistant
species may be affected by UV-B radiation in the event of ozone reduction if such
species possess long-lived leaves which are exposed to full sunlight.

The effects of UV-B irradiance on photosynthetic capacity have been shown to be
associated with the nonstomatal component of photosynthesis [36]. That is, reduction of photosynthesis was not due to a simple increase in the diffusion resistance
to CO_2 entering the leaf through the stomates. Instead, the damage appeared to be
associated with the photosynthetic apparatus itself and related studies have shown
that electron transport associated with the photosystem II may be most affected.

Since stomatal diffusion resistance did not change, the capacity for transpirational water loss would remain the same; therefore, the water use efficiency of photosynthesis (photosynthesis/transpiration ratio) would also become less favorable for a sensitive plant such as *Rumex patientia* exposed to UV-B radiation.

As there is no more relevant experimental data available, the following points of view need to be carefully considered for an evaluation of the consequences of an increasing level of UV radiation on utilized plants:

- Eliminating of one seemingly insignificant member of an agricultural or natural ecosystem might result in serious alterations in such balanced communities. For timber production natural systems are utilized which are not completely controlled by man and therefore might be indirectly influenced by UV in an unexpected way. Long-term development of a highly UV resistant tree might be limited by UV damage of some inconspicuous but essential organisms of the ecosystem.

- If it turns out that certain effects of UV damage accumulate over a long time, as has been supposed by Caldwell (Book of Abstracts VII, International Congress on Photobiology, 58, Rome 1976) for the repression of net photosynthesis, even a slight increase in UV will become serious for plant productivity or maybe for survival by changing the competitive balance between species.

- As a rule, cultivated plants are grown under conditions far from those of their natural habitat. That means plants will often not be adapted to UV loads to be tolerated in the field. For example, wind conditions may expose certain sensitive phyto-elements to solar UV radiation that ordinarily would remain sheltered under calm conditions in another region. Distances in crop stands might be of importance for UV-tolerance.

- It has to be considered that cultivated plants might have lost UV protective mechanisms by selection.

- There are some indications that UV might be much more damaging when plants are under competitive stress.

- Productivity of certain useful plants depends on effective pollination by insects. Many insects are known to utilize UV for orientation. In case of altered UV intensities orientation might be disturbed and, therefore, the effectiveness of insect pollinators and maybe also of predatory insects be reduced.

- It should be pointed out that knowledge about UV effects on aquatic (especially of marine) plants is of greatest importance, because primary production of the ocean depends on photosynthetic activity of these organisms. Clean water is not effective in absorbing UV-B radiation. Under laboratory conditions some single celled algae proved to be extremely sensitive to UV damage [33, 34, 38]. There is no information at all about potential UV-effects upon planktonic organisms living in UV-transparent natural environments.

- Although the relationship between ozone reduction and UV-B_{BE} has been calculated to show an approximate 2% increase in UV-B_{BE} per 1% decrease in ozone thickness, the physiological responses of plants reflecting the net unrepaired radiation damage is not necessarily a linear function of UV-B_{BE}. Thus, the question of the basic dose-response relationships must be addressed. Several related questions are also pertinent: Is there a threshold level of UV-B_{BE} to which plants must be exposed before physiological manifestations of unrepaired damage appear? Do plants respond to the total accumulated exposure of UV-B_{BE} or are the physiological manifestations better correlated with the level of UV-B_{BE} irradiance during rather brief periods of time such as during critical developmental stages.

Experimental data thus far available are not strong enough to predict severe damage to plants in response to increased solar UV irradiance. Even if plants will survive, however, there is good evidence that certain useful plants under unfavourable

growth conditions will be influenced by UV in such a way as to result in reduced
productivity.

PROTECTIVE MECHANISMS AGAINST UV DAMAGE IN PLANTS

Plants vary widely in the extent of their response to UV radiation, depending on
morphological, biochemical or physiological variables. Plants have evolved adaptive
mechanisms for utilizing certain beneficial photoreactions at the same time protecting themselves against others. UV damages biological systems primarily through
destruction of specific cellular constituents rather than alteration of broad
physiological processes. It must be considered that the UV portion ($\lambda < 320$ nm) most
influenced by changes in ozone concentration is also that most effectively absorbed
by nucleic acids and proteins. Plants have developed three principle protective
mechanisms to withstand UV damages:

- Plant organs or cell organelles are able to avoid high intensity light by orientation and UV sensitive tissues (meristematic and generative cells) are located
  several cell layers below the screening plant surface. Susceptible tissues like
  young vegetative buds, only covered by embryonic leaves, may be protected if
  growth takes place at night or in periods of light when the UV portion is low.
  Basic to this concept of avoidance would be plant processes altered in a positive
  manner by UV itself. Plants can respond to UV, selectively inducing physiological
  processes that might be linked to survival. Base curvature of Avena coleoptiles
  [14] and root elongation of several Umbelliferae seedlings [39, 40, 41] were
  found to be specifically mediated by UV with a maximum response at about 295 nm.
  This type of UV effects have not yet been analysed in other plants.

- Plants are able to shield themselves by coverings and pigments, mechanisms which
  at the same time often serve other purposes such as protection from infections
  or (from) oven transpiration. UV is absorbed and reflected by extracellular
  compounds such as cuticular substances and cell wall materials. Optical
  properties of such extracellular material vary considerable within single plant
  species or even within different parts of one plant, and will only in exceptional
  cases be a main factor in UV resistance [2, 42, 43]. The thickness of an epidermis appears to influence the amount of UV radiation that penetrates cells
  lying below, but the biochemical potential of the epidermal cell will have a
  greater influence on the transmittance of UV. Transmittance through the epidermis of most higher plant leaves is on the order of 80 or 90% in the visible
  part of the spectrum and typically decreases to less than 10% in the UV-B portion
  of the spectrum [12, 44]. Although the epidermis itself must be considered a
  physiological target of some importance, this selective absorbance of UV-B
  radiation by the epidermis undoubtedly reduces the impact of UV-B radiation on
  mesophyll tissues. Epidermises, when exposed to radiation in some cases quickly
  become impervious for UV due to the formation of anthocyanins or other flavonoid
  or phenolic compounds [2, 12]. There are a few cases where flavonoid synthesis
  can be induced by UV itself [2, 12]. A very sensitive UV effect on flavonoid
  formation was observed in parsley seedlings [39-41, 45] with a maximum quantum
  efficiency at wavelengths near 300 nm. A linear relationship was demonstrated
  between UV dose and the amount of pigments. General distribution in the plant
  kingdom of highly effective defense mechanisms like this needs to be examined.

- Reactions which reconstitute UV damaged nucleic acid molecules (esp. DNA) to the
  original state have been termed "repair mechanisms". Such processes have been
  shown in a wide variety of species and can be presumed to operate in all organisms [2, 4, 28, 38, 46]. There are several different mechanisms of "dark
  repair" [47] (not molecularly understood in higher organisms) that might be of
  importance in reversing UV damage. "Photoreactivation" is of special interest
  with aspect to potential UV effects on plants, and must be considered as an
  interfering factor in all irradiation experiments. Photoreactivation is the
  reversal of UV damage to DNA the basic genetic constituent of cells and is

brought about by radiation of longer wavelengths (blue-violet spectral range) of sufficient intensity. This energy is needed for the activation of the specific enzyme responsible for the repair process.

Photoreactivation has been demonstrated in a variety of higher plants for several physiological manifestations of UV irradiation at 254 nm [2, 46]. A few studies also have demonstrated that some UV-B induced damage can be photoreactivated [12]. It is quite likely that these molecular repair systems such as photoreactivation are of survival value to higher plants under present-day ozone conditions rather than being simply a vestigial element lingering from earlier stages of the evolution of plant life during that time when the earth's ozone layer was only just forming.

This importance of photoreactivation for plants growing under a normal solar irradiation regime was implied in earlier experiments where leaves of several alpine plant species were exposed to 297 and 300 nm radiation with intensities controlled to closely simulate UV-B$_{BE}$ on a summer day at alpine elevations [12]. During and following this irradiation, the leaves were not exposed to any other wavelengths. Some of these leaves did exhibit severe lesions indicating UV-B-induced damage which had never been observed under field conditions. The absence of such lesions in nature might suggest the importance of photoreactivation of UV-B-induced damage as an effective protective mechanism. Photoreactivability of these UV-B-induced lesions was also shown in these studies.

It should be pointed out that photoprotective mechanisms [48] and photoreactivation respond to visible light. Since UV and visible are under present conditions highly correlated, the avoidance responses to one serve as mechanisms for avoiding the other. The increase in UV, however, which would result from a decrease in ozone would be disproportionately large in relation to total solar radiation and therefore protective reactions would not be adapted to the increased UV level.

ECOLOGICAL IMPLICATIONS OF OZONE REDUCTION

Although there is little doubt as to the destructive potential of solar UV-B radiation for at least senstive higher plant species, the probable ecological impact of increased UV-B radiation resulting from decreased mean atmospheric ozone concentrations is difficult to assess. For agricultural or forest species which are sensitive to UV-B radiation, it is very conceivable that even the increase of UV-B radiation that would result from moderate ozone reduction may have effects on productivity. Since both photosynthesis and leaf expansion are critical to plant productivity and are processes readily affected by UV-B radiation, impacts on plant production and yield may be expected if the species in question are sensitive to UV-B radiation.

Because higher plant species do appear to be differentially sensitive to UV-B radiation another consequence of changes in UV-B radiation may be a shift in the competitive balance among species in nature.

With ozone reduction, even small depressions of photosynthesis, growth, or water use efficiency of photosynthesis caused by exposure of sensitive species to increased UV-B radiation may under some circumstances shift the competitive balance in favour of more UV-B-resistant species. Recent experiments in M.M. Caldwell's laboratory by F.M. Fox (personal communication) suggest that the competitive balance between species can be changed with even modest supplements of UV-B radiation.

Although it has never been documented or perhaps even much contemplated, some shifts in competitive balance between higher plant species may be taking place during the course of natural fluctuations in atmospheric ozone. Natural fluctuations of the ozone column occur over periods of days, months, and years. During the course of frontal storm system movements in mid-latitudes, the atmospheric ozone column over

a given locality can change by a magnitude of 20 to 40% in a few days time [49].
However, such short-term changes are probably of little influence since in most
cases plants would be responding to the average UV-B flux over a longer period of
time. (As was discussed earlier plants appear to be subject to an accumulative UV
dose.) Seasonal variations of 20 to 40% also take place in middle latitudes, the
high concentrations occurring in the spring and low concentrations in the fall [49].
However, because these changes are phased with the seasonal change of prevailing
solar angles from the zenith, they merely modify the present variations in monthly
average UV-B radiation at different times of the year. Although seasonal changes
of ozone probably exert little change on the competitive balance of species, the
seasonal timing of major periods of photosynthetic activity and growth of UV-B
sensitive plant species may have adjusted in the course of evolution to avoid, to a
certain extent at least, periods of particularly high UV-B flux during the year.

In terms of changes in competitive balance among plant species, it is the variation
of ozone over the course of several years which would be most influential. Long-
term records of atmospheric ozone concentrations are unfortunately very limited.
At only a couple of geographical locations do records covering even the past four
decades exist. Nevertheless, such records bear some witness to changes in mean
annual concentration for about 10% which appear to follow an approximate 10-to
12-year cycle [49, 50]. These changes would result in roughly a 20% change in
UV-B_{BE} over periods of half a dozen years which might result in some shifts in
competitive balance of species. Though evidence for changes over longer periods of
time do not exist, they might be expected. For example, if the periods of low ozone
during the 10- to 12-year variation are indeed linked to periods of low sunspot
activity, with appropriate lags, as has been suggested by Christie [50], periods
when there was an extended lull in sunspot activity such as has been suggested to
have occurred during the latter part of the 17th century [51] may have resulted in
an extended period of low ozone concentrations. Unfortunately, a retrospective
surmise as to the biological consequences of such extended periods of low ozone
concentration, or even during the course of the 10- to 12-year fluctuations, has
not been attempted and would undoubtedly be difficult, if not impossible.

The natural fluctuations of ozone concentration, particularly over the longer term,
do, however, suggest that higher plants, and ecosystems in general, must display a
certain resilience and ability to accommodate some change in UV-B flux. Although
the basic character of an ecosystem might show little change during the course of
long-term ozone fluctuations, it would not be unreasonable to expect some shifts in
species prevalence during these periods.

The magnitude of change over the 10-to 12-year cycle, i.e. 10%, is at least equal
to the ozone reduction anticipated from several man-induced depletions of atmospheric ozone. Nevertheless, the anthropogenic reduction of mean atmospheric ozone
would be superimposed on natural fluctuations. Therefore, periods of low ozone
concentrations in the course of natural variations, if compounded by a man-induced
reduction in mean ozone concentration, could result in exposure of ecosystems to
UV-B flux not yet experienced in particular regions.

Mean ozone concentrations also vary considerably with geographical location. This
is primarily a function of latitude, with lower concentrations toward the equator.
Although this natural ozone gradient has been useful in deducing possible trends in
human skin cancer, analogous inferences concerning higher plant response to UV-B
radiation are not so feasible. Plant response to this UV-B gradient is compounded
by the influence of other environmental factors such as temperature and moisture
supply on plant growth and success.

Since the ozone depletion question as it relates to ecological consequences is
restricted to a very narrow waveband of the solar UV spectrum (see Fig. 3.1), it is
perhaps not surprising that a background of scientific information upon which to

base an assessment does not exist. Due to the lack of particular scientific attention to this narrow waveband, compounded with the problems of generating UV-B flux
of appropriate intensities and wavelength composition, few studies dealing with UV-
B had been undertaken until the questions of man-induced ozone depletion came to the
fore. A concerted effort to place some perspective on the ecological consequences
of ozone depletion on the order of 15% or less is, however, urgent in the face of
current trends in stratospheric flight and the release into the troposphere of
agents which can be calculated to have potential ultimate consequences for the
global stratospheric ozone layer.

SUMMARY

The biological impacts of an increase in solar UV radiation as would result from
ozone destruction are evaluated preferably in consideration of potential effects
on economically important plants.

Although there are natural fluctuations in the ozone layer any man-induced reduction
of the ozone layer would be superimposed on the natural fluctuations and would be
justifiable cause for concern.

Proteins and nucleic acids are the compounds in living beings most susceptible to
UV radiation from the spectral range below 320 nm (UV-B). Where these compounds are
the UV receptors, biological effectiveness of UV will increase exponentially with
decreasing wavelengths. The effectiveness of atmospheric ozone in filtering out UV
is highest at lowest wavelengths of solar radiation. These are the two reasons why
even minor increases in the range of shortest wavelengths of sunlight in response
to ozone destruction might result in distinct - as a rule damaging - biological
effects.

For each 1% reduction in the atmospheric ozone column there would be a predictable
increase of approximately 2% in solar UV radiation when weighted for biological
effectiveness.

Plants are able to protect themselves very effectively from UV radiation
- by movement of organs (e.g. leaves) or cell organelles to avoid intense irradiation;
- by surface coverings (cuticula, cell walls, layers of dead cells); or
- by pigments accumulated in the outer cell layers (epidermis).

There are a few examples where pigment formation can be adapted to UV levels. In
many cases, however, plants respond to total radiation and not UV selectively for
protective reactions.

Finally all organisms are considered to possess a variety of molecular systems in
order to repair UV mediated disruptions of nucleic acids. Photoreactivation should
be very important for plants in preventing UV damage to DNA.

It has been found that plant species differ significantly in sensitivity. UV levels
which can no longer be compensated for protective mechanisms are unknown for plants
in natural or agricultural habitats.

Functioning of an ecosystem depends on the well balanced interrelationships between
different species and UV damage to one organism might finally result in dramatic
effects upon another one which is UV resistant. In this indirect way timber production might be negatively influenced.

Since both photosynthesis and leaf expansion are critical to plant productivity and are processes readily affected by UV-B radiation, impacts on plant productivity and yield may be expected if the species in question are sensitive to UV-B radiation.

So far there is no convincing evidence that any plant species would not withstand an increase in biologically effective UV corresponding to a 50% decrease in stratospheric ozone. Minor UV effects (e.g. on growth or photosynthesis) potentiated by additional stress factors might result in long-term consequences for the survival of certain plants and very likely for productivity of cultivated plants.

REFERENCES

[1]   Green, A.E.S., T. Sawada and E.P. Shettle 1974. Photochem. Photobiol. 19, 251.

[2]   Caldwell, M.M. 1971. Solar ultraviolet radiation and the growth and development of higher plants. Photophysiology (A.C. Giese, ed.) 6: 131-171. Academic Press, New York.

[3]   Bener, P. 1972. Technical Report, U.S. Army, London, Contract No. DAJA 37-68-C-1017.

[4]   Giese, A.C. 1964. Studies on ultraviolet radiation action upon animal cells. Photophysiology (A.C. Giese, ed.) 2: 203-245.

[5]   Nachtwey, D.S. 1975. In Impacts of Climatic Change on the Biosphere: Part 1 Ultraviolet Radiation Effects (D.S. Nachtwey, M.M. Caldwell and R.H. Biggs, eds.), pp. 3-50. U.S. Dept. Transportation, Wash,. D.C.

[6]   Knapp, E., A. Reuss, O. Riesse and Schreiber H. 1939. Naturwissenschaften 27, 304.

[7]   Hollaender, A. and C.W. Emmons, 1941. Cold Spring Harbor Symp. Quant. Biol. 9, 179.

[8]   Bell, L.N. and G.L. Merinova 1961. Biofizika 6, 21.

[9]   Glubrecht, H. 1953. Z. Naturforsch. B 8 17.

[10]   Stadler, L.J. and F.M. Uber 1942. Genetics 27, 84.

[11]   Lockiesch, M. Applications of Germicidal, Erythemal, and Infrared Energy. Van Nostrand-Reinhold, Princeton, New Jersey.

[12]   Caldwell, M.M. 1968. Solar ultraviolet radiation as an ecological factor for alpine plants. Ecol. Monogr, 38: 243-268.

[13]   Kirby-Smith, J.S. and D.L. Craig 1957. Genetics 42, 176.

[14]   Curry, G.M., K.V. Thimann and P.M. Ray 1956. The base curvature response of Avena seedlings to the ultra violet. Physiol. Plant 9: 429-440

[15]   Caldwell, M.M. 1972. Ecologic considerations of solar radiation change. In Proc. Second Conf., Climatic Impact Assessment Program. Cambridge Mass. (A.J. Broderick ed.). U.S. Dept. of Transportation, Rep. No. DOT-TSC-OST-73-4, pp. 388-393.

[16]   Cline, M.G. and F.B. Salisbury 1966. Effects of ultra-violet radiation on the leaves of higher plants. Radiat. Bot. 6: 151-163.

[17]   Cline, M.G. and F.B. Salisbury 1966. Effects of ultra-violet radiation

alone and simulated solar ultraviolet radiation on the leaves of higher plants. Nature 211: 484-485.

[18] El-Mansy, M.I. and F.B. Salisbury 1971. Biochemical responses of Xanthium leaves to ultraviolet radiation. Radiat. Bot. 11: 325-328.

[19] Lockhart, J.A. and U.B. Brodführer-Franzgrote 1961. The effect of ultraviolet radiation on plants. In Encyclopedia of Plant Physiol. 16 (W. Ruhland, ed.) pp. 532-554. Springer-Verlag, Berlin.

[20] Ambler, J.E., D.T. Krizek and P. Semenink 1975. Influence of UV-B radiation on early seedling growth and translocation of 65 Zu from cotyledons in cotton. Physiol. Plant. 34: 177-181.

[21] Benda, G.T. 1975. Some effects of ultraviolet radiation on leaves of French bean (Phaseolus vulgaris L.). Ann. Appl. Biol. 43: 71.

[22] Mantai, K.E., J. Wong and N.I. Bishop 1970. Comparison studies on the effects of ultraviolet irradiation on photosynthesis. Biochim. Biophys. Acta 197: 257-266.

[23] Foley, R.F. 1965. Some solar radiation effects on vegetable crops in Idaho: Proc. Amer. Soc. Hot. Sci 87: 443-338.

[24] Nilsen, K.N. 1971. Plant responses to near-ultraviolet light. Hort. Science 6: 26-29.

[25] Sisson, W.B. and M.M. Caldwell 1975. Lamp/filter systems for simulation of solar UB irradiance under reduced atmospheric ozone. Photochem. Photobiol. 21: 453-456.

[26] Anderson, R. and M.J. Kasperbauer 1973. Chemical composition of tobacco leaves altered by near-ultraviolet and intensity of visible light. Plant Physiol. 51: 723-726.

[27] Bridge, M.A. and W.L. Klarman 1973. Soybean phytoalexia, hydroxyphaseollin, induced by ultraviolet irradiation. Phytopathology 63: 606-609.

[28] Jagger, J. 1976. Effects of near-ultraviolet radiation on microorganisms. Yearly Review. Photochem. Photobiol. 23: 451-454.

[29] Maddison, A.C. and J.G. Manners 1973. Lethal effects of artificial ultraviolet radiation on cereal rust uredospores. Trans. Br. Mycol. Soc. 60: 471-494.

[30] Young, S.Y. and W.C. Yearian 1974. Persistence of Heliothis NPV on foliage of cotton, soybean and tomato. Environ. Entomol. 3: 253-255.

[31] Murphy, T.M. 1973. Inactivation of TMV (tobacco mosaic virus) RNA by ultraviolet radiation in sunlight. Int. J. Radiat. Biol. 23: 519-526.

[32] Murphy, T.M. and M.P. Gordon 1971. Ultraviolet irradiation of tobacco leaves: Inhibition of tobacco mosaic virus - ribonucleic acid photoreactivation. Photochem. Photobiol. 14: 721-731

[33] Climatic Impact Committee (1975). Environmental Impact of Stratospheric Flight. National Academy of Sciences, Wash., D.C.

[34]   Climatic Impact Assessment Program. Monograph 5. 1975. U.S. Dept. of Trans-
        portation, Wash., D.C.

[35]   Krizek, D.T. 1975. Influence of ultraviolet radiation on germination and
        early seedling growth. Physiol. Plant. 34: 182–186.

[36]   Sisson, W.B. and M.M. Caldwell 1976. Plant Physiol. 58, 563.

[37]   Biggs, R.H., W.B. Sisson and M.M. Caldwell 1975. In Impacts of Climatic
        Change on the Biosphere: Part 1 Ultraviolet Radiation Effects (D.S.
        Nachtwey, M.M. Caldwell, and R.H. Biggs, eds.), pp. 4–34. U.S. Dept
        Transportation, Wash., D.C.

[38]   Halldal, P. and O. Taube 1972. Ultraviolet action and photoreactivation in
        algae. Photophysiology (A.C. Giese, ed.) 7: 163–188. Academic Press,
        New York.

[39]   Wellmann, E. 1974. Regulation der Flavonoidbiosynthese durch ultraviolettes
        Licht und Phytochrom in Zellkulturen und Keimlingen von Petersilie
        (Petroselinum hortense Hoffm.) Ber. Deutsch. Bot. Ges. 87: 267–273.

[40]   Wellmann, E. 1976. Der Einflub physiologischer UV-Dosen auf Wachstum und
        Pigmentierung von Umbelliferenkeimlingen. In Industrieller Pflanzenbau
        5, Wien.

[41]   Wellmann, E. 1976. Specific UV effects in plant morphogenesis. Yearly
        Review. Photochem. Photobiol. 23.

[42]   Allen, L.H., Jr, H.W. Gausman and W.A. Allen 1975. Solar ultraviolet radia-
        tion in terrestrial plant cummunities. J. Environ. Qual. 4: 285–293.

[43]   Gausman, H.W., R.R. Rodriguez and D.E. Escobar 1975. UV radiation reflec-
        tance, transmittance and absorptance by plant epidermises. Agron. J.
        67: 720–724.

[44]   Lautenschlager - Fleury, D. 1955. Ber. Schweiz. Bot. Ges. 65, 343.

[45]   Wellmann, E. 1975. UV-dose dependent induction of enzymes related to
        flavonoid biosynthesis in cell suspension cultures of parsley. FEBS
        Letters 51: 105–107.

[46]   Cline, M.G., G.I. Connor and F.B. Salisbury 1969. Simultaneous reactivation
        of ultraviolet damage in Xanthium leaves. Plant Physiol. 44: 1674–1678.

[47]   Howland, G.P. 1975. Nature 254, 160.

[48]   Mohr, H. 1972. Lectures on photomorphogenesis. Springer, Berlin-Heidelburg-
        New York

[49]   Dütsch, H.U. 1974. Can. J. Chem. 52, 1491.

[50]   Christie, A.D. 1973. Pure Appl. Geophys 106, 1000.

[51]   Eddy, J.A. 1976. Science 192, 1189.

PAPER 4

Aircraft Engine Emissions

Presented by the International Civil Aviation Organization

THE NATURE AND RESPONSIBILITIES OF ICAO

The International Civil Aviation Organization (ICAO) is the specialized agency of the United Nations which deals with aviation matters. Its Member States at present number 135. The Chicago Convention on International Civil Aviation (1944) agreed on the principles by which international civil aviation could be developed in a safe and orderly manner and ICAO was the permanent body established to administer these principles.

The Organization's principal means of discharging this responsibility in the technical and operational areas of aviation is the publication of annexes to the Chicago Convention. These annexes contain internationally agreed specifications covering such subjects as operation of aircraft, rules of the air, personnel licensing, communications, and airworthiness. The specifications are in the form of Standards and Recommended Practices. All Contracting States to the ICAO Convention undertake to incorporate Standards into their national codes. However, if they do not wish to do so in a particular case, they must file a difference with ICAO, which must then be published for the information of the other States.

In addition to the annexes to the Chicago Convention, ICAO also publishes many other kinds of documents. Some of these are directly connected to specific annexes and take the form of supporting guidance material to assist States in implementing the intentions of the Standards and Recommended Practices.

In the environmental field, Annex 16 dealing with aircraft noise was published in 1972, culminating several years of intense effort. In addition, circulars have been published on such subjects as "Noise Assessment for Land-Use Planning" (Circular 116-AN/86) and "Guidance Material on SST Aircraft Operations" (Circular 126-AN/91).

LIAISON WITH WMO AND WHO

ICAO has been actively supporting the World Meteorological Organization (WMO) programme to investigate stratospheric pollution, including the effects of man's activities on the ozone layer. As the organization responsible for international civil aviation, ICAO stands ready to supply WMO with authoritative operational data as required by that Organization for its studies, such as the information already provided on supersonic aircraft exhaust emissions.

With regard to the World Health Organization (WHO), ICAO has established liaison with that Organization's section for Control of Environmental Pollution and Hazard and maintains contact with it on programmes of mutual interest - particularly those relating to the effects of aircraft engine exhaust emissions in the vicinity of airports.

ICAO ACTIVITY IN ENVIRONMENTAL MATTERS TO DATE

Annex 16 to the Chicago Convention contains specifications relating to aircraft noise. Continuing work on this subject, including the updating of this annex, is carried out by an advisory body of ICAO, the Committee on Aircraft Noise. ICAO has also published a circular, "Noise Assessment for Land-Use Planning" (Circular 116-AN/86), which provides descriptions of various methods used in the assessment of noise exposure and a comparison of these methods.

A circular entitled "Guidance Material on SST Aircraft Operations" (Circular 126-AN/91) has been published which contains, among other things, information regarding sonic boom and its avoidance. This subject remains under consideration by another ICAO advisory body, the Sonic Boom Committee.

To consider the question of possible atmospheric pollution by aircraft, a study was initiated in 1973 by the Secretariat, assisted by a group of experts from States. These efforts were directed towards aircraft engine emissions in the vicinity of airports and consisted of:

- fact finding on relevant technical and regulatory developments in States;

- the development of specifications for units and methods of measurement of engine emissions;

- a study of the need for a certification scheme to control engine emissions; and

- the formulation of proposals for implementation of such a scheme if it were found necessary.

It was concluded that, although at present aircraft engine emissions do not pose a serious pollution problem except at a few particular airfields, a certification scheme for control of such emissions was desirable. This was because other sources of emissions at airports, chiefly ground vehicles, were being controlled and consequently aircraft would contribute an increasingly significant proportion of pollution if left uncontrolled. It was also recognized that early action before a real problem developed would enable engine designers to plan for low emission engines and avoid any further need for retrofitting with "cleaner" engines.

Work has therefore been continuing on proposals for a certification scheme for the control of engine emissions, including smoke, vented fuel and gaseous emissions as well as full specifications for instrumentation and measurement procedures. Substantial progress has been achieved in this work and it can be expected that the initial results will see publication in the near future. This material will initially be restricted to turbo-jet and turbo-fan engines used for propulsion of aircraft at subsonic speeds, but will likely be extended to other classes of engines in due course.

Although the work has been specifically directed towards emissions at low altitudes, the existence has been recognized of a potential high altitude pollution problem which should be kept in sight when the low altitude question is being addressed. Investigation into upper atmosphere pollution is of great interest to ICAO, and should it confirm that aircraft might be a significant contributory factor, ICAO would wish to take steps to alleviate the problem at the earliest possible moment.

Besides the work on engine emissions outlined above, ICAO has also been considering the need and possibility of controlling emissions at airports by operational means. Such procedures as the delayed start-up of engines before take-off and early shut-down after landing are being considered as ways of reducing taxiing time and there-fore emissions.

FUTURE ACTIVITY BY ICAO

In fulfilling its role in the development of international civil aviation, ICAO is conscious of the adverse environmental impacts which may stem from aircraft oper-ations. It is also conscious of its responsibility and that of its Member States to achieve the maximum compatibility between the safe and orderly development of civil aviation and the quality of the human environment.

In discharging this responsibility ICAO will continue to cooperate with States and other international organizations in studying questions, as they arise, regarding the impact of civil aviation on the environment.

It is hoped that if potential environmental problem areas are identified at an early stage, then designers can be given time to incorporate technology into new equipment, which will avoid or minimize such problems. ICAO's work to minimize the disruptive and costly effect of retroactive regulations is also mentioned in the paper submit-ted by the United Nations Department of Economic and Social Affairs.

PAPER 5

Effects of Ultra-violet Radiation on Human Health*

Presented by the World Health Organization

INTRODUCTION

Solar radiation is a very important component of the environment although the wide scope of its biological effects is often not fully appreciated. The fact that solar energy conversion makes life possible is widely known, but it is not so generally understood that solar radiation may also have adverse effects. Most people are aware that excessive exposure to the sun will result in painful sunburn, but sunlight may have even more subtle effects on living cells including the production of mutations, and skin cancer may develop as the result of long-term exposure to sunlight.

The discussion of the effects of ultra-violet radiation (UVR) contained in this paper should not be taken as a definite evaluation. A detailed study, in collaboration with some 20 Member States, is in progress and will be presented in a WHO environmental health criteria document on UVR that should be completed by the end of 1977 or early 1978.

BIOLOGICAL EFFECTS OF EXCESSIVE UVR

The shortwave portion of the solar spectrum is potentially highly detrimental to plant and animal cells. A layer of ozone in the upper atmosphere filters out these harmful wavelengths of UVR and thus prevents most of such radiation from reaching the surface of the earth. However, even with the presence of this ozone layer, a biologically significant amount of UVR does reach the surface of the earth.

The photobiological effects involving the region of the ultra-violet spectrum under consideration (280-320 nm, referred to as UV-B) have received little attention.** Most UVR research has related to the 254 nm radiation because it is easily generated

*The first draft of this paper was prepared by Professor F. Urbach of the Temple University Health Sciences Center, Philadelphia, USA, and was then reviewed and revised by a group of experts - Dr. Y. Skreb of the Institute for Medical Research and Occupational Health, Zagreb, Yugoslavia, Dr. V. Stenback, Assistant Professor of Pathology, Medical Faculty of Oulu University, Finland, Dr. O.M. Jensen, Unit of Epidemiology and Biostatics, International Agency for Research on Cancer, Lyon, France, and WHO staff. In reviewing the first draft, use was made of the USSR national review on ultra-violet radiation, prepared by members of the Institute of General and Community Hygiene, Moscow, of the Academy of Medical Sciences of the USSR, under the general guidance of Professor N.M. Dancig.

**UV-A: 400-320 nm   UV-B: 320-280 nm   UV-C: 280-200 nm

by low-pressure mercury vapour lamps and is efficiently absorbed by nucleic acids.
Since wave-lengths shorter than 285 nm are effectively absorbed even by an atmosphere severely depleted of ozone, this 254 nm photobiology is of little value in
quantitatively assessing the impact of increased solar UV-B irradiation [1].
Only in the last 3 years have experiments been initiated that relate specifically
to the biological effects of increased UV-B radiation that could occur in the event
of stratospheric ozone depletion [2].

These preliminary experiments and the earlier photobiological literature indicate:

- that most observed biological effects of UV-B radiation are decidedly detrimental;

- that many organisms have developed protective mechanisms against excessive solar
  UV-B radiation, such as the cuticular waxes and pigments of plants, the feathers,
  fur, pigments etc. of animals that absorb radiation before it reaches sensitive
  physiological targets, behavioural patterns for avoiding excessive exposure, or
  some tolerance to UV-B radiation stress, e.g. by molecular repair processes;

- that, nevertheless, the capacity of most organisms to attenuate, avoid or
  tolerate UV-B radiation is very limited, and that for some micoorganisms and
  plants any increase in UV-B radiation could be deterimental [3].

Although skin cancer has become a sensitive issue, it should be recognized that the
potential effects of increased solar UV-B radiation on plants, animals and microorganisms of both terrestrial and aquatic ecosystems may be of even greater importance for man's health and well-being. Even subtle effects, such as slightly
depressed photosynthesis of phytoplankton in the ocean or of higher plants on land,
or increased mutation rates of airborne pollen or microorganisms, could have far-
reaching consequences for agriculture, forestry and fisheries. Considerably more
effort must be devoted to identifying these potentially important consequences of
increased UV-B radiation [3].

                        BENEFICIAL EFFECTS OF UVR

Sunbathing is popular, and there is a widespread feeling that "sunlight is good for
you", but the physiological benefits that presumably underlie the feeling of well-
being have not been adequately explained or studied.

Various systemic beneficial effects on the development of children, on work performance, psychological well-being etc. have been reported, but further scientifically
controlled studies on such positive health effects are needed, as well as on the
adverse effects of UVR deficiency [4,5].

A thoroughly established beneficial effect of UVR on the skin is the conversion of
7-dehydrocholesterol to vitamin D_3. Throughout most of human existence, 7-dehydro-
cholesterol was converted in the skin to vitamin D_3 by UVR [6]. Only in northerly
climates, and particularly with the advent of cities, was the sun-produced
vitamin D insufficient to prevent rickets in growing children and osteomalacia in
adults, both diseases being produced by defective bone calcification [4,5]. An
association of UVR deprivation and dental caries has also been reported in northerly
climates [7].

Some anthropologists have long looked upon white skin as an adaptation to low levels
of UVR that occurred in poorly insolated areas, associated with glacial periods
[8-10]. On the other hand, there is evidence that excessive dietary intake of
vitamin D (almost always vitamin D_2) can produce vitamin D intoxication. Whether
this is due to the difference in route of administration or to a difference between
vitamin D_2 and D_3 is not clear from the literature, and is an important area for

investigation. We know that vitamin D intoxication can be produced by oral administration, but we cannot at the present time say whether an increase in environmental UVR could cause vitamin D intoxication in healthy man or animals [10].

DETRIMENTAL EFFECTS OF UVR

Acute Effects

Acute effects on the skin. Solar erythema or, more commonly, sunburn consists, in
its mildest form, of a reddening of the skin that appears one to six hours after
exposure to UVR and gradually fades in one to three days. In more severe form,
solar erythema causes inflammation, blistering and peeling, followed by tanning of
the skin that becomes noticeable within two to three days following irradiation.
Although sunburn would occur with shorter exposures if stratospheric ozone were
reduced, the protective measures now available (clothing, chemical sunscreens, etc.)
could prevent this from becoming a serious limiting factor for human activities [2].

Acute effects on the eye. Although more energetic than the visible portion of the
electromagnetic spectrum, UVR is not detected by the visual receptors in mammals,
including man. Thus, exposure to UVR may result in ocular damage before the
recipient is aware of the potential danger. Many cases have been reported of
keratitis of the cornea and cataract of the lens due to exposure to UVR produced by
welding arcs, high-pressure pulsed lamps, and the reflection of solar radiation
from snow and sand.

In the UVR region reaching the earth from the sun, which is biologically the most
effective (290-320 nm UV-B), the action spectrum for photokeratitis is almost
identical to that for skin erythema. Thus, all the conclusions regarding the increased hazard from a change in this UV band that apply to acute effects on the
skin apply equally to the eye [11].

Chronic Effects

UVR and skin aging. Solar or actinic changes in human skin (farmer's skin, sailor's
skin) are shown by atrophy, freckling with hyper- and hypo-pigmentation, dilated
blood vessels, and a yellow discolouration due to increases in abnormal elastic and
collagenous tissue. Wrinkling is also a prominent feature. Of all the hazards of
increased UVR in man's environment, most attention has centred on skin cancer [12],
but for every patient who actually develops cancer, there are probably thousands
who develop readily visible actinic changes that may lead to serious psychological
problems in some individuals.

Experimental induction of skin cancer by UVR. Much of our information on the
causation of human skin cancer has come from extensive studies on laboratory
animals, particularly mice. All photobiological responses to UVR and visible light
show a dependence on the energy of the incident photons, with a maximum response at
a fairly well-defined and limited photon energy range beyond which the lower photon
energies are very much less effective. Primarily this depends on absorption of UVR
by specific biologically important molecules, such as deoxyribonucleic acid (DNA),
and hence the biological action of UVR is restricted to the absorption spectrum of
these molecules. Skin carcinogenesis by UVR is generally assumed to be the result
of damage to DNA molecules, followed by inaccurate repair. Thus the skin carcinogenesis action spectrum and the DNA absorption spectrum may correspond. The
erythemal spectrum compares well with the DNA absorption spectrum in the spectral
region from 295 to 330 nm, which appears to cause nearly all the effects. Hence,
the DNA absorption spectrum and the erythema action spectrum will give nearly the
same effects. Furthermore, the action spectrum for skin carcinogenesis in mice
covers the same spectral range as the human erythema action spectrum [13].

The evidence, based primarily on animals studies, is consistent with the concept
that UV-induced photodamage to the skin is the main causative factor in skin cancer,
and that there is practically no threshold for this effect. Thus a relationship
should exist between skin-cancer incidence and cumulative dose. In mice, the
quantitative relationship between UV dose and the production of skin cancer has been
thoroughly explored [14].

This photobiological model is based on a number of simplifying assumptions, e.g.
monochromatic UV-B (280-320 nm UVR) input, synchronized cell population, and constant environmental conditions. In practice, these conditions are seldom so clear.
An estimate of the increase in response (e.g. skin-cancer incidence) can be made
from estimates of increased energy input (UV-B) only if the mathematical relationship between the two variables is precisely known. This relationship is known
moderately well for animals, where direct experimentation is possible [15]. The
UV-B skin-cancer relationship in man is much less clear, because the evidence must
be laboriously gathered from epidemiological data [16,17].

An understanding of the etiology of human skin cancer is the ultimate goal of
investigations into the carcinogenic effects of UVR. As noted, human skin is not an
appropriate experimental model, not only because of the obvious moral and socio-
logical objections, but also because of the long tumour-development time. The
experimental production of tumours has therefore been confined to appropriate
laboratory animals, and skin cancers have been induced in a variety of mouse strains,
in hamsters and in rats with artificial UVR as well as with sunlight.

Beginning in 1940, Blum and his co-workers initiated and carried out a series of
experiments concerning the energy requirement, dose-rate and time relationships that
remain milestones in the field of experimental UV carcinogenesis [14]. These
authors used set exposure procedures monitored by a stable photoelectric cell and an
integrating meter, and confined their observations to tumour formation on the ears
of an inbred strain of albino mice. Though the exposure was not exceptionally
efficient, the reproducibility of their results confirms the significance of their
findings. In addition, they used very large numbers of mice, further enhancing the
validity of their results. They first examined the problem of dose-rate dependence,
and found that reciprocity held within the accuracy of the experiments when dose-
rates greater than 3.5 to 4.0 Jm^{-2} sec^{-1} were used. Below this dose-rate, UV became less efficient in producing tumours [18].

In subsequent studies, they reported that by increasing the dose and/or shortening
the interval between subsequent doses, tumour development could be accelerated, but
the shape of the incidence curve remained unchanged. Even if exposure was discontinued prior to the appearance of visible tumours, cancer was nevertheless produced if sufficient energy had been applied over a sufficiently long period of
time [15].

These early experiments on photocarcinogenesis did not include detailed descriptions
of the light sources used, or histological confirmation and description of the
tumours produced. However, more recent studies, with rigorous control of light
sources and other experimental factors, have confirmed the essential findings of
these older experiments that UV-induced carcinogenesis is a continuous cumulative
process that begins with the initial exposure [14].

Studies concerning the influence of different physical stimuli on UV-induced carcinogenesis have been sparse, with the exception of those on temperature and air
movement. In contrast, there have been a number of studies evaluating the effects
of various chemicals on UV-induced carcinogenesis [19]. The effects of repeated
exposure to non-carcinogenic UVR of longer wavelength (UVA 320-400 nm), in the
presence of a photosensitizer with the proper action spectrum, in producing skin

cancer have now been well established. However, when polycyclic aromatic hydrocarbons were used as photosensitizers, a variety of conflicting results were
reported. The relationship between UVR and chemical factors in the development of
skin cancer has been reviewed by Epstein [20] and Emmett [21].

Human skin cancer and UVR. The role of the sun in the production of human skin
cancers does not lend itself to direct experimentation. However, extensive clinical
observations, beginning with Dubreuilh and Unna in the 1890s, have strongly
suggested the etiological significance of light in the induction of such tumours.
Skin cancers in white-skinned individuals are, in general, most prevalent in geographical areas of highest insolation and among the most highly-exposed persons,
i.e. those working outdoors [22]. Such cancers are rare in Indians, Japanese,
Africans and similarly pigmented groups who are well protected from UVR injury.
Further, individuals with the fairest complexions, such as those of Scottish or
Irish descent, appear to be most susceptible to skin cancer formation when they
live in areas of high UV exposure [16,22,23]. When skin cancers do occur in the
darkly pigmented populations, they are not distributed primarily in the sun-exposed
areas as in light-skinned persons but are most commonly stimulated by trauma, such
as chronic leg ulcers, irritation due to going barefoot, the use of a Kangri (an
earthenware pot that is filled with burning charcoal and strapped to the abdomen
for warmth), the wearing of a Dhoti (loin-cloth), and so on. In contrast, the
distribution of skin cancer in the Bantu albino and in patients with xeroderma
pigmentosum follows sun-exposure patterns [22].

Blum [14], Urbach et al. [17,24] and most recently Emmett [21] have reviewed the
evidence supporting the role of sunlight in human skin-cancer development. Briefly,
the main arguments are:

- Superficial skin cancers occur most frequently on the head, neck, arms and hands,
  i.e. those parts of the body habitually exposed to sunlight.

- Pigmented individuals, who sunburn much less readily than the white-skinned,
  have very much less skin cancer, and when it does occur, it affects areas less
  frequently exposed to sunlight.

- Among white-skinned individuals, there appears to be a much greater incidence of
  skin cancer in those who spend much time outdoors than in those who work predominantly indoors.

- Skin cancer is more common in white-skinned individuals living in areas of high
  insolation.

- Hereditary characteristics (albinism, xeroderma pigmentosum) that result in
  greater sensitivity of the skin to the effects of solar UVR are associated with
  a marked increase in, and premature development, of skin cancer.

- Superficial skin cancers, particularly squamous cell carcinomas, occur predominantly on those areas of the body receiving maximum amounts of solar UVR, and
  where histological changes due to chronic UV damage are most severe.

- Skin cancer can be produced readily on the skin of mice and rats with repeated
  doses of UVR, and the upper wavelength limit of the most effective cancer-
  producing radiation is about 320 nm, i.e. the spectral range that produces solar
  erythema in human skin.

Though these arguments do not constitute absolute proof, there is good qualitative
epidemiological evidence supporting the role of sunlight in three types of skin
cancer: basal cell carcinomas, squamous cell carcinomas, and malignant melanomas.

Basal cell carcinoma. Basal cell carcinoma is the most common human skin cancer
that occurs in white-skinned people. It originates in epidermal or adnexal (or

both) basal cells and frequently presents itself as a slow-growing, shiny nodule
composed of masses of cells with darkly staining nuclei that simulate epidermal
basal cells. These lesions invade locally, but rarely metastasize. Epidemiological
studies in many countries indicate that these tumours are found primarily on the
head and neck - that is, in sun-exposed areas [22]. They are exceedingly rare in
persons with pigmented skin. There is a statistically significant tendency for
patients with basal cell carcinoma to have blue or blue-gray eyes, light hair and
light complexions, to sunburn easily, and to spend more hours outdoors than a control group without such tumours [25]. These findings add further support to the
importance of sunlight in basal cell carcinoma formation. However, although 80% of
these cancers occur on skin areas highly exposed to UVR, they often occur around the
eye, behind the ear and other relatively protected sites, so that factors in addition to exposure to the sun must play a role in their induction [17,24,26].

Squamous cell carcinoma. Squamous cell carcinoma is the second most common skin
cancer in white-skinned people and is found more often than basal cell carcinoma in
persons with pigmented skin. It generally is manifest as a relatively slow-growing
keratotic tumour composed of masses of cells that simulate epidermal cells and tend
to form keratin. However, the individual cells frequently show bizarre forms and
shapes as well as malignant dyskeratosis. These lesions do metastasize at times and
can be fatal.

Evidence of the role of sunlight in producing squamous cell carcinomas in white-
skinned people, albinos, and patients with xeroderma pigmentosum is even more convincing than it is for basal cell carcinoma formation. These growths are distributed primarily over the head and neck and, to a lesser extent, the exposed areas of
the upper extremities. As noted for basal cell carcinomas, squamous cell carcinoma
are found most frequently in men working outdoors in areas of high insolation [21].
In addition, though both squamous cell and basal cell carcinomas are more prevalent
in areas of high insolation, there is a relatively greater increase in squamous cell
carinoma incidence with decreasing latitude and increasing insolation. Chemical
carcinogens as well as ionizing radiation can cause squamous cell carcinoma; the
proportion varies from country to country [26,27].

Squamous cell cancers are uncommon in persons with deeply pigmented skin, and when
they do occur, they are found primarily on the lower extremities and on the abdomen,
depending on sites of chronic irritation and injury but not of solar damage [22].
Further, in contrast to basal cell carcinomas, squamous cell carcinomas in white-
skinned persons occur almost exclusively at anatomical sites that receive maximal
UV irradiation, such as the rim of the ear, so that the role of the sun is fairly
clear. The geometry of sunlight exposure on the human body further supports the
association of sunlight with basal cell and squamous cell cancer. The areas receiving most UVR when the sun is high are the head (and since the top of the head is
usually covered with hair and is thus protected, the exposed areas are the rim of
the ear, the nose, forehead, cheeks, lower lip and chin), the shoulders, back of
neck and upper arms and, to a lesser extent, the outer surface of the arms and hands.
The chest, back, abdomen and legs are exposed only when the sun is low, and little
UV-B is present in the solar radiation. This concept gains additional support from
experimental studies in which squamous cell carcinomas can be produced in appropriate
animals by UV irradiation [17,28].

Incidence of non-melanoma skin cancer in man. Surveys of the incidence of skin
cancer have been performed with varying success in the recent past. In the USA,
the major surveys have been those carried out by the National Cancer Institute (NIH)
in 1947-48 (the "Ten-City Survey") [22], the recent Third National Cancer Survey
(1971-72) [29] and the M.D. Anderson Institute Texas Survey (1962-72) [30]. Data
are also available from Iowa (1950) [22], Minnesota (1963) [28] and from McDonald's
study in Texas (1960-70) [30], as well as from studies carried out in Australia

(1960-70) [12], in the USSR [16,23], in Sweden, and from Vol. III of Cancer Incidence in 5 Continents (IARC) [31]. World incidence data on skin cancer are given in Table 5.1. From all these data, an inverse relationship with latitude is apparent [31]. For example, the Third National Cancer Survey (USA) shows that the annual incidence rates for skin cancer (excluding malignant melanoma) vary in the USA from 379/100,000 in Dallas-Fort Worth to 124/100,000 in Iowa [29].

However, the enumeration of skin-cancer incidence in a population is made very difficult by the relative benignity of the disease, which allows for curative treatment in physicians' offices rather than hospitals. Surveying all physicians in any one area is difficult and expensive, so that most skin-cancer incidence studies have seriously underestimated actual conditions.

Comparing the older and more recent studies, there is a strong suggestion that skin cancer has increased in the last decades. For instance, in Minnesota, if one compares the metropolitan areas, an apparent two-fold increase has occurred since 1963, and in Texas a three-fold increase from 1944 to 1966, perhaps a reflection of the trend for more people to expose more of themselves more often to the sun's rays along with the changes in life styles, clothing and leisure time.

The best estimate of the present annual incidence of non-melanoma skin cancer in the USA (in white-skinned people) is 165/100,000 population. This means that about 300,000 cases of skin cancer occur in the USA each year, or that about one-third to one-half of all cancers are skin cancers [29]. This very high rate certainly does not apply to other countries; for example, in the USSR, skin cancer amounts to 15-26% of all cancers in the south, and 9-14% in the north [32]; in Oxford, England, 12% of all male cancers are skin cancers, in Miyal Prefecture, Japan, 1%, and in Cuba, 10% [31].

Sunlight and the etiology of malignant melanoma. The influence of latitude of residence on the incidence of and mortality from melanoma is the original and strongest evidence that exposure to sunlight of white-skinned individuals is a factor in the etiology of malignant melanoma. The gradient with latitude of the death rate from malignant melanoma is not as great as for other cancers, but it is substantial. Both the incidence of, and mortality rates from, malignant melanoma are rising rapidly in all countries where they have been studied. Mortality rates are rising by about 3-9% annually, so that they have doubled in about the last 15 years [33]. All this is unlikely to be due to chance. Where exposure of particular sites differs between the sexes because of conventional dress and hair styles (ears and neck in males, lower limbs in females), there is a higher incidence and mortality rate for a site exposed in one sex than for the same site that is unexposed in the opposite sex [34].

Despite this evidence, the importance of sunlight as an etiological factor in malignant melanoma has only been recognized in recent years and tends to be minimized in the older literature. Malignant melanomas are not common tumours and, until the Sixth Revision of the International Statistical Classification of Diseases (WHO, 1948), they were not separated from other skin tumours, and this prevented recognition of their variation from one population to another. However, more important was the fact that malignant melanoma does not appear mainly on the face and neck, in obvious contrast to squamous and basal cell carcinomas. It is of interest, too, that there is a close relationship between the death rate from squamous cell carcinoma and malignant melanoma in the USA and the Canadian provinces, suggesting that a common environmental factor is involved [34].

The interesting features of melanoma, i.e. the increasing incidence with exposure and decreasing latitude, the particular sites of lesion according to sex and occupation and, at the same time, the lack of clustering at sites of exposure, have

TABLE 5.1 Incidence per 100,000 (World Standard Population) of Malignant Melanoma of the Skin (ICD 172) and other Skin Cancer (ICD 173) in Different Areas of the World. (Waterhouse, Muir, Correa & Power, 1976) [31]

Registry	Males		Females	
	Malignant Melanoma (ICD 172)	Other Skin Cancer (ICD 173)	Malignant Melanoma (ICD 172)	Other Skin Cancer (ICD 173)
Nigeria: Ibadan	0.9	1.2	2.2	1.6
Bulawayo (Africans)	3.0	9.9	2.6	0.9
Brazil: Recife	1.6	41.6	1.2	36.7
Sao Paulo	2.2	41.4	1.9	37.4
Canada: Alberta	2.2	56.2	2.7	41.6
B.C.	3.6	101.3	4.8	67.6
Manitoba	2.6	54.9	3.4	39.3
Mar.Prov.	2.0	68.4	2.7	39.1
N'Fdland	1.6	47.9	1.9	30.6
Quebec	1.4	30.2	1.8	19.8
Saskatch.	2.8	90.2	3.4	59.5
Colombia: Cali	2.1	39.8	2.0	39.5
Cuba	0.5	19.4	0.3	13.3
Jamaica:				
Kingston	1.4	13.2	1.1	11.4
Alameda (White)	5.4		5.9	
(Black)	0.9		0.5	
Bay Area (White)	6.3		6.6	
(Black)	0.9		0.6	
(Chinese)	0.7		0.0	
USA:				
Connecticut	4.5		4.3	
Iowa	3.2		2.7	
Detroit (White)	2.7		3.1	
(Black)	0.5		0.6	
N.Mexico (Spanish)	0.8	16.5	0.9	9.9
(Other White)	4.8	96.7	5.3	42.4
(Amer.Indian)	0.7	9.6	1.2	9.2
New York State	3.4		3.0	
El Paso (Spanish)	0.0	30.9	1.0	24.0
(Other White)	3.8	144.9	4.8	73.3
Puerto Rico	0.7	42.6	0.8	40.3
Utah	5.5		5.1	
India: Bombay	0.2	2.1	0.2	1.2
Israel:				
Jews	3.4		4.5	
(Born Israel)	4.7		7.0	
(Born Eur.Amer.)	3.4		4.8	
(Born Afr.Asia)	0.9		0.6	
Non-Jews	0.7		0.1	

TABLE 5.1 contd

Registry	Males		Females	
	Malignant Melanoma (ICD 172)	Other Skin Cancer (ICD 173)	Malignant Melanoma (ICD 172)	Other Skin Cancer (ICD 173)
Japan: Miyagi	0.3	1.3	0.1	1.3
Okayama	0.1	1.9	0.0	1.9
Osaka	0.3	0.8	0.2	0.5
Singapore (Chinese)	0.6	6.6	0.3	4.9
(Malay)	0.4	4.6	0.0	3.3
(Indian)	0.4	2.7	0.9	4.8
Denmark	2.9	24.7	4.9	15.4
Finland	2.9	25.3	2.8	22.5
German Dem.Republic	2.1	31.1	2.3	19.7
Fed.Rep.Germany: Hamburg	2.4	4.8	1.8	2.2
Saarland	2.3	8.5	2.3	5.6
Hungary: Szabolcs	1.6	20.5	1.4	22.4
Vas	1.8	28.4	2.1	21.0
Iceland	1.6	2.7	3.6	1.7
Malta	0.5	32.7	0.9	12.1
Norway	5.4		5.7	
(Urban)	6.7		6.5	
(Rural)	4.5		5.1	
Poland: Cieszyn	0.6	13.1	2.5	11.4
Cracow	1.7	13.7	2.0	11.2
Katowice	1.2	12.9	1.5	9.8
Warsaw (City)	2.3	10.4	2.1	9.0
" (Rural)	0.8	6.2	0.8	6.1
Romania: Timis	1.5	14.2	1.4	12.2
Spain: Zaragoza	0.3	19.7	0.3	10.0
Sweden	4.1	6.3	4.9	3.3
Switzerland: Geneva	3.9	24.9	1.8	14.7
UK: Birmingham	1.3	29.8	2.2	18.8
Oxford	2.2	29.9	3.1	17.5
Sheffield	1.1	23.3	1.5	15.9
SMCR 1963-66	1.3	30.7	2.4	16.8
SMCR 1967-71	1.6	27.4	2.8	14.7
South West	1.6	29.6	4.0	18.0
Liverpool	0.9	29.7	2.0	17.3
Ayrshire	2.0	33.7	2.9	23.5
Yugoslavia: Slovenia	1.8	15.8	2.9	15.4
Hawaii (Hawaiian)	0.9	0.0	1.0	0.8
(Caucasian)	6.8	1.9	5.7	2.2
(Chinese)	0.0	0.0	0.0	0.0
(Filipino)	0.3	0.0	0.0	0.0
(Japanese)	0.3	0.3	0.3	0.3
New Zealand (Maori)	1.5		1.5	
(Non-Maori)	7.4		11.7	

been most clearly presented by the Australian clinical studies. There is a broad
similarity in the anatomical distribution of the malignant melanomas reported from
white populations with widely differing incidence rates, but direct questioning of
Australian melanoma patients failed to elicit a history of specific exposure at
the primary site of lesion.

TABLE 5.2 Percentage Distribution of Malignant Melanoma of the Skin
and other Skin Cancers, by Anatomical Site, Males and
Females, Denmark 1943-57. (Clemmesen, 1965) [35]

	Head & Neck Percentage	Trunk Percentage	Upper Limb Percentage	Lower Limb Percentage	Multiple Sites Percentage	Total
Malignant Melanoma						
male	32	25	9	22	12	100
female	24	19	14	33	10	100
Other Skin Cancers						
male	81	7	4	6	2	100
female	78	7	8	4	3	100

One simple hypothesis has been proposed to account for the occurrence of melanoma
at non-exposed sites, namely, that malignant melanomas have two origins - one being
independent of sunlight and occurring wherever there are melanocytes (perhaps
related to pre-existing naevi), the other related to exposure to sunlight [33].

The incidence of malignant melanomas is increasing in many industrialized countries.
For example, mortality from this disease has now reached 6-10 per 100,000 in the
USA [31] despite improvements in treatmeat in the last 20 years. As a result, the
death rate from all forms of skin cancer has shown no decline in recent years, and
the increase in the death rate from malignant melanoma in the younger age groups
has offset the decrease in the death rate from squamous cell carcinoma in the
elderly. There seems to be at present no reason to associate the increasing incidence of melanoma with anything other than change in cultural habits.

The potential effect of reduced stratospheric ozone on skin-cancer incidence (basal
and squamous cell carcinoma). Because of the association between skin-cancer incidence in man and effective UVR, any long-term increase in UVR may be expected to increase the incidence of skin cancer, and it is clearly important to quantify this
association. The most straightforward answer should result from a comparison of
the ozone thickness over the areas for which epidemiological data on skin-cancer
incidence are available, and extrapolation of this information to other areas taking into account the influence of factors, other than ozone, thought to affect
skin-cancer incidence. This presupposes, however, accurate data on local ozone
conditions, not usually available at the present time. Existing ozone maps are
based on measurements made at various places (at none of which are there epidemiological data on skin-cancer incidence), at various times and by various methods
[36,37]. Furthermore, there are known marked seasonal, longitudinal and latitudinal changes in the ozone layer. Since, however, UVR (as modified by stratospheric
ozone thickness and atmospheric conditions) is considered to be causally related to
skin-cancer incidence, it should be possible to measure UVR with some degree of
accuracy in some places and compare these measurements with those of ozone [38,39].

Wavelength-to-wavelength spectroradiometric measurements of global and diffuse solar
UVR have been performed by Bener [38] (World Meteorological Center, Davos,

Switzerland) and by scientists of Moscow State University [40]. These measurements are very exact but could not be carried out daily on a long-term basis because of the technical difficulties and the expenses.

Long-term measurements with analog integrating dosimeters have been carried out in various places in the USSR by Moscow State University [40], and in 10 areas in the USA by Temple University in cooperation with the National Oceanic & Atmospheric Administration. In each case, the solar UVR flux is weighted by an action spectrum (i.e. a suitable response function), which purports to be representative of the effect expected - skin erythema, DNA damage, or cutaneous carcinogenesis.

In practice, however, none of the integrating analog UVR field meters accurately represents even the skin erythema action spectrum, although their response function is reasonably parallel and covers the approximate wavelengths known to elicit erythema. But the ability of such instruments to indicate some parameter of actual UVR reaching the detector (despite alterations that may occur owing to varying ozone thickness, cloud, haze, aerosol etc.) and their continuous long-term operational stability is an advantage that far exceeds their spectral-match limitations.

A number of authors have used such measurements for their model computations. UV dose estimates have then been calculated on the basis of certain assumptions, and these calculations have been found to correspond reasonably well with actual measurements [39,41].

RELATIONSHIP OF OZONE THICKNESS, UVR DOSE AND THE INCIDENCE OF SKIN CANCER

Using the figures for the incidence of basal cell and squamous cell skin cancer in the USA, obtained by the Third National Cancer Survey, and a variety of assumptions on the relationship of UVR and skin cancer, several models of the potential effects of reduced stratospheric ozone have been proposed. Underlying these models are several assumptions:

- A decrease in stratospheric ozone will result in an increase in UVR with wavelengths shorter than approximately 320 nm.

- An increase in UVR with wavelengths shorter than approximately 320 nm will result in an increased skin-cancer incidence in susceptible human populations.

- The observed increase in skin-cancer incidence with decreasing latitude is due to several interacting factors. Among these factors are ozone thickness, local atmospheric conditions, genetic background of the populations, and the type, length and kind of outdoor exposure. There are also some as yet unspecified factors, e.g. chemical agents.

On average, according to these models, the result of reduced ozone layer thickness could be the appearance of skin cancer in susceptible persons at an earlier age. As skin cancer is predominantly a disease of the elderly, this shift to younger persons, when integrated over the entire population, would be seen as an increase in the total number of skin-cancer cases.

The various models [42-45] indicate two separate "amplification" factors. One of these is physical and related to the UV absorption spectrum of ozone and the thickness of the ozone layer. A quantitative relationship between ozone thickness and UVR that is erythemogenic for untanned white skin has been calculated by a number of investigators [42,46]. The difference in estimates (increases in UVR effectiveness ranging from 7 to 11.5% for a 5% ozone increase) are in part due to the non-linearity of the rate of change of the biological effectiveness of solar UVR with the change in the assumed original ozone thickness. The average

calculated increase in UVR effectiveness for a 5% decrease in ozone is 9.25% (or
a multiplication factor of 1.83) [2]. Comparisons of UVR meter readings at Hilo
(Hawaii) and Bismarck (North Dakota) with Dobson meter measurements of ozone at
these same stations gave multiplication factors of 2.15 and 2.0 respectively [41].
This gives an approximate physical amplification factor of about 2 (range 1.4 – 2.5).

The second amplification factor is biological. To derive this factor, an estimate
of the dose-response relationship of UVR to skin carcinogenesis in man is needed.
This estimate has been based on comparisons of measured UVR doses with skin-cancer
incidence data. At this time, such data appear to be available only at four
locations in the USA which are not ideally suited for the purpose. The models
suggest that this biological amplification factor is not linear with dose. Furthermore, the actual skin dose received by the population at risk had to be estimated,
as the measurements represented the total dose received by the measuring instrument and not the exposure of the population which is related to their habits. As
the models developed, different amplification factors were reported. At present,
they appear to range from 0.5 to 3.5, but for working purposes a biological
amplification factor of 2 is currently being used in the USA.

The estimation of a potential increase in skin-cancer incidence from possible
changes in ozone concentration in the stratosphere is subject to even greater
variation. The currently used *total* amplification factor ranges from 0.7 to 8.76,
with a working value of 4 (applicable to the USA conditions for which the models
were developed).

It cannot be overemphasized that all the calculations, by all the authors who
attempted them, are subject to great, and at this time immeasurable uncertainty,
owing to inadequate knowledge of the growth of cancer cell populations [47].
Therefore, until far better data become available , at least on the effects of
shifts in the spectral distribution of UVR and on the effects of the flux, time-
dose relationships and seasonal cycles of irradiation, and until we have more reliable information on the true incidence of skin cancer, all the present numerical
estimates must be considered as very preliminary and subject to significant
correction as new data become available. These estimates will certainly vary
considerably from country to country.

The skin cancers discussed above are principally basal cell and squamous cell
carcinomas; the melanomas have not been included. The incidence of malignant
melanoma, though small, also shows a marked latitude gradient, and death rates from
the two broad groups of primary skin neoplasms are highly correlated in the
individual states of the USA and the provinces of Canada. The importance of malignant melanomas lies in the high mortality (with current methods of treatment about
40%, i.e. comparable with breast cancer) [34].

Furthermore, present observations on the relationship between exposure to sunlight
and skin-cancer incidence (basal cell and squamous cell carcinoma, and melanoma)
refer largely to a static situation. There is considerable evidence that the
incidence of skin tumours of all types is rising [29,30,34]. Thus, if we are to
make projections into the future, we should estimate the likely incidence of human
skin tumours, assuming the continuation of present trends, probably in relation to
changes in human behaviour.

Finally, it must be emphasized that all estimates concerning the effects of stratospheric ozone depletion on the incidence of human skin cancer represent conditions
that can be expected after a *new steady state* of *both* UVR and skin-cancer incidence
has been reached. Since it is assumed that the development of skin cancer is related to the accumulated lifetime UVR dose, a new steady state in skin-cancer
incidence for a population will not be reached until that time when all members of

that population have been exposed to the new increased UVR levels for a considerable part of their lifetime. Cutchis [48] has estimated the manner in which skin-cancer incidence may be expected to build up from an initial to a new equilibrium rate. Assuming a sudden increase in UVR, which is unlikely, Cutchis shows that one-quarter of the expected increase in skin-cancer incidence would develop about 10 years afterwards and that, if UVR were to return to its original level at that time, it would take about 15 years for the skin-cancer incidence to return to its base-line level.

CONCLUSIONS AND PROPOSALS FOR FURTHER ACTION

An increase in UVR could cause acute effects on human skin and on the eyes, which it should be possible to prevent. However, the effects of chronic exposure to increased solar UVR would be insidious and could result in accelerated skin "aging" and in an increased incidence of non-melanoma and melanoma skin cancer. This would affect primarily the white-skinned portion of the world's population, particularly those living in areas of high insolation, and would be accentuated by cultural habits (especially sunbathing and the use of artificial UV sources for "health" purposes). However, all presently available quantitative estimates of these effects are subject to great uncertainty and are applicable only to conditions for which the models were developed.

The proposals for further action in each of the following sections are presented in order of priority.

Evaluation of Present Knowledge

All available data on the health and other biological effects of UVR should be collected and evaluated with a view to establishing dose-effect and dose-response relationships summarized in criteria documents. These are needed to obtain a balanced and critical evaluation of existing information that is available at national level. This would be of direct benefit to countries not in a position to carry out the necessary studies themselves.

Recommendations for Further Research

Monitoring of solar UVR. An internationally coordinated network of ground-based stations for the measurement of UVR with uniform internationally standardized methods should be established. Some of these UVR measurements should be made at sites where detailed epidemiological studies of skin-cancer incidence are in progress, so that corresponding UVR data can be obtained for assessing the dose-response relationship of UVR and skin cancer.

Epidemiological studies. An epidemiological programme is needed for the monitoring of melanoma and non-melanoma skin-cancer incidence in various parts of the world. So that valid comparisons may be made, any such programme would have to be carried out in accordance with a uniform protocol covering not only UVR and genetics but other pertinent environmental factors as well. Data thus collected on UVR intensity and distribution, and on population skin-doses (from personal dosimeters that should be developed) could also be used as the basis of prospective cohort studies.

Development of criteria and methods for identifying populations at greatest risk. Populations studies on the prevalence of skin cancer strongly suggest that persons with certain skin characteristics (fair skin, light eyes, freckles) sunburn easily, tan poorly and run a greater risk of developing skin cancer, and that this is a genetic trait. Simple screening methods should be developed to identify the most susceptible members of a population.

Beneficial effects of UVR. Beneficial effects of UVR have been reported but are insufficiently documented. It is therefore suggested that potential beneficial effects of UVR, as well as the adverse effects of UVR deficiency be further investigated.

Experimental research. More extensive studies are needed on the fundamental effects of UVR at molecular and cellular level, in particular on the photochemistry of macromolecules (including DNA damage and repair), metabolic and cytogenetic effects on cells and cell constituents, on the nucleus and cytoplasm, and the significance of such effects for the induction of photobiological effects, including cancer.

More extensive animal studies are needed to clarify the relationship of UVR exposure and photobiological effects including cancer. Animal models, particularly for malignant melanoma, need to be developed.

It has already been shown that there is an interaction of UV-B and the rest of the solar spectrum in relation to skin-cancer development. Further investigations of this aspect of UVR carcinogenesis are needed.

Preliminary experiments have shown that protracting the delivery of a dose of UVR has a significant influence on skin carcinogenesis. Further studies are necessary because lower dose rates may be more effective under certain circumstances. Similarly, the influence of the length of intervals between exposures needs clarification; this may be very important because, although any possible change in the ozone layer will affect the flux and spectral distribution of UVR, exposure times depend on the individuals concerned.

Further experimental work on animals is needed on the interactions of UVR and other physical, chemical and biological factors in skin carcinogenesis. Particular attention should be paid to the modifying effects of various classes of chemicals, both those purposely applied, such as cosmetics and drugs, and the others, including endogenous photosensitizers such as porphyrins, on UVR-induced skin carcinogenesis.

Lastly, further research is needed on natural and artificial protection of the skin against UVR, including studies on systemic sunscreening substances.

REFERENCES

[1]   Smith, K.C. (ed.) (1973), *Biologic impact of increased intensities of solar ultra-violet radiation*. Environmental Studies Board, National Academy of Sciences, Washington DC.

[2]   Nachtwey, D.S. et al. (ed.) (1975), *Impact of climatic changes on the biosphere* (Monograph 5, Parts 1 & 2, Climatic Impact Assessment Program, US Dept. of Transportation, CIAP). US Govt Prtg Office, Washington DC.

[3]   Brooker, H.G. (ed.) (1975), *Environmental impact of stratospheric flight*. National Academy of Sciences, Washington DC.

[4]   Talanova, I.K. & Sabalueva, A.P. (1972), Evaluation of ultra-violet requirements in children in the north and middle USSR. In: *Problems of experimental and clinical health resort studies and physiotherapy, Vol. XX*. In Russian.

[5]   Petrova, N.N. (1971), Study of ultra-violet deficiency in the population of Western Siberia during the spring. In: *Ultra-violet radiation*, Moscow. In Russian.

[6]   De Luca, A. (1971), Vitamin D: A new look at an old vitamin. *Nutr. Rev.*,

29: 179.

[7]   Dancig, I.N. (1974), Dental caries and ultra-violet deficiency. *Stomatologija,*
       No. III. In Russian.

[8]   Blum, H.F. (1968), Vitamin D, sunlight and natural selection. *Science, 159:*
       652.

[9]   Blum, H.F. (1969), Is sunlight a factor in the geographical distribution of
       skin colour? *Geogr. Rev., 69:* 557.

[10]   Loomis, W.F. (1967), Skin pigmentation regulation of vitamin D biosynthesis
       in man. *Science, 157:* 501.

[11]   Pitts, D.G. & Tredici, J.G. (1971), The effect of ultra-violet on the eye.
       Am. Indust. Hyg. Assoc. J., 32: 235.

[12]   Silverstone, H. & Gordon, D. (1966), Regional studies in skin cancer, Second
       Report. *Med. J. Austr., 2:* 733.

[13]   Setlow, R.B. (1974), The wavelengths in sunlight effective in producing skin
       cancer: a theoretical analysis. *Proc. Nat. Acad. Sci. US, 71:* 3363.

[14]   Blum, H.F. (1959), *Carinogenesis by ultra-violet light.* Princeton Univ.
       Press, Princeton NJ.

[15]   Blum, H.F. (1950), On the mechanism of cancer induction by ultra-violet
       radiation. *J. Nat. Cancer Inst., II:* 463.

[16]   Vasil'iev, Ju. M. (1974), The cancer-inducing factors. In: *Reference Book*
       on Oncology, Moscow. In Russian.

[17]   Urbach, F. (1969), Geographic pathology of skin cancer. In: Urbach, F. (ed.),
       The biologic effects of ultra-violet radiation, p. 635. Pergamon Press.

[18]   Blum, H.F. (1943), Effects of intensity on tumour induction by ultra-violet
       radiation. *J. Nat. Cancer Inst., 3:* 533.

[19]   Epstein, J.H. (1965), Comparison of the carcinogenic and co-carcinogenic
       effects of ultra-violet on hairless mice. *J. Nat. Cancer Inst., 34:*
       741.

[20]   Epstein, J.H. (1970), Ultra-violet carcinogenesis. In: Giese, A.C. (ed.),
       Photophysiology, Vol. 5, p. 235. Academic Press, New York.

[21]   Emmett, E.A. (1974), Ultra-violet radiation as a cause of skin tumours. *CRC*
       Critical Reviews in Toxicology, No. 22.

[22]   Urbach, F. (ed.), *The biology of cutaneous cancer* (National Cancer Institute
       Monograph No. 10), US Govt Prtg Office, Washington DC.

[23]   Caklin, A.B. (1974), Epidemiology of cancer. In: *Reference Book on Oncology,*
       Moscow. In Russian.

[24]   Urbach, F. et al. (1972), Genetic and environmental interaction in skin
       carcinogenesis. In: *Environment and Cancer,* p. 355. Williams &
       Wilkins, Baltimore.

[25] Urbach, F. et al. (1974), Ultra-violet carcinogenesis, experimental, global
       and genetic aspects. In: Fitzpatrick, T.B. (ed.), *Sunlight and Man*,
       Tokyo Univ. Press, Tokyo.

[26] Department of the Environment (Central Unit on Environmental Pollution) (1976),
       Chlorofluorocarbons and their effects on stratospheric ozone (Pollution
       Paper No. 5). HMSO, London.

[27] Bogovskij, P.A. (1960), *Occupational skin cancers induced by products of
       fossil fuel processing*. Leningrad. In Russian.

[28] Lynch, F.W. et al. (1970), Incidence of cutaneous cancer in Minnesota.
       Cancer, 25: 83.

[29] Scotto, J. et al. (1974), Non-melanoma skin cancer among Caucasians in four
       areas in the US. *Cancer, 34:* 1333.

[30] Macdonald, E. (1973), *Cancer of the skin in five regions in Texas, 1962-1966*.
       Dept of Epidemiology, Univ. of Texas, Houston.

[31] Waterhouse, J., Muir, C., Correa, P. & Powell, J. (eds.) (1977), *Cancer
       incidence in 5 continents, Vol. III.* IARC, Lyon.

[32] Cerkovinij, G.F. et al. (1975), Malignant neoplasm morbidity among the popu-
       lation of the USSR. *Vop. Oncol., 21:* (1) 3. In Russian.

[33] Lee, J.H.H. & Carter, A.P. (1970), Secular trends in mortality from malignant
       melanoma. *J. Nat. Cancer Inst., 45:* 91.

[34] Elwood, J.M. et al. (1974), Relationship of melanoma and ultra-violet radia-
       tion in the United States and Canada. *Int. J. Epidemiol., 3:* 325.

[35] Clemmesen, J. (1965), *Statistical studies in the aetiology of malignant
       neoplasms II*, p. 408, Munksgaard, Copenhagen.

[36] London, J. (1974), Global trends in atmospheric ozone. *Science, 194:* 987.

[37] Kohmyr, W.D. et al. (1973), Total ozone increase over North America during
       the 1960s. *Pure & Appl. Geophys., 106-108:* 981.

[38] Schulze, R. (1974), Increase of carcinogenic ultra-violet radiation due to
       reduction in ozone concentration in the atmosphere. In: *Proc. Intl
       Conf. on Structure and Composition and General Circulation of the Upper
       and Lower Atmosphere and Possible Anthropogenic Perturbation, Vol. 1,*
       p. 479.

[39] Belinskij, V.A. & Garadza, M.P. (1972), Some results of measurement of ultra-
       violet radiation on Eastern Pamirs. Institute of Geology & Geography
       of the Academy of Sciences of Lithuania SSR Sci. Reports, II, XIII.
       In Russian.

[40] Belinskij, V.A. & Andrienko, L.M. (1976), *Ultra-violet radiation of sun and
       sky* (atlas, maps, etc). Moscow Univ. Press. In Russian.

[41] Berger, D. et al. (1975), Field measurements of biologically effective ultra-
       violet radiation. *CIAP, Vol. 5*, Appendix D, DOT.

[42] Scott, E.L. et al. (1975), *Environmental impact of stratospheric flight*

Appendix C, p. 177. National Academy of Sciences, Washington DC.

[43] Urbach, F. & Davies, R.E. (1976), Estimate of effect of ozone reduction in the
      stratosphere on the incidence of skin cancer in man. In: Hard, J.M. &
      Broderick, A.J. (ed.), *Proc. IV Conference on the Climatic Assessment
      Program*, Springfield, Va.

[44] Green, A.E.S. et al. (1976), The ultra-violet dose dependence of non-melanoma
      skin cancer incidence. *Photochem. & Photobiol., 24:* 353.

[45] Rundel, R.D. & Nachtwey, D.S. (1977), Skin cancer and ultra-violet radiation
      (NASA - Johnson Space Center, Houston, Texas). *Photochem. & Photobiol.*

[46] Mo, T. & Green, A.E.S. (1974), A climatology of solar erythema dose.
      Photochem. & Photobiol., 20: 483.

[47] Blum, H.F. (1974), Uncertainty of growth of cell populations in cancer.
      J. Theor. Biol., 46: 143.

[48] Cutchis, P. (1975), *Ultra-violet, ozone and skin cancer,* Appendix C, p. 7.
      Monograph No. 5. Dept of Transportation, CIAP, Washington DC.

PAPER 6

Atmosphere Ozone - A Survey of the Current State of Knowledge of the Ozone Layer

Presented by the World Meteorological Organization

INTRODUCTION*

The Problem

In recent years there has been increasing concern about the effects of the pollution of the stratosphere, and in particular the possibility of a reduction in the amount of ozone present at these levels due to photochemical reactions involving certain trace substances, the concentration of which may be increased due to human activities. This concern arises from the danger of increased solar ultra-violet radiation penetrating to the Earth's surface and from the possibility of induced changes in the global climate.

Some of WMO's Activities in Connexion with Atmospheric Ozone

WMO was first involved in this field by collecting and publishing world ozone data for the International Geophysical Year (IGY) and International Geophysical Co-operation (IGC). Following a decision by the Third World Meteorological Congress in 1959 to extend WMO's sphere of action to include responsibilities in international ozone work, Members were urged to establish total-ozone observing stations and programmes of ozonesondes to measure the vertical ozone distribution. WMO regional associations were asked to designate at least one of their ozone stations as a regional standard.

Following the IGY, WMO has sponsored, by arrangement with the Atmospheric Environment Service of Canada, the continued publication of world ozone data. These were issued annually for the years 1960 to 1964, and since 1965 the data have been published bi-monthly in order to get them to research workers more quickly. A catalogue containing a list of stations measuring atmospheric ozone and giving the programme of observations was first published in 1962 and is continuously updated and incorporated periodically in the bulletins issued by the World Ozone Data Centre at Toronto.

In response to increasing public concern as to the possibility of a depletion of the stratospheric ozone due to pollutants resulting from human activities, and the attendant risk of increased amounts of solar ultraviolet radiation reaching the Earth's surface, some nationally sponsored studies of the problems have been

---

*Readers for whom a summarized form of this paper is sufficient may confine their attention to pp. 75-80, 106-121.

undertaken, notably the Climate Impact Assessment Program (CIAP) of the U.S.A.

The Seventh World Meteorological Congress (May, 1975) agreed that "there was an urgent need for more studies (and for a definitive review of these studies) to determine the extent to which man-made pollutants might be responsible for reducing the quantity of ozone in the stratosphere". Additionally, Congress "stressed the need to determine the role played by chlorofluoromethanes (used in refrigerants and aerosol cans) in destroying ozone. As part of these studies, the global ozone-monitoring system should be strengthened and in this connexion the Secretary-General was requested to examine the possibility of collaborating with UNEP within the context of its Global Environment Monitoring System (GEMS)".*

Some four months later, the ICSU International Union of Geodesy and Geophysics (IUGG) at its sixteenth General Assembly in Grenoble (France) in turn called for greater efforts in research to increase our knowledge of progesses affecting the upper atmosphere, with special emphasis on the stratosphere and possible consequences of pollution therein, and urged co-operation in international monitoring and other activities aimed at determining any long-term trends in the stratosphere in so far as these might affect environmental quality. IUGG recommended that appropriate intergovernmental organizations, especially WMO, provide encouragement, support and co-ordination to such endeavours.

At a meeting in September 1975 the WMO Commission for Atmospheric Sciences Working Group on Stratospheric and Mesospheric Problems, together with some of the world's most eminent scientists in the field, discussed the subject of stratospheric ozone. Representatives of UNEP and ICSU took part in the meeting during which an official WMO statement entitled "Modification of the ozone layer due to human activities and some possible geophysical consequences" was drafted and subsequently released with the agreement of the WMO Executive Committee at the end of 1975. It was at this session also that the outline of a Global Ozone Research and Monitoring Project was formulated, further detailed elaboration being undertaken by a smaller ad hoc group of experts in January 1976 (with UNEP and ICSU again being represented).

WMO Global Ozone Research and Monitoring Project

At its twenty-eighth session in June 1976, the WMO Executive Committee adopted Resolution 8 (EC-XXVIII) and thereby approved the proposed Global Ozone Research and Monitoring Project. The resolution is reproduced in full in Appendix A.

The aim of the Project is to enable WMO to provide advice to Members and to the United Nations and other appropriate international organizations concerning:

- the extent to which man-made pollutants might be responsible for reducing the quantity of ozone in the stratosphere, with particular emphasis on the role played by chlorofluoromethanes and nitrogen oxides;
- the possible impact of changes in the stratospheric content of ozone on climatic trends and on solar ultra-violet radiation at the Earth's surface;
- the establishment of the basis and identification of the needs for strengthening the long-term monitoring programme of the ozone system for determination of trends and of future threats to the ozone shield.

The means by which the objectives may be achieved are as follows:

- to gather and evaluate existing knowledge on atmospheric ozone;
- to provide information on the atmospheric concentrations of substances of

---

*Abridged report of Seventh Congress, general summary paragraph 3.2.1.3.

of relevance to the N_2O, NO_x, HO_x, ClO_x, BrO_x cycles leading to a better appreciation of their currently-known important sources and sinks and their impact on the ozone budget;

- to extend and improve analyses and circulation studies of the stratosphere based on actual data, with a view to clarifying and predicting the exchange processes between the troposphere and the stratosphere;

- to assess the impact on the atmospheric environment of possible ozone depletion, recognizing that this involves a variety of interdisciplinary interactions requiring input from the fields of meteorology, oceanography, agriculture and microbiology;

- to arrange for an exchange of knowledge through reports, newsletters, scientific meetings and exchange of personnel.

The Global Ozone Research and Monitoring Project was welcomed by the International Ozone Commission of ICSU's International Association of Meteorology and Atmospheric Physics (IAMAP) which went on to recommend that all measures be taken to implement the proposals as completely and as quickly as possible, especially the monitoring of trace gases at ozone measuring stations.

A Brief Review of Scientific Progress

Ozone is formed at stratospheric levels as a result of the photodissociation of molecular oxygen. It is of major importance in the meteorology of the stratosphere, since by virtue of its absorption of solar ultra-violet radiation and the resulting heating effect, it largely determines the basic temperature structure and general circulation of this region. Moreover, since it behaves essentially as an inert tracer in the lower stratosphere, observations of its distribution provide considerable information on transport mechanisms in this region and also transfer between the stratosphere and troposphere.

If the ozone formation and destruction processes were determined solely by the photochemical reactions involving only the oxygen species (the Chapman reactions) no effects resulting from human activities would be expected. However, improved observations of the vertical distribution of ozone obtained during and after the International Geophysical Year made it clear that the pure oxygen photochemistry did not completely explain the actual processes found to be taking place, and subsequently the importance of stratospheric trace gases was discovered, especially the oxides of nitrogen (NO_x), in controlling the ozone balance. This led to the conclusion that ozone might be less stable against outside man-made influences than had previously been believed.

The main natural source of NO_x in the stratosphere is thought to be the oxidation of nitrous oxide (N_2O) (of biological origin at the surface of the Earth and the sea) by excited oxygen atoms which themselves are a product of the photodissociation of ozone. Other natural sources of NO_x whose magnitude is not yet well enough established are cosmic rays and solar proton events and downward transport from the thermosphere in high latitudes of the winter hemisphere. The possibility of some direct supply of NO_x from the troposphere below cannot be ruled out.

A theory has recently been propounded that increased use of agricultural fertilizers and/or of nitrogen-fixing vegetation might affect the nitrogen cycle and result in an increase in the amounts of nitrous oxide (N_2O) released from the surface into the atmosphere. This would then lead to an increase in NO_x in the stratosphere and hence a decrease in the ozone amount. This source of N_2O might also be stimulated by increases in the acidity of rain.

There has already been considerable research conducted into the direct injection of NO_x into the stratosphere by fleets of supersonic aircraft and the consequent effect on the ozone layer. Although there are considerable uncertainties in both measurements and theory (in the latter case being of about a factor of two), the role of NO_x is sufficiently well established to be able to state with reasonable confidence that:

- currently planned SSTs due to their lower flight altitudes of 17 km and their limited numbers (fewer than 50 projected) are not predicted to have an effect that would be significant or that could be distinguished from natural variations;

- a large fleet of supersonic aircraft flying at greater altitudes is predicted to have a noticeable effect on the ozone layer, and permissible total emission levels may have to be defined by international agreement.

Recently it has been shown that ozone is also destroyed by the catalytic cycle of oxides of chlorine ($Cl-ClO_x$). In this case, ozone destruction is accelerated by increased HO_x, but retarded by the presence of methane and nitric oxide. The amount of naturally produced ClO_x in the stratosphere seems to be small and its effect on ozone considerably smaller than that of NO_x. The role played by the bromine $Br-BrO_x$ cycle could be similar to that of the $Cl-ClO_x$; but given equal amounts of Br and Cl in the stratosphere the former would be the most effective in destroying ozone.

The increase in the manufacture and release into the atmosphere of chlorofluoro-methanes, especially $CFCl_3$ (F-11) and CF_2Cl_2 (F-12), see Table 6.1, is predicted to result in a rapidly increasing amount of ClO_x in the stratosphere. The removal rate of ClO_x is slow; it is to be expected that the stratosphere ClO_x concentration will continue to increase for several years even were all emissions of chlorofluoro-methanes into the atmosphere to cease, due to the slow diffusion rates into and through the stratosphere. Thereafter the recovery rate would be very slow (a few decades).

Methods of Measuring Atmospheric Ozone

At this point it may be useful to summarize briefly the kinds of measurements of atmospheric ozone that are currently being made.

Total ozone (surfaced-based). The total amount of ozone in the column of the atmosphere above the observing station is measured by a spectrophotometer. The instrument most generally used is the *Dobson ozone spectrophotometer* which, although bulky and delicate, gives estimates of total ozone to within 2% if carefully operated and maintained. Certain other recently-developed spectrophotometers are being tested against the Dobson instrument with satisfactory results. WMO has periodically organized intercomparisons of Dobson ozone spectrophotometers, the latest having been at Aspendale (Australia) in 1972 and at Belsk (Poland) in 1974. The next is planned for the summer of 1977 in Boulder (Colorado, U.S.A.).

Total ozone (satellite). Observations by a total ozone measuring device carried by a polar-orbiting satellite have the obvious advantages of quasi-global coverage by one and the same sensor (thus being readily comparable). The Backscatter Ultra-violet (BUV) instrument carried by a satellite operated by the U.S.A. gives values of total ozone which are compatible to within 6% of those obtained by Dobson spectrophotometers.

Surface ozone. Since this involves a simple in-situ chemical analysis process, relatively inexpensive automatic instruments are available for measuring the amount of ozone at or near the surface.

Vertical distribution of ozone. Balloon-borne ozonesondes are probably the most

TABLE 6.1   Estimates of the Global Production of Chlorofluoromethanes $CFCl_3$ and CF_2Cl_2 - in millions of
pounds. (Data from the Manufacturing Chemists'
Association, 1 March 1976)

$CFCl_3$ Annual	$CFCl_3$ Cumulative	Year	CF_2Cl_2 Annual	CF_2Cl_2 Cumulative
.0	.0	1931	1.2	1.2
.0	.0	1932	.3	1.5
.0	.0	1933	.7	2.2
.1	.1	1934	1.5	3.7
.1	.2	1935	2.2	5.9
.2	.4	1936	3.8	9.7
.3	.7	1937	6.8	16.5
.2	.9	1938	6.2	22.7
.2	1.1	1939	8.7	31.4
.4	1.5	1940	10.0	41.4
.6	2.1	1941	13.8	55.2
.7	2.8	1942	13.1	68.3
.9	3.7	1943	18.1	86.4
.8	4.5	1944	36.9	123.3
.8	5.3	1945	44.3	167.6
1.6	6.9	1946	36.7	204.3
2.9	9.8	1947	44.4	248.7
6.6	16.4	1948	54.6	303.3
9.9	26.3	1949	57.6	360.9
14.6	40.9	1950	76.2	437.1
20.0	60.9	1951	79.9	517.0
29.9	90.8	1952	82.1	599.1
38.1	128.9	1953	102.5	701.6
46.1	175.0	1954	108.3	809.9
58.0	233.0	1955	128.0	937.9
71.7	304.7	1956	151.9	1089.8
74.9	379.6	1957	164.2	1254.0
65.4	445.0	1958	162.8	1416.8
78.8	523.8	1959	194.6	1611.4
110.1	633.9	1960	221.4	1832.8
133.8	767.7	1961	242.4	2075.2
173.2	940.9	1962	287.4	2362.6
207.2	1148.1	1963	329.8	2692.4
246.4	1394.5	1964	385.0	3077.4
273.8	1668.3	1965	432.0	3509.4
312.2	1980.5	1966	499.6	4009.0
359.9	2340.4	1967	567.2	4576.2
411.7	2752.1	1968	609.7	5185.9
491.0	3243.1	1969	685.4	5871.3
541.0	3784.1	1970	737.9	6609.2
604.2	4388.3	1971	783.0	7392.2
698.4	5086.7	1972	877.0	8269.2
809.1	5895.8	1973	970.2	9239.4
880.5	6776.3	1974	1041.9	10281.3
786.0	7562.3	1975	915.9	11197.2

most reliable method at present of measuring the vertical distribution of ozone, but the highest altitude reached is usually about 30 km, or somewhat above the level of maximum ozone concentration. Fortunately, remote sensing from satellites is currently capable of providing useful information on the vertical ozone profile down as far as the level of maximum concentration, so that the two techniques are complementary in producing a complete profile. These sondes and satellite sensors each give a precision of 5 to 10% above 30 km. Rocket-borne ozonesondes are a means of obtaining a homogeneous profile. WMO has organized intercomparisons of ozone-sondes and comparisons with satellite-derived data.

As a back-up for the ascending/sensing systems mentioned above, it is possible to obtain some indication of the vertical distribution of ozone using an ozone spectrophotometer near sunrise and sunset, the so-called Umkehr technique.

REVIEW OF CURRENT KNOWLEDGE RELEVANT TO ATMOSPHERIC OZONE

Ozone Photochemistry: Facts and Theories

Solar radiation, in particular in the UV region, is the driving force in the ozone chemistry at all heights. The first theory of ozone generation and destruction, which was generally accepted for many years, was that propounded by Chapman in the early 1930s. Only since the mid 1960s was knowledge made more refined, as will be seen from the following paragraphs. Any assessment of the mechanisms leading to the existing ozone distribution, as well as predictions of anthropogenic perturbations in ozone, rest upon adequate knowledge of (a) the solar spectrum, (b) the attenuation of the solar flux caused by absorption and scattering, and (c) the resulting capability to dissociate various gases. These factors are reflected in the recommendations in the section "Studies of photochemistry".

In order to provide a somewhat more fundamental background in ozone photochemistry for the benefit of those intimately involved in the problem, as well as to inform on the more recent developments, the following represents the very basic photo-chemistry in symbols and in words.

The Chapman theory. Photodissociation of molecular oxygen (O_2) by solar ultra-violet radiation ($h\nu$) at wavelength (λ) less than 242 nm

$$O_2 + h\nu \longrightarrow O + O$$

results in the formation of ozone below 80 km

$$O + O_2 + M \longrightarrow O_3 + M$$

where M is any body (mainly N_2). The ozone in turn can be dissociated by solar radiation ($\lambda < 320$ nm) and to some extent also by visible radiation

$$O_3 + h\nu \longrightarrow O_2 + O*$$

where O* is an oxygen atom in an electronically excited state ($O* \equiv O(^1D)$), and can enter into a destructive reaction with atomic oxygen

$$O_3 + O* \longrightarrow 2 O_2$$

The latter (Chapman) reactions destroy about 20% of the ozone formed.

Hydrogen system. The hydrogen system, while of only minor importance in the direct destruction of ozone (it destroys about 10% of the ozone formed, although becomes dominant in this regard above 40 km), is of great indirect importance in ozone photochemistry.

Several reaction schemes have been suggested to account for this influence. One involves the radicals OH and HO_2 that arise mostly from the photochemistry of water

$$H_2O + O^* \longrightarrow 2HO.$$

The following reactions are particularly important above 40 km:

$$HO + O_3 \longrightarrow HO_2 + O_2$$

$$HO_2 + O \longrightarrow OH + O_2$$

$$\text{Net: } O + O_3 \longrightarrow 2 O_2$$

whereby ozone and one atomic oxygen atom are transformed into two molecular oxygen atoms, and the catalyst (OH) is renewed to repeat the process. In this way one OH radical can be responsible for the destruction of many ozone molecules. OH radicals could be supplied to the stratosphere also when methane (CH_4) (produced principally by bacteria in marsh lands and lakes) diffuses upward and reacts with O^* in the stratosphere to produce the hydroxyl radical (OH)

$$CH_4 + O^* \longrightarrow CH_3 + OH$$

which thereafter follows the above scene. Hydrogen (H), the perhydroxyl radical (HO_2) and OH interconvert very rapidly by reaction with O, so that all three tend to be in a steady state. Depletion of the radicals OH and HO_2 is accomplished principally by

$$OH + HO_2 \longrightarrow H_2O + O_2$$

resulting in the formation of water which can drift down out of the stratosphere. The latter reaction, and its rate, is one of the most important (and sensitive) reactions in the photochemistry of the stratosphere, and hence in the photochemistry of ozone.

Nitrogen system. The ozone balance of the stratosphere is maintained principally by the oxidation of nitrous oxide (N_2O) released at the Earth's surface. N_2O is produced by bacterial action of micro-organisms in the ocean and soil (denitrification) and diffuses upward through the troposphere to the stratosphere. It is relatively inert. In the stratosphere

$$N_2O + O^* \longrightarrow 2 NO$$

$$NO_2 + h\nu \longrightarrow NO + O$$

and the nitric oxide (NO) thus produced catalyzes ozone by

$$NO + O_3 \longrightarrow NO_2 + O_2$$

$$NO_2 + O \longrightarrow NO + O_2$$

$$\text{Net: } O_3 + O \longrightarrow 2 O_2$$

where once again ozone and atomic oxygen are transformed to molecular oxygen. The reaction of nitrogen dioxide (NO_2) with the hydroxyl radical OH

$$OH + NO_2 + M \longrightarrow HNO_3 + M$$

forms nitric acid which is eventually washed out of the troposphere and hence represents the major sink of the system. Almost 60% of the ozone formed is believed to be destroyed by this natural system. Other natural, but less important, stratospheric sources of NO include galactic cosmic rays, solar proton events, meteoroides, lightning in the troposphere (with some upward transport of NO through the tropical tropopause), and a downward flux from the winter thermosphere. Furthermore, it should be noted that increased use of fertilizers and increased acidity of rain could enhance the N_2O release from the ground (see Appendix D). Since NO is formed at temperatures exceeding 2000°K by the reactions

$$O_2 + M \longrightarrow O + O + M$$
$$O + N_2 \longrightarrow N + NO.$$

NO is formed in nuclear explosions and also in jet engines, and the latter is thought to be a serious problem in the case of large numbers of high-flying aircraft.

Chlorine. Natural chlorine makes only a minor (few percent) contribution to ozone destruction, but the chlorine included in man-made chlorofluoromethanes (CFMs) released into the atmosphere may have a serious effect. In particular, the chlorofluoromethanes $CFCl_3$ and CF_2Cl_2 (also known as Freon 11 and 12 respectively) which are used in refrigeration and as a propellant in various aerosol sprays, are inert in the troposphere but are dissociated at a height of about 30 km (near the height of maximum ozone formation) by ultra-violet radiation in the 180-225 nm "window" between the Schumann-Runge and Hartley absorption bands. The chlorine released by the photolysis destroys ozone by the catalytic chain.

$$Cl + O_3 \longrightarrow ClO + O_2$$
$$ClO + O \longrightarrow Cl + O_2$$
$$\overline{\text{Net: } \quad O + O_3 \longrightarrow 2 O_2}$$

with the same consequences as previously. The ClO_x catalytic efficiency is reduced by the reaction

$$ClO + NO \longrightarrow NO_2 + Cl$$

followed by

$$NO_2 + h\nu \longrightarrow NO + O.$$

The chlorine sink is HCl (hydrogen chloride, also known as hydrochloric acid) which is formed primarily through the reaction of chlorine with methane and the perhydroxyl radical

$$CH_4 + Cl \longrightarrow HCl + CH_3$$
$$HO_2 + Cl \longrightarrow HCl + O_2.$$

The balance between the chlorine radicals (Cl and ClO) and HCl is determined by the above two reactions in competition with the recycling step

$$HCl + OH \longrightarrow H_2O + Cl.$$

The latter reactions, and its rate, is another very important (and sensitive) reaction in ozone photochemistry. It might be noted that whereas the chlorine catalysis of ozone is about six times as efficient as the nitric oxide catalysis of ozone, the conversion to "inert" hydrogen chloride is also more efficient than the conversion to "inert" nitric acid, so that the two processes are in fact comparable in overall efficiency. Interaction of the nitrogen and chlorine systems comes about through

$$ClO + NO \longrightarrow ClNO_2$$

forming nitryl chloride and

$$ClO + NO_2 \longrightarrow ClNO_3$$

Chlorine nitrate effect. The U.S.A. National Academy of Sciences (NAS) report of 1976, based on recent studies of the chlorine nitrate reactions, estimated that this effect reduces the degree of ozone depletion due to other reaction from a projected 14% to 7.5%, i.e. by a factor of 1.85. Model calculations and a few measurements, however, suggested that the chlorine nitrate concentration is less than previously thought, apparently because the reaction

$$ClO + NO_2 + M \longrightarrow ClONO_2 + M$$

is much slower than the reaction

$$OH + NO_2 + M \longrightarrow HNO_3 + M.$$

Recent measurements and calculations suggest that in fact the chlorine nitrate chemistry reduces the degree of ozone depletion from 14% to 10-11% (rather than 7.5%). It must be stressed that there is still great uncertainty concerning the magnitude of the chlorine nitrate effect, and more measurements and investigations are needed.

Bromine. Other halogens, bromine in particular, react in a manner similar to chlorine and thus have the same ozone-destroying possibilities, although being perhaps even more efficient than chlorine in this regard. Bromine could be delivered through the decomposition of methyl bromide (CH_3Br), most of which is believed to be produced by marine algae. However, a small but growing man-made source of CH_3Br exists as a result of fumigation processes in agriculture, although the latter is believed to be small compared with the chlorofluoromethane release.

Relevant processes in the troposphere. The amount of ozone in the troposphere is small by comparison with the stratosphere, but nevertheless constitutes a not insignificant part of the total ozone. Consequently changes in tropospheric ozone will result in changes in total ozone to an extent which cannot be ignored.

Sources of tropospheric ozone are twofold, firstly the downward transport from the stratosphere, in particular in middle and high latitudes (this is a natural source which is not likely to undergo systematic changes in the future). Secondly, there is the photochemical production of ozone in the troposphere, in "clean" air as well as heavily polluted air. Whereas some information on these two sources is available, our knowledge about tropospheric ozone is incomplete, particularly as regards the photochemical production.

Photochemical production of ozone in the troposphere is caused partly naturally and partly as a result of human activities. As in the stratosphere, OH and HO_2 play important roles in the ozone chemistry. In clean air hydroxyl originates from the decomposition of water vapour by excited atomic oxygen

$$H_2O + O* \longrightarrow OH + OH.$$

In turn, HO_2 is formed when naturally occurring methane is decomposed through reaction with OH. HO_2 is of particular importance since it converts NO to NO_2

$$NO + HO_2 \longrightarrow NO_2 + OH$$

whereby the production of odd oxygen through photodissociation of NO_2 is increased.

Under certain conditions, the production of ozone in the lowest layers may be drastically increased in heavily polluted urban and industrial areas, where a new source of odd hydrogen, the hydrocarbons, is emitted as a result of consumption of fossil fuels. The HO_2 content (mainly responsible for the conversion of NO to NO_2 in the "clean" atmosphere) is increased by orders of magnitude as a result of oxidation of hydrocarbons. In addition, other radicals, of the type RO_2, which also promote the NO $\longrightarrow NO_2$ conversion, are produced. Ozone produced locally in this way may be transported over long distances and mixed with the surrounding air.

Some recent assessments. The calculated ozone reduction is extremely sensitive to any tropospheric sink for the chlorofluoromethanes. For example, only about 1% of the CFMs in the atmosphere is dissociated each year, so that if there were an inactive sink that removes 2% of the CFMs each year without destroying ozone, then the ozone reduction would be only one-third of the estimate without such a sink. On this basis, the NAS report reduced the eventual ozone depletion from 7.5% to 6.0% largely on the assumption of some CFM absorption by sea water, followed by an unknown degradation process. The presence of such an inactive sink could be verified if accurate estimates of CFM release and the amount of CFM in the atmosphere could be obtained. The latter parameters are particularly uncertain, with considerable controversy regarding the existing gradient of CFM between the northern and southern hemispheres. More work needs to be done on this topic too.

The importance of carbon tetrachloride (CCl_4) in the ozone depletion problem also remains controversial because of the uncertainty as to how much is natural and how much is man-made. Even though the anthropogenic release of CCl_4 should be decreasing, the Adrigole (Ireland) surface measurements continue to show that it is increasing in the atmosphere at the same rate as $CFCl_3$. Until the CCl_4 problem is resolved, uncertainty is bound to surround the ozone depletion problem.

A speculative positive feedback process involves an increase in the upwelling of water vapour through the tropical tropopause due to warming of the tropopause because of redistribution of the ozone to lower levels owing to destruction of ozone at around 30 km by dissociation of CFMs. This would lead to more OH radicals

$$H_2O + O* \longrightarrow 2\ OH$$

and hence to catalytically active chlorine through

$$HCl + CH \longrightarrow H_2O + Cl.$$

The only long-term stratospheric water vapour measurements available at Washington, D.C., indicate a 30% increase in stratospheric (14-26 km) water vapour between 1964 and 1969, little change between 1970 and 1974, and a sharp drop since 1974 back to almost the 1964 level.

A few additional facts which became available at the Logan (Utah, U.S.A.) conference just at the time of release of the NAS report in September 1976 put in question several of the assumption made in the NAS report and as already noted (see Chlorine nitrate effect above) have tended to support an even greater reduction in ozone than the projected figure of 7.5% given in the NAS report.

For instance, measurements by Dr Rasmussen of Washington State University of chlorine compounds in the atmosphere showed a greater level of CFMs carbon-tetrachloride, and other potential ozone-reducing compounds. These findings were consistent with the major release of these substances by the industrialized nations. The findings also strongly suggested that much of the atmospheric load of carbon-tetrachloride is due to human activity. This represents an additional source of ozone depletion by the human race.

Of similar import were the findings of Dr Anderson of the University of Michigan, who presented the first simultaneous measurements of chlorine atoms and chlorine

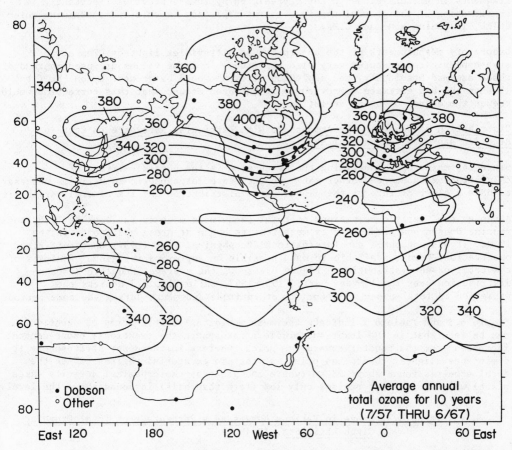

Fig. 6.1. Global average total ozone distribution for the years 1957–1967, expressed in m-atm-cm (Dobson units). The dots and small circles indicate station locations. (London and Bojkov)

oxides at different altitudes in the stratosphere. The findings of these chemical fragments in amounts close to those predicted by theory provides important confirmation of the general theory of ozone depletion by chlorine-containing mole-cules, including man-made CFMs.

Laboratory measurements of the absorption of ultraviolet light by CFMs at stratospheric temperatures were also reported. If these values are used instead of those assumed previously, the depletion of ozone would be greater by one-half to 1% than the NAS estimate. Previously it had been argued that this correction would reduce the degree of ozone depletion.

Distribution of Ozone in the Atmosphere with Particular Reference to Certain Recorded Man-Made and Natural Events

Since the main topic underlying the discussion is the possible effect of chlorofluoromethanes and other man-made species on atmospheric zone, it is necessary to have a clear picture of the global ozone distribution and its natural variability.

Although some early observations of total ozone were made as far back as 1908, and routine Dobson spectrophotometer measurements began at Arosa and Tromso in the 1930s, it was not until the International Geophysical Year (1957) that sufficient observations became available to make possible an approximative global analysis of the total ozone distribution. Figure 6.1 shows the average annual total ozone distribution over the 10-year period 1957-1967, while Fig. 6.2 depicts the variation in total ozone as a function of latitude and month during the same period.

Figure 6.3 and Table 6.2 indicate the mean meridional distribution of ozone. It can be seen that in the lower and middle stratosphere the fractional ozone content increases from the Equator toward the poles, whereas in the upper stratosphere it diminishes slightly toward the poles, so that in equatorial regions 38% of the total ozone is found above 28 km (where the greatest photochemical activity takes place) whereas in polar regions only 18% (less than half) is found above the level.

TABLE 6.2. Ozone in Various Layers as a Percentage of Total Ozone at Given Latitudes. (Bojkov)

Layer (km approx.)	0-10	10-15	15-19	19-24	24-28	28-33	33-38	Above 38 km
Latitude: polar	10	17	18	19	17	10	5	3.5
middle	9	11	14	20	20	14	7	4.5
subtropical	8	6	9	20	25	17	9	5.4
equatorial	4	3	5	19	30	21	10	7.3

Figure 6.4 shows the variability of the ozone partial pressure in terms of its standard deviation. The variability is largest in the low stratosphere in middle and high latitudes, and decreases to a relatively small value at all levels in the tropics. In other studies it is indicated that the ozone partial pressure turns out to be most highly correlated with the total ozone in the column at heights between 10 and 20 km in middle and high latitudes, i.e., changes in total ozone mainly reflect changes in ozone amount in this layer. In the upper stratosphere these correlations are very weak, and in middle latitudes there is even a region where the correlation is negative, that is, if the total ozone increases, the ozone amount in the region near 50 km tends to decrease.

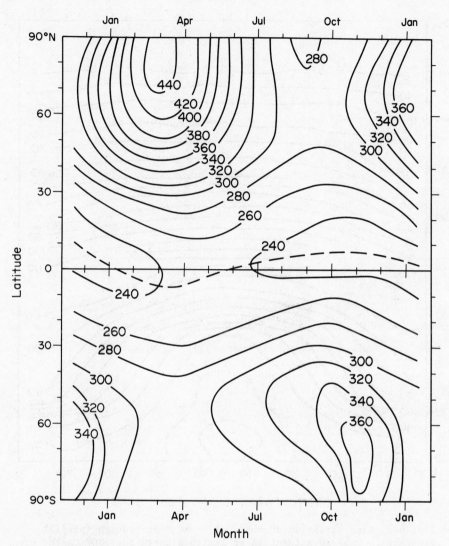

Fig. 6.2. Average variation in total ozone (m-atm-cm) by latitude
and month, based on observations during 1957-1967. (London and Bojkov)

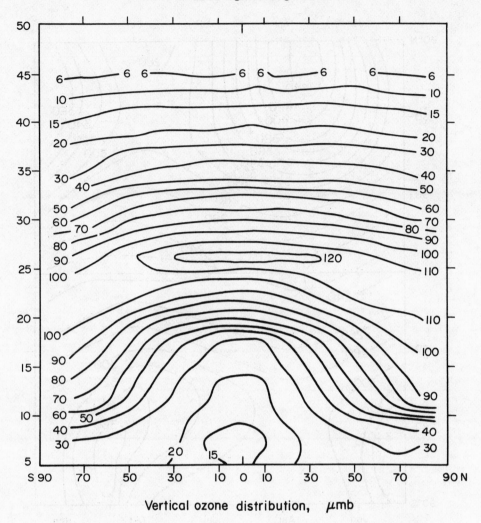

Vertical ozone distribution, μmb

Fig. 6.3. Mean variation of vertical ozone distribution (partial
pressure in μmb) as a function of latitude based on 8,500 Umkehr profiles and ozone soundings taken between 1956 and 1966.   (Bojkov)

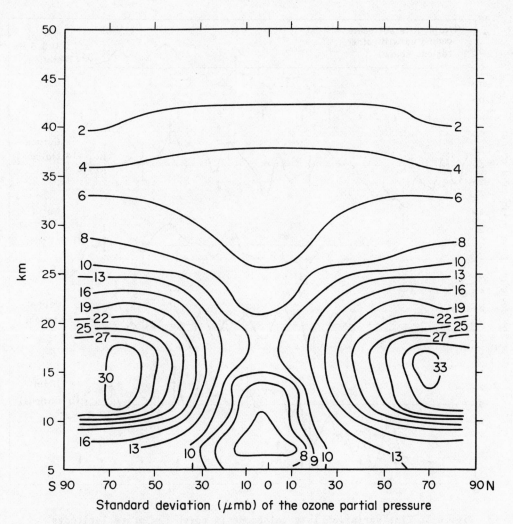

Fig. 6.4. Standard deviation of ozone partial pressure (μmb) as a function of latitude and height. Large values indicate high variability. (Bojkov)

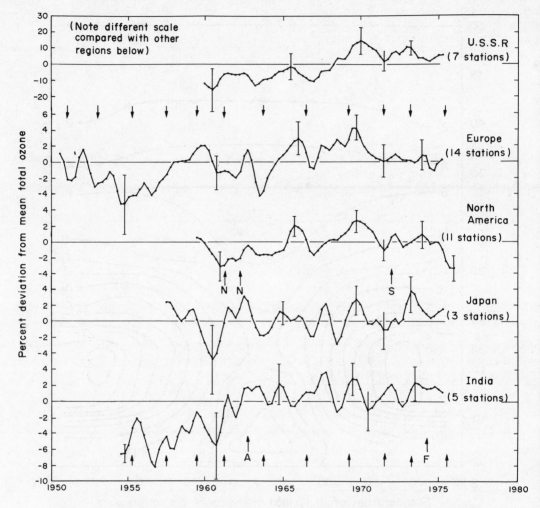

Fig. 6.5. Time variation in total ozone in north temperate latitudes
expressed as a percentage deviation from the mean for the total length
of record (the annual oscillation has been removed). A 1-2-1 smoothing
(divided by four) has been applied twice to the successive seasonal
values.   (Angell and Korshover)

Vertical bars represent two standard errors of estimate based on
individual station values within the regions. Single-shafted arrows
indicate occurrence of quasi-biennial west wind maximum at 50 mb
in the tropics;

A = Eruption of Mt Agung (Indonesia)   F = Eruption of Mt Fuego
                                                 (Guatamala)

N = Large nuclear explosions   S = Large solar proton event.

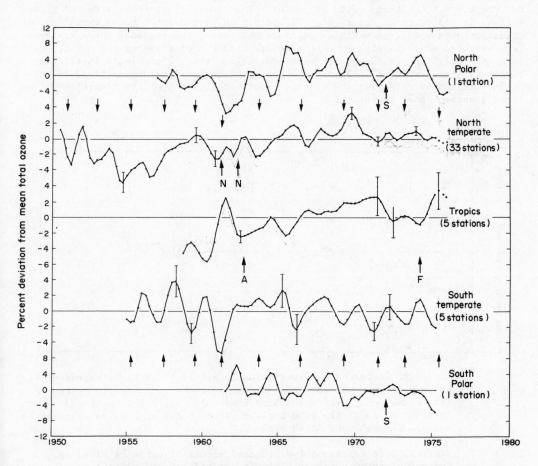

Fig. 6.6. Time variation in total ozone in global climatic zones
expressed as a percentage deviation from the mean for the total length
of record (the annual oscillation has been removed). A 1-2-1 smoothing (divided by four) has been applied twice to the successive seasonal
values. (Angell and Korshover)

Vertical bars represent two standard errors of estimate based on
individual station values within the regions. Single-shafted arrows
indicate occurrence of quasi-biennial west wind maximum at 50 mb
in the tropics;

A = Eruption of Mt Agung (Indonesia)   F = Eruption of Mt Fuego
                                             (Guatamala)

N = Large nuclear explosions   S = Large solar proton event.

Figure 6.5 shows the time variation in total ozone for those regions in north temperate latitudes possessing adequate data. The annual oscillation has been eliminated and a 1-2-1 smoothing (divided by 4) applied twice to successive seasonal values. Note that the ordinate is precentage deviation from the mean with the scale for the U.S.S.R. stations quite different. The number of stations with the given regions is indicated at the right. Figure 6.6 shows the total ozone variation for climatic zones where, in north temperature latitudes, the regional averages have been weighted subjectively according to area, number of stations and quality of data (the weighting is Europe 3, North America 3, Japan 1, and India 1, filter ozonometers are excluded). Figure 6.7 shows the derived world-wide trend in total ozone obtained by weighting the climatic zones of Fig. 6.6 by the area of the Earth's surface they represent.

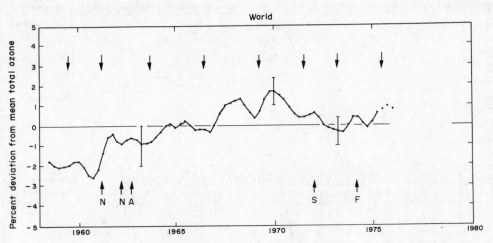

Fig. 6.7. Estimated time variation in global total ozone expressed as as percentage deviation from the mean for the total length of record (the annual oscillation has been removed). A 1-2-1 smoothing (divided by four) has been applied twice to the successive seasonal values. (Angell and Korshover)

Vertical bars represent two standard errors of estimate based on individual station values. Single-shafted arrows indicate occurrence of quasi-biennial west wind maximum at 50 mb in the tropics;

A = Eruption of Mt Agung (Indonesia)   F = Eruption of Mt Fuego
                                             (Guatamala)

N = Large nuclear explosions   S = Large solar proton event.

In general, based on the period of available observation (which is really *much too short* for the accurate delineation of trends), the total ozone increased during the 1960s, peaked about 1970, then declined about half as much between 1970 and 1973 as it had risen during the 1960s, and since 1973 has remained essentially unchanged. Note, however, that Japan and India do not exhibit this variation, and in the Southern Hemisphere there is evidence of a slight decrease in total ozone over most of the period of record. The single-shafted arrows in Figs. 6.5 - 6.9 indicate the time of quasi-biennial west wind maximum at 50 mb (20 km) in the tropics, and it can be seen that whereas this oscillation is most obvious in total ozone in south temperate latitudes, it also permeates most other climatic zones. It is emphasized, however, that these variations and trends must be accepted with great caution since the total ozone stations are mostly confined to the continents so that changes in circulation patterns could affect the results.

There is evidence, particularly in Europe, that the amount of ozone peaked about 1960 as well as in 1970, and both times are near the time of sunspot maximum. It is not at all certain that there is a real relation here, but *if* there is, this would explain the levelling-off of the ozone in recent years (the present sunspot minimum is unusually prolonged) and we should expect the ozone to increase slightly in the near future.

The double-shafted arrows labeled N in Figs. 6.5 - 6.9 indicate the time of large nuclear explosions at Novaya Zemlya (75ºN) in the autumn of 1961 and 1962. It is known that nitric oxide (NO) is produced by the high temperatures in a nuclear explosion, but on the other hand as the fireball cools by radiation emission, that radiation between 185 and 242 nm would dissociate oxygen and create ozone in a region surrounding the explosion. Hence, for a few days following the explosion one might even find an enhancement rather than reduction in total ozone downwind of the detonation site. After a week or two, though, the nitric oxide effect should dominate. Whilst the magnitude of the above explosions suggests a large production of nitric oxide in the low and middle stratosphere, the nitric oxide would not (basically) have been able to react with, and decrease, ozone for a few months due probably to circulation specifics. The model calculation of Chang indicated a 3-4% decrease in hemispheric total ozone due to these explosions, but such an effect is not obvious from Fig. 6.5 - 6.7. Even though the picture is clouded by quasi-biennial oscillation, it is difficult to see how the nuclear-induced reduction in total ozone in northern latitudes could have exceeded 1%. The discrepancy between observation and theoretical model has not yet been explained.

The double-shafted arrow labeled A in Figs. 6.5. - 6.9. indicates the time of the eruption of Mt Agung in Bali, Indonesia (8ºS) in the spring of 1963. At the time of the eruption the strong quasi-biennial oscillation in total ozone in south temperate latitudes ceased, and the ozone amount remained at a relatively high level until 1965, when the oscillation resumed, though not with its original vigour. It is tempting to relate the breakdown of the quasi-biennial oscillation in total ozone to the eruption, but the quasi-biennial oscillation in zonal wind in the tropical stratosphere increased from 2 to nearly 3 years at this same time, which seems a more likely cause for the breakdown in the total ozone oscillation. In any case, there is no evidence that the hydrogen chloride and water vapour emitted by Mt Agung reduced the ozone column; indeed no hydrogen chloride was observed in the stratosphere from the eruption of Mt Fuego in Guatemala in 1974, which may also have been the case with the Mt Agung eruption.

High energy solar proton events have the capability of producing nitric oxide (NO) through ionization of nitrogen by means of secondary electrons formed by energetic protons. The largest such event in the last 15 years (there apparently was an even larger event in 1959) occurred on 4 August 1972 (double-shafted arrow labeled S in Figs. 6.5 - 6.8. Figure 6.6 indicates no obvious effect of the solar proton event on total ozone at the north polar region (Resolute), very close to the magnetic pole where any effect on the ozone amount should be maximal (Resolute is the only high latitude Arctic station).

Figure 6.8 shows the time trend in ozone for the height intervals 32-37, 37-42, and 42-47 km as derived from Umkehr measurements in the 4 regions (Europe, Japan, India, Australia) presently reporting such measurements. The ozone trend at these elevations is generally similar to the trend in total ozone, although the reduction in ozone after 1970 is not so obvious at the high levels. It is not certain whether the tendency for the percentage variation to increase with height is real, or

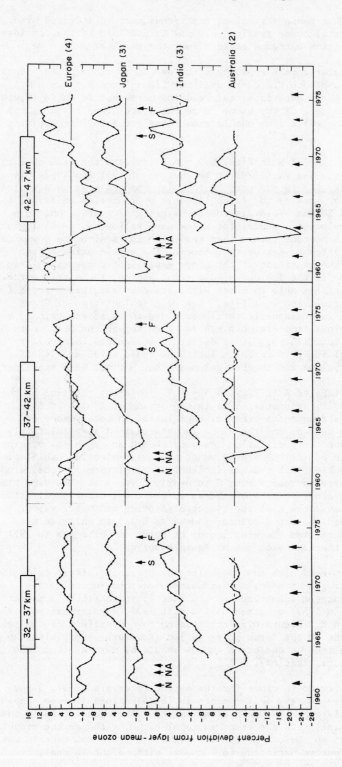

Fig. 6.8. Time variation of ozone (percentage deviation from the mean) in different layers as determined from Dobson spectrophotometer Umkehr measurements at stations in Europe, Japan, India and Australia. (Angell and Korshover)

The number of stations is indicated in brackets. Single-shafted arrows indicate occurrence of quasi-biennial west wind maximum at 50 mb in the tropics;

A = Eruption of Mt Agung (Indonesia)   F = Eruption of Mt Fuego (Guatamala)
N = Large nuclear explosions   S = Large solar proton event.

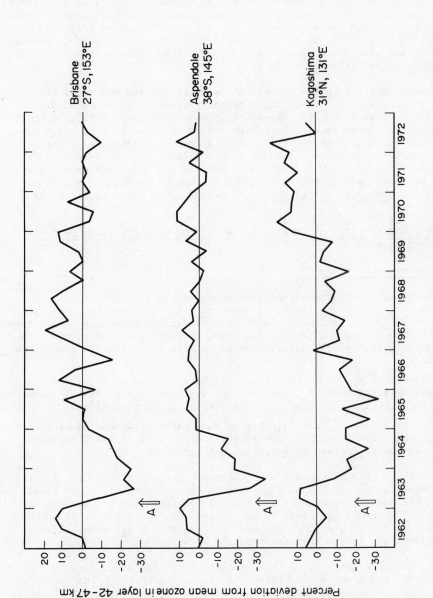

Fig. 6.9. Indicated variation (percentage deviation from the mean) of ozone in the 42-47 km layer at Brisbane, Aspendale and Kagoshima. The annual oscillation has been eliminated, but the seasonal data are unsmoothed. Double-shafted arrow "A" marks the time of eruption of Mt Agung (8°S, 115°E). (Angell and Korshover)

inherent in the analysis technique. A striking feature is the indicated decrease
in ozone above 35 km at Australian stations following the eruption of Mt Agung in
1963. Figure 6.9 shows that the decrease occurred slightly later at sites further
removed (latitudinally) from the eruption site, as though a dust cloud were spread-
ing meridionally. Since the height of the volcanic dust cloud could hardly have
exceeded 25-30 km, and yet the magnitude of the ozone reduction is indicated to
increase with height up to 50 km, it is probable that the indicated ozone reduction
is fictitious rather than real, demonstrating that Umkehr observations in a turbid
atmosphere should be scrutinized carefully.

The Nimbus 4 ozone data are in the process of being re-reduced and until this is
accomplished, and new analyses made, it is inappropriate to review these data.
Nevertheless, it must be noted that the BUV instrument on Nimbus 4 (Fig. 6.10) has
suggested a dramatic consequence of the solar proton event of 1972 on total ozone
amounts above the level of 4 mb (40 km) at high geomagnetic latitudes. The
percentage decreases in ozone is indicated to be at least 15% at these high levels
(4 August is day 218), but only slightly more than 1% for the total ozone column,
a value which would be lost in the noise level of the Dobson data at Resolute.
Thus, we appear to have here the first "hard" evidence for ozone reduction by
nitric oxide.

SOME CONSEQUENCES OF INCREASING AMOUNTS OF MAN-MADE TRACE GASES IN THE ATMOSPHERE

Consequences on the Ozone Layer

There has already been considerable research conducted into the direct injection
of NO_x into the stratosphere by fleets of supersonic aircraft and the consequent
effect on the ozone layer. Although there are considerable uncertainties in both
measurements and theory (in the latter case being of about a factor of two), the
role of NO_x is sufficiently well established to be able to state with reasonable
confidence that:

- currently planned SSTs, due to their lower flight altitudes of 17 km and their
  limited numbers (fewer than 50 projected) are not predicted to have an effect
  that would be significant or that could be distinguished from natural variations;

- a large fleet of supersonic aircraft flying at greater altitudes is predicted to
  have a noticeable effect on the ozone layer, and permissible total emission
  levels may have to be defined by international agreement.

A theory has recently been propounded that increased use of agricultural fertilizers
and/or of nitrogen-fixing vegetation might affect the nitrogen cycle and result in
an increase in the amounts of nitrous oxide (N_2O) released from the surface into
the atmosphere. This would then lead to an increase of NO_x in the stratosphere
and hence to a decrease in the ozone amount. This source of N_2O might also be
stimulated by increases in the acidity of rain.

Because of the extreme complexity of this problem, involving as it does the whole
global nitrogen cycle, uncertainties regarding the consequences on the ozone layer
are still very great. Although there is no likelihood of a significant change in
the ozone layer in the near future as a result of changing agricultural practices,
the matter deserves thorough study because of the importance of fertilizers in food
production and the substantial long-term effects which some scientists consider to
be possible (see also Appendix B).

Prediction of the degree of ozone depletion resulting from a given rate of release
of chlorofluoromethanes (CFMs) into the atmosphere must be made using atmospheric
models, and so far the model generally employed has been a 1-D model which, although

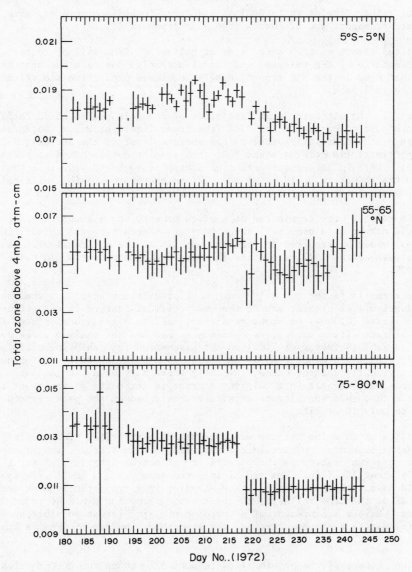

Fig. 6.10. Time variation of total ozone above 4 mb (above 40 km) as a function of latitude, as measured by the BUV instrument on Nimbus 4. A large solar proton event occurred on day 218. (Angell)

permitting rather complex photochemical reactions to be incorporated, simplifies atmospheric transport and diffusion (see also Appendix C).

The predicted effect on the ozone layer of release of CFMs will of course depend upon the scenario of CFM release, and for illustrative purposes the one chosen here is that used in the NAS report, namely continued production and release of CFMs at the 1973 rate.

On this basis, the ultimate "peak" depletion of ozone is predicted in the NAS report to amount to between 6 and 7.5% (the lower figure assuming that a small percentage of the CFMs is absorbed by the surface layer of the oceans); 95% confidence limits lie between 2 and 20%. The predicted time taken to reach half the peak (i.e. 3 to 3.75 percent reduction of ozone) is 40 to 50 years, so that by the year 2000 the reduction would be at most 2% and therefore possibly difficult to detect from the total-ozone observation network.

If the CFM release were terminated at a given moment, the maximum reduction in ozone would not be attained until about 10 years thereafter, and an additional 60-70 years would be required to recover one-half of the maximum loss. Each year of CFM release at the 1973 rate would increase the peak reduction in ozone by about 0.07%.

The above scenario is not the only one. Most predictions suggest a somewhat larger ozone reduction. At present one of the most critical factors is the effect of chlorine nitrate ($ClONO_2$) on ozone reduction. As has already been stated (see chlorine nitrate effect above), the NAS report reduced the maximum ozone reduction from 14% to 7.5% (a factor of 1.85) on the assumption that chlorine nitrate is relatively stable and hence ties up an appreciable fraction of the chlorine radicals which catalytically destroy ozone. Recently there have been both measurements and calculations that suggest that chlorine nitrate is not quite so important in this regard, and that chlorine nitrate chemistry should reduce the peak ozone depletion from 14% to only 10 or 11%.

The stability of chlorine nitrate was somewhat of a surprise. The possibility must be considered that other photochemical surprises are in store. Probably the one other most critical factor is the possibility of a tropospheric sink for the CFMs. Since only about 1% of the CFMs are dissociated each year, a small tropospheric sink would have tremendous impact. For example, if a tropospheric sink accounted for only 2% of the CFMs released, the peak ozone reduction would be reduced by two-thirds. Direct measurement of a tropospheric sink is not possible because both the CFM release to the atmosphere and the atmospheric burden of CFMs is imprecisely known.

An important aspect of the hypothesized reduction in ozone due to CFM release is shown in Fig. 6.11. This diagram presents the expected percentage reduction in ozone, as a function of height, for the years 1972, 2000 and 2025. The reduction is strongly peaked near a height of 40 km, i.e. near the region where the CFMs dissociate most rapidly and where ozone is preferentially formed. According to this diagram, the direct measurement of a reduction in ozone due to CFMs could best be carried out near a height of 40 km. This is too high for most ozonesondes, but BUV satellite data are good and Umkehr measurements are relatively precise if correctly taken. Some of the available Umkehr measurements (Fig. 6.8) show only slight evidence of a recent ozone decrease at these heights, but considerable evidence of an ozone *increase* at these heights between about 1963 and 1973.

Consequences on Climate

Any consideration of the possible influence on the climate of an eventual ozone depletion or of CFMs released in the atmosphere must take account of the fact that

there have obviously been naturally-caused climate changes in the past, and it
seems likely that there will be similar patterns of change in the future. The long-
term record of the Earth's climate has been reconstructed from studies of ocean and
lake sediments, the location of ancient glacial moraines, the chemistry of ice
cores taken from Greenland and the Antarctic, and so forth, and it now appears that
the most pronounced changes in the past million years or more might have been
associated with changes in the Earth's axis of rotation and its orbit around the
sun. These quasi-periodic alternations between major glaciations and interglacials
(such as the present) occur with periods ranging from 25,000 to 100,000 years.
While there are apparently periodicities with shorter time scales, their amplitudes
are small relative to the larger and more or less random fluctuations of the weather
and climate, and it is not possible at this time to predict naturally-induced
climate changes for the next year, decade, or century with any useful degree of
confidence.

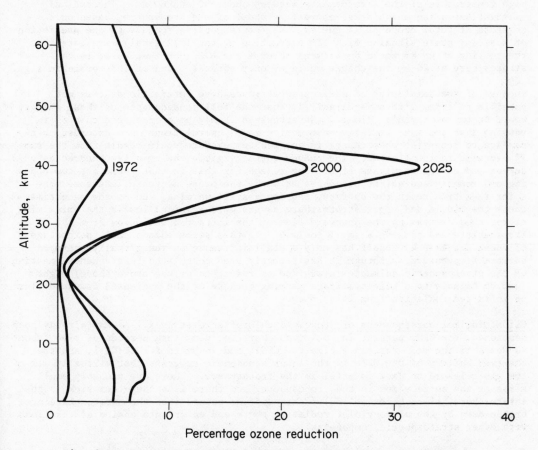

Fig. 6.11. Percentage ozone reduction resulting from the release
of chlorofluoromethanes at the 1973 rate, as a function of height.
Indicated are the model results for the years 1972, 2000 and 2025.
(After Crutzen)

Climatological consequences of possible ozone depletion. Because solar ultra-violet
radiation is very strongly absorbed by ozone, the temperature of the stratosphere
is largely maintained by a balance between absorption of solar radiation by ozone
and emission of atmospheric infra-red radiation by ozone, carbon dioxide and water
vapour. Any change in the stratospheric heating rates will have a direct influence
on the temperature distribution in the stratosphere and possibly in the troposphere,
and these temperature changes will have an effect on the patterns of atmospheric
circulation and hence on weather and climate. Such thermodynamic and dynamic
changes will themselves have significant interactions or feedback on the stratospheric composition.

A depletion of ozone due to its reaction with the dissociation products of CFMs or
with NO_x will result in a decrease in the temperature *of the stratosphere,* since
ozone absorbs strongly the solar ultra-violet radiation in the 200 to 300 nm part
of the spectrum, as well as some visible radiation, thus causing the relatively
high temperature of the stratosphere between about 30 and 60 km. Theoretical
calculations using a radiative-convective model of the atmosphere, based on a
decrease of total ozone of 7% due to CFMs (corresponding roughly to one prediction
of a steady-state situaion with CFM production at the 1972 level), indicate that
the cooling at 40 km would be between about 4 and $8^{o}C$, but that lower in the
stratosphere at 20 km the change would be only about $1^{o}C$ and of indeterminate sign.

Because of the complexity of stratosphere-troposphere interactions it is not
possible to infer with any reliability what the full consequences of these changes
would be on the Earth's climate. Nevertheless it may be of interest to note in
passing that the same radiative-convective model calculations were extended to the
surface to determine what change in surface temperature would result from the same
7% decrease in total ozone. The change was negligible, being a few hundredths of a
degree and of indeterminate sign. The reason for this is that there are two oppos-
ing and roughly equivalent processes at work; the ozone decrease lets some more
solar radiation reach the surface, causing a small warming, and at the same time
cools the middle and upper stratosphere as has been seen and lowers the emissivity
there, thus decreasing the downward flux of the infra-red radiation from the
stratosphere and causing a small cooling. It thus seems that a small percentage
of ozone decrease by itself has only a small influence on the globally averaged
surface temperature, although it has recently been postulated that enhanced heating
of the stratospheric sulphate layer (due to reduction of the ozone above) might
lead on balance to a slight surface warming because of the increased downward flux
of infra-red radiation from this layer.

Climatological consequences of increased chlorofluoromethanes. As has already been
mentioned, the CFMs persist in the troposphere for very long periods of time (about
40 years is the mean residence time for $CFCl_3$ and 70 years for CF_2Cl_2), and the
observed buildup of the CFMs in the lower atmosphere suggests that virtually all of
the gas released to date is still in the troposphere. There are probably small
sinks at the surface and in the troposphere, and there is a long-term sink in the
stratosphere, since those molecules that diffuse upward into the stratosphere are
broken down by the ultra-violet radiation there and enter into chains of reactions
with other stratospheric components.

The present mean tropospheric concentration of total CFMs is about 0.4 parts per
billion (10^{-9}) by volume (ppbv), and calculations by Crutzen and others indicate
that, if CFMs continue to be produced at the 1973 production rates, by 2000 AD the
$CFCl_3$ would reach about 0.32 ppbv and the CF_2Cl_2 about 0.58 ppbv and would level
off in the middle of the next century at about 0.7 and 1.9 ppbv, respectively (for
example, see NAS report, 1976).

The molecules CF_2Cl_2 and $CFCl_3$ have strong infra-red absorption bands in the

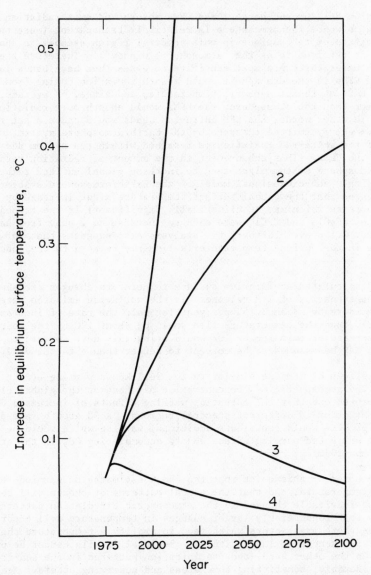

Fig. 6.12. Computed equilibrium surface temperature increase due to chlorofluoromethanes in the atmosphere according to various scenarios. (After J.M. Mitchell Jr)

Curve 1 – at 1975 emission rate until 1979, thence annual five per cent growth

Curve 2 – at 1975 emission rate throughout

Curve 3 – at 1975 emission rate until 1979, thence annual five per cent reduction

Curve 4 – at 1975 emission rate until 1979, thence total cessation of emissions

spectral region 8-12 μm and hence these CFMs absorb and emit radiation in this
spectral region where the atmosphere is relatively transparent (more than 75% of the
radiation emitted by the surface in this spectral region escapes to space, which for
this reason is referred to as the "atmospheric window"). Therefore the atmospheric
and surface temperatures are most sensitive to gases that have bands in the 8-12 μm
region. The CFMs in the atmosphere would absorb radiation coming from the Earth's
surface and emit at the atmospheric temperature, and since, on average, the surface
is much warmer than the atmosphere, the CFMs would absorb more radiation energy than
they emit. In other words, the CFM infra-red bands would cause a net reduction in
the radiative energy emitted to space by the Earth/atmosphere system, which implies
that more of the infra-red radiation is retained within the system due to the
presence of the CFMs. This enhancement in the amount of radiation retained within
the Earth/atmosphere system might tend to increase global surface temperatures.
Adopting a simple one-dimensional model of the Earth/atmosphere system, Ramanathan
(1975) indicated that the global surface temperature might increase by about
$0.9^{\circ}K$ (an increase which may be climatically significant) if the tropospheric concentrations of CF_2Cl_2 and $CFCl_3$ were each to increase to 2 ppbv from an assumed
level of about 0.1 ppbv. Further, his analysis also suggests that for small changes
the increase in the surface temperature is linearly related to the concentration
of CFMs.

In Fig. 6.12 calculated equilibrium surface temperature changes are shown for a
few different scenarios of CFM release. If the worldwide emission rate of CFMs
continues to increase by about 5% per year (or half the rate of increase up to the
present time), then the temperature rise would be about $1^{\circ}C$ by the year 2025. A
continued increase of emission by 10% would bring this date forward to 2010. The
calculations may be assumed to be correct to better than a factor of 2.

The warming effect of CFMs is similar to the well-known warming effect that an
increase in the atmospheric CO_2 concentration may have on the global climate. It
should be pointed out that the potential warming effects of increased CO_2 and
increased CFMs occur in different spectral regions. A 5% growth rate per year in
the release of CFMs would contribute additional warming which would be of the same
range as the projected warming effect due to accumulating CO_2 in the atmosphere
(about $1.4^{\circ}C$ by 2030).

It should be noted of this point that the above estimates of warming refer to a
global mean temperature, and that the actual patterns of change will be greatly
influenced by inevitable changes in the atmospheric circulation patterns. Thus,
it is likely that considerably larger changes in temperature will occur in polar
regions where there are factors which tend to amplify a temperature change, with
relatively less changes at low latitudes. Regional details cannot be predicted at
this time. On the other hand, the precipitation patterns in the sub-tropics may be
altered considerably, benefitting some areas and depriving others. Again, it is
at present impossible to predict exactly how these patterns of change will appear.

Most of the estimates of climate change due to mankind's activities are based on
one-dimensional models of the climate system, and there is a need to take better
account of the dynamical interactions within the atmosphere and between the
atmosphere, the oceans, the ice masses of the polar regions, and the biosphere.
Thus, the development of more complete models of the complex system that determines
the global climate must be developed before we shall be in a position to give
satisfactory and detailed answers to the many problems being raised.

Advances in our knowledge of climate mechanisms over the next two years will improve
our assessment of both climatic effects due to CFMs (through ozone reduction and
displacement and through infra-red absorption), but these advances cannot be
expected to make our assessment of the climatic effects as precise as our assessment

of ozone reduction. If the increased infra-red absorption and emission due to the presence of CFMs in the atmosphere were to alter our climate by small amounts, the most important effects would be on agriculture, particularly through the boundaries of the regions in which particular crops can be grown effectively, although other agricultural effects are possible.

Biological Effects

A reduction in the ozone column would result in an increase of UV-B (290-320 nm) radiation received at the Earth's surface, and theoretical calculations using both absorption and scattering by ozone and other atmospheric constituents have indicated an average magnification factor from slightly less than 1.5 to slightly more than 2.0 between percentage changes in the ozone column and percentage changes in UV-B radiation in clear sky conditions. In other words, a reduction of 10% in ozone would result in an increase of about 20% in UV-B.

However, an important point to note is that whereas an ozone depletion would result in increased penetration of UV-B, in respect of the damaging potential the corresponding increase is even greater than the increase in total UV-B, because the increase is proportionally greater at shorter wavelengths. Relative effectiveness of one quantum for altering DNA (the material carrying the genetic information of all living cells) is roughly as follows:

320 nm	0.03
315 nm	0.1
310 nm	0.6
305 nm	2.6
300 nm	15
295 nm	60
290 nm	160

(these define the action spectrum of UV-B radiation for DNA, which is here normalized to a value of 1000 at 265 nm, the wavelength for peak sensitivity. At still shorter wavelengths the sensitivity drops off.)

Whereas the problem of the biological consequences of increased solar UV radiation reaching the Earth's surface is not within the sphere of WMO, the fact that occurrences of sunburn, cancer and eye damage are directly related to the intensity of UV radiation, and the possibility that an increase in UV may also reduce agricultural yields and disturb ecosystems, means that it is essential that the competent bodies undertake appropriate studies. It should be the role of WMO to ensure that the relevant data and information obtained through the Global Ozone Research and Monitoring Project are readily available for these studies.

It is assumed that a detailed discussion on the biological effects of ozone depletion will be submitted to the UNEP meeting in Washington, D.C. Nevertheless, the following actions are suggested as being suitable for an interdisciplinary effort in support of studies into the effects on the biosphere of an increase in solar UV radiation:

(a) to document the world climatology (including trends) of the ultra-violet radiation received at the surface of the Earth, particularly in the "active" wavebands for effects on man and the biosphere;

(b) to determine more precisely than is presently known, the "active" wavebands causing effects on plants and micro-organisms;

(c) to determine quantitatively the dose-response relationships for various exposure times;

(d) to develop epidemiological relationships from the information in (a) and (c), the ultimate objective being to provide alternative management strategies.

In view of the complexity of the problems involved in these fields, but bearing in mind the seriousness of the possible consequences of an attenuation of the ozone layer, it is proposed that UNEP consider establishing a scientific committee or working group on which the following bodies might be represented:

Governmental: FAO, UNEP, UNESCO (MAB), WHO and WMO

Non-governmental: ICSU and following affiliated unions: UBS, IUGG, IUPAB, IUPS.

The purpose of the committee would be to recommend a suitable study programme to investigate the impact on the biosphere of any ozone changes established under the WMO Global Ozone Research and Monitoring Project. It is to be expected that the appropriate UN specialized agencies would accept responsibility for developing programmes within their respective spheres of competence.

MODELLING THE ATMOSPHERIC OZONE PROCESSES

In order to make reliable estimates of future changes in ozone amount, it is necessary to rely upon the ability to simulate the physical and photochemical processes in the atmosphere. As the full complexity of the atmospheric processes cannot be simulated on a laboratory scale, the most efficient tool available is numerical simulation, or modelling, of such processes with the aid of fast electronic computers. It is important to design models of the atmosphere that, on an average basis, simulate or describe the photochemical or physical processes taking place on all scales. Since many models already exist which simulate the atmosphere dynamical-photochemical system in varying degrees of complexity, special effort needs to be made to co-ordinate studies using these models and to establish standard cases and comparative model runs.

One-Dimensional Models

These models maximize the photochemical complexity and representation while requiring that all other atmospheric processes be represented by simple diffusion. They are essential means of determining the importance of the many chemical reactions with their variability in the vertical dimension, and of deriving efficient parameterization schemes for more complex atmospheric models. The major task is to derive a set of diffusion coefficients that are applicable to the problem under study. Efforts must continue to obtain the most reliable eddy exchange coefficients for those models using only diffusive transport (1-D and some 2-D models). A plausible approach would be through studies based on available data of the transport of relatively stable tracers, when assumptions about chemical reactions in the stratosphere would be unnecessary. Special adaptation of these models is required to support the design, intercomparison and interpretation needs of experimental programmes.

Intermediate Models

Two-dimensional models (latitude, vertical) can represent the major north-south variations in the atmosphere whilst retaining complete interaction between the various photochemical components and the atmospheric circulation. They must, however, represent the synoptic- and planetary-scale waves in the atmosphere by some form of diffusion parameterization, and as yet there is no completely satisfactory formulation of this for all levels. Tests of the models on the ozone/NO_x problems suggest that they give fairly reliable results and a reasonable description of the north-south variations. Development work on parameterization in such models should be encouraged. Since the ozone formation, destruction and distribution

processes, as well as those of the associated trace constituents, are highly
latitude-dependent, it follows that these models have an important role to play in
simulation problems and in interpreting experimental data. More research work needs
to be done to develop simplified models which can include the major planetary forcing functions without requiring the full complexity of the three-dimensional general
circulation models.

In so doing, it should continually be recognized that in these models a major
uncertainty results from the necessity of predicting the transport by eddies in
terms of zonal mean (longitude-independent) quantities using the diffusion assumption that all net local eddy transfers occur in a downgradient fashion and with a
magnitude determined by appropriately chosen exchange coefficients. This assumption
is often violated in nature because the meridional circulations and eddies are a
tightly coupled pair of processes. In a two-dimensional sense the transport by
eddies of heat and quasi-conservative trace constituents can, in general, be
considered as acting downgradient, though the one-dimensional (vertical or meridional) components may be countergradient. It is understood that some 2-D models
incorporating both diffusion and organized circulation are being developed.

Three-dimensional Models

Because only the three-dimensional numerical models can adequately represent the
synoptic-scale wind systems which are responsible for much of the transport in the
atmosphere, and the planetary-scale systems which dominate the winter stratospheric
flow, it is essential that they be used to determine and verify many aspects of the
ozone problem. As a minimum these models should include complete interaction
between the atmosphere's thermodynamics and ozone, as well as the action of other
specified trace constituents on ozone. The three-dimensional models should then be
used to establish and verify the transport parameterization in simpler models and to
test depletion mechanisms wherein the major concentrations determined by similar
models are perturbed by variations in an input parameter. Special attention should
thus be given to models which can be run in simpler modes or in conjunction with
simpler models. It should be understood that even in these complex 3-D models it
is assumed that there would be no changes in the way the stratosphere transports
trace substances in the event of large ozone reductions due to CFMs; a reasonable
assumption since it is not currently known precisely what the response of the
atmosphere would be in such circumstances.

Specialized Models

In addition to regular modelling activities, particular treatment is required for
selected parts of the atmospheric ozone problem.

- Many sub-sets of the atmospheric chemistry problem must be handled by the
  elimination of very short-lived or transient species. Also reactions which vary
  rapidly at particular times (e.g. at sunrise or sunset) must usually be modelled
  using slower averaged rates.

- There are specific problems of strongly interdisciplinary nature which must be
  resolved. Of prime importance in the ozone problem is an improved model for the
  complete nitrogen cycle, its possible changes due to man's activities and the
  implications for atmospheric chemistry.

- As regards the climate aspects, studies aimed at assessing the long-term impact
  of ozone changes on climate should be undertaken, even using highly parameterized
  models based on currently accepted hypotheses, which should take account of the
  progress achieved so far in this field.

NEED FOR CO-ORDINATED MONITORING PROGRAMMES OF OZONE AND OF RELEVANT TRACE GASES AND ATMOSPHERIC PARAMETERS

From the foregoing discussion it should be apparent that research is now urgently needed in order to advance knowledge in the field of atmospheric ozone (including the impact of changes in the amount of ozone). This work could not be successfully undertaken on the basis of the present observing network and there is a need for a limited period of intensive observations of ozone (particularly its vertical distribution) and also of those other atmospheric trace gases which have a direct or indirect influence on ozone. A real effort is needed over a period of at least 3-4 years in order to permit many of the assumptions and working hypotheses used in current ozone studies to be confirmed and refined so that future research would be built on solid foundations.

The WMO Commission for Atmospheric Sciences Working Group on Stratospheric and Mesospheric Problems has stressed that the generally wide spacing of the observing network and the nature of the parameters to be monitored means that nothing less than global integration would be required, and that it is essential therefore that data from the network be uniform and of very high quality and that they be made widely available. It is hoped that mutual collaboration and assistance between Members in respect of the ozone investigation would be forthcoming in a spirit of international co-operation. Following the initial period of intensive efforts, requirements could be relaxed to a level commensurate with a continuing system of monitoring ozone and other relevant constituents whose main objective would be to detect trends in the global zone concentration and contribute to better understanding of the stratospheric environment.

Monitoring is the process of repetitive measurement, for defined purposes, of one or more elements or indicators of the environment according to prearranged schedules in space and time. There are a number of reasons why it is important to operate a global ozone monitoring system:

- to define the ozone climatology of the stratosphere in sufficient detail to determine hemispheric and world trends;

- to provide data for research and help in making predictions of future states;

- to provide, on a synoptic basis, "ground truth" for many of the critical experiments.

A programme of ozone measurements would have limited value in future research without a parallel programme to measure the concentration in the atmosphere of those substances responsible for the destruction of ozone. The most important "parent" substances of the reactive radicals are nitrous oxide (N_2O) and the chlorofluoro-methanes (CFMs). They should preferably be monitored both near the ground *and* in the stratosphere. Since these substances are relatively inert and fairly uniformly distributed in the troposphere, apart from experimental stratospheric profiles of their vertical distribution to define diffusive and decay properties, and occasionally to construct meridional profiles, it would be sufficient if these substances were measured at the baseline air pollution stations to establish trends. Thus relevant geophysical and chemical parameters should be monitored for the following reasons:

- to assist in explaining the observed ozone distribution and trends, and to provide essential data for relevant studies;

- to document the climatology (including trends) of the ultra-violet radiation received at the surface of the Earth, particularly in those wavelengths causing sunburn.

Scientific Design Criteria

A number of scientific criteria should be followed when recommending modifications and additions to existing systems for monitoring ozone and related geophysical/chemical parameters.

- Optimization of network densities requires a knowledge of:

    • the variability in time and space of the elements being sampled;

    • a conceptual model of the relevant physical-chemical processes;

    • the objectives of the monitoring system.

- Notwithstanding the above points, a monitoring network should not be tied too closely to a particular model of the atmosphere. A new model may arise later, and data sets may be used for entirely different purposes than originally contemplated. Thus if some additional measurements can be made at relatively small supplementary cost, then this should be done.

- The design of a monitoring system often requires a pilot field study over a limited period of time. It is important that such a study be identified and carried out before the monitoring system becomes operational.

Existing Monitoring Programmes

The WMO world ozone network. The existing network of stations provides the following data:

- total amounts of ozone on particular days and hours at about 80 locations;
- Umkehr observations, yielding estimates of vertical distribution of ozone at about 15 of the stations mentioned above but only sporadically taken;
- ozonesonde observations at 5-10 stations, once weekly at the most;
- rocket-borne vertical ozone measurements at one station less than once a month.

The existing network contains gaps, particularly over the oceans and in the Southern Hemisphere, and it is difficult, if not impossible, to separate regional trends (due to shifts in the positions of the long-wave troughs) from global trends in ozone concentrations. The Umkehr technique permits the vertical distribution of ozone to be estimated in a crude fashion using the Dobson spectrophotometer, and knowledge of the ozone variation at heights exceeding 30 km is particularly valuable since CFMs dissociate in this region and should preferentially attack the ozone there.

Satellite measurements of ozone. Observations of ozone by a polar-orbiting satellite have the obvious advantages of quasi-global coverage by one and the same instrument (so that observations are readily comparable) and availability of the data in quasi-real time. Moreover, *if* individual total ozone observation errors using the BUV instrument are random and of an order of 5%, then the hemispheric and global mean monthly ozone content can be estimated to within a fraction of 1%. (Much of this uncertainty is due to the fact that observations in polar latitudes during the winter season are not obtainable at present.)

The BUV and LR1R type instruments can obtain layer-average ozone amounts at heights above the level of ozone maximum more precisely than any other methods currently available, and thus should be capable of early detection of any reduction of ozone due to photodissociation of CFMs at 30 to 40 km, especially in the tropics.

However, the limitations of satellite monitoring systems must also be recognized. The time between the proposal and launch is long (7 years or more), so that such systems are somewhat inflexible for quick turnaround programmes. They are also limited in vertical and horizontal resolution, and often in accuracy, making them in general not very suitable for studies of detailed chemical mechanisms.

The following satellite observations of relevant atmosphere parameters have been made so far:

Nimbus 4:   Temperature to 50 km (SIRS, IRIS, SCR)
                                       Ozone total, 30-50 km (BUV, IRIS)

Nimbus 5:   Temperature to 50 km (SCR)

Nimbus 6:   Temperature to 80 km (LRIR, PMR)
                                       Ozone 15-50 km (LRIR)
                                       Water vapour 15-45 km (LRIR)

Atmospheric Explorer E: Ozone total 30-50 km (BUV)

The following are planned for the future:

Block 5D-1 System (1977):   Total ozone, vertical temperature and moisture
                                       profiles

Nimbus-G (4th Quarter 1978):   Temperature up to 80 km (LIMS, PMR)
                                       Ozone 10-50 km (LIMS, SBUV)
                                       H_2O, NO_2, HNO_3 (LIMS)
                                       H_2O, CO, NO, CH_4, N_2O (SAMS)
                                       Aerosols (SAM II)

SAGE: Aerosols
                                       Ozone

Shuttle (after 1980):   A large number of active and passive instru-
                                       ments for remote probing of the upper atmo-
                                       sphere are planned for various shuttle missions.
                                       The AMPS payload has a portion dedicated to
                                       study of the atmosphere from the troposphere to
                                       120 km.

The backscatter ultra-violet (BUV) instrument on polar-orbiting Nimbus 4 is still in operation after 6 years, although a power shortage now allows it to operate rather less than half the time. This instrument provides an estimate of total ozone, as well as estimates of ozone amounts in layers above the ozone maximum. The data from Nimbus 4 are being re-reduced because of uncertainties in satellite positioning and orientation , and it is not expected that the data reduction will be completed until early in 1978. Data are available only for those parts of the Earth's atmosphere in sunlight; there is no information from polar winter regions.

The Limb Radiance Infra-Red Radiometer (LRIR) on Nimbus 6 (launched June 1975) had a lifetime of six months due to the use of cryogenic cooling. This instrument provides a more detailed picture of ozone variation with height than the BUV instrument, but its ability to assess total ozone is uncertain and may be less accurate.

The Atmospheric Explorer E (AE/E) was launched in December 1975 into a low-incli-nation orbit (± 20° of Equator) and provides BUV ozone data in the tropics. This satellite permits a recalibration of the Nimbus 4 BUV instrument.

The polar-orbiting Nimbus G, scheduled for launch in late 1978, will have a two mode step-scanning BUV instrument (SBUV) and a Continuous Scan Mode which will scan the Earth across the orbital track and provide nearly complete areal coverage of the globe each day. Nimbus G will also include an instrument similar to the LRIR (the LIMS).

Consideration is being given to incorporating either the BUV or SBUV ozone-measuring instrument into the Tiros N operational satellite series. However, incorporation would probably not take place until the third or fourth member of the series.

The World Weather Watch. This is the system organized by WMO to provide synoptic information on the three-dimensional structure of the atmosphere. It is the largest and most reliable environmental monitoring system in the world.

The WMO Background Air Pollution Station Network. This network includes 110 identified regional air pollution stations proposed by 65 countries, and about 15 baseline air pollution stations proposed by 9 countries. About half of these are already functioning. An Operations Manual for Sampling and Analysis Techniques for Chemical Constituents in Air and Precipitation has been published (WMO No. 299, 1974).

As indicated in paragraph 1.5 of Part II of the Operations Manual, the minimum programme for baseline stations "shall include measurements·of CO_2 in addition to turbidity, radiation and chemical composition of precipitation. Measurements of total ozone would also be highly desirable unless they are already being made at a nearby site". Baseline stations and regional stations with expanded programmes are encouraged to monitor other trace gases, and the Operations Manual contains recommended methods for sampling ozone, SO_2, CO, NH_3, NO, NO_2 and N_2O.

Baseline stations are fully or partially operational at Moana Loa (Hawaii), South Pole (Antarctica), Point Barrow (Alaska), Samoa, Ocean Weather Station "Papa" (50^ON, 145^OW), Alert (Canada), Sable Island (Canada), Mount Hobart (Australia), Patagonia (Argentina), Huancayo (Peru), Minamitorishima (Japan), Kislovodosk (U.S.S.R.). In addition, financial support has been secured to undertake a site-selection survey of Mount Kenya. Ocean Weather Station "Papa" may have a limited lifetime.

The UV Monitoring Network. Dr. D.F. Robertson of Australia has designed a simple instrument which measures global ultra-violet solar radiation with a response function similar to that of human skin sunburning. About half a dozen of these instruments have been operated in and around Australia during recent years. The U.S.A. has installed modified versions of this instrument (the Robertson-Berger Sunburning UV Meter) at some 17 locations within its own country and similar instruments also are used at Davos (Switzerland), Warsaw (Poland) and Aspendale and Brisbane (Australia).

RECOMMENDATIONS

Global Ozone Monitoring System

It is recommended that because global monitoring systems are costly, they should meet the needs of multiple users wherever possible. Thus the world ozone network should be fully co-ordinated with, and sometimes integrated with, the observing systems of the World Weather Watch, the WMO background air pollution network, and the UV monitoring network.

An intensive effort should be carried out over a period of 3-4 years with the understanding that following this initial period of intensive effort, requirements could be relaxed to a level commensurate with a continuing monitoring system of ozone whose main objective would be to detect trends in the global ozone concentration.

Every effort should be made to ensure that global ozone data are uniform, of high quality, and will be made widely available, and that mutual collaboration and assistance between Members in respect of the ozone investigation is carried out in a spirit of international co-operation.

Detailed recommendations for improving the global ozone monitoring system follow later.

Total ozone - ground-based systems. In spite of the complexity, the bulk and the expense of the Dobson ozone spectrophotometer, it is recommended that a campaign be launched to bring back into operation as many of the existing Dobson instruments as possible, and to transfer some of these to countries in areas lacking any ground-based means of measuring total ozone. One or two countries, preferably the U.S.A. and Canada, with the necessary facilities for overhauling and recalibrating Dobson ozone spectrophotometers have already accepted some commitments in this respect. An intercomparison and intercalibration test between Dobson instruments will be carried out at Boulder (U.S.A.) in 1977.

Urgent attention should be given to the development of new surface-based instruments with the capability of automatically measuring total ozone, and that the forth-coming comparison between the new Canadian (Brewer) instrument and the Dobson be carried out for at least a year, with the hope that a similar comparison can be carried out soon with the New Zealand (Mathews) total ozone instrument. Consider-ation should also be given to automating Umkehr measurements, including a computer programme to unscramble observations when clouds are present.

All the existing, and reactivated, total ozone measuring stations should be main-tained for at least 20 to 25 years, and arrangements should be made for a qualified expert on ozone instruments to visit the stations periodically to advise on observ-ing procedure and instrument maintenance and possibly to effect intercomparisons with a travelling standard instrument. This would be additional to systematically conducted intercomparisons between key regional instruments to ensure stability of the quality of collected data.

In view of a recent suggestion that near 300 nm the radiation may change by as much as 20% during the 11-year sunspot cycle, whereas it is implicit in the Dobson measurements that this radiation be constant, extraterrestrial ultra-violet (UV) radiation should be monitored as precisely as possible for a few solar cycles (using both an extrapolation to zero air mass of surface UV measurements and direct measurement by satellite).

A concensus should be reached concerning the possible influence of atmospheric aerosols on Dobson total ozone measurements, and in particular on the Dobson Umkehr measurements. The problem of the temperature dependence of ozone absorption should also be clarified.

A detailed comparison should be made of total ozone estimates obtained by a Dobson instrument under cloudy and clear skies, since there is some evidence that these observations are not completely compatible.

Advanced models such as the General Circulation Model at the Geophysical Fluid Dynamics Laboratory (NOAA, U.S.A.) should be used immediately to estimate errors in climatic or hemispheric areal means of total ozone resulting from ozone observations being only at certain Dobson stations, and within certain regions. Comparisons on a daily basis and any trends that may occur should be used for this purpose.

Satellite stations. It is recommended that the best available ozone measuring sensors be considered a primary requisite among satellite instrumentation.

Member countries operating meteorological satellites should be encouraged to include atmospheric ozone sensors of proven capability aboard future satellites, with the hope that continuous ozone measurements by satellite can be obtained for at least one solar cycle (11 years).

Efforts should be made to ensure that the BUV data become available to the user within one year of the observations.

The BUV data should periodically be compared with high quality data from Dobson ozone spectrophotometers and balloon-borne ozonesondes. Moreover, data from rocket-borne ozondesondes should be improved to a degree at which they may be used also for intercomparison with satellite-borne sensors.

Efforts should continue to develop limb-radiance instruments capable of observations for several years (these instruments can measure ozone profiles better than the BUV instrument and to lower altitudes, but the total ozone amount derived thereby may not be quite so accurate).

As soon as the reduction of the Nimbus 4 total ozone data up to the end of 1976 has been completed, a comparison should be made between the trend (since 1970) determined from these data and from the surface-based Dobson network, thus indicating the error in hemispheric and global ozone trends deduced from the Dobson network.

Climatic monitoring of stratospheric temperatures should be carried out in concert with the monitoring of stratospheric ozone so that trends can be compared, since ozone equilibrium values may be partially a function of temperature. (Note: while an attempt is being made in the U.S.A. to determine the worldwide temperature trend in the lower stratosphere from radiosonde data, this is much more logically done from satellite information, particularly at the higher elevations, where radiation effects on radiosonde temperatures are severe).

Umkehr measurements. It is recommended that all stations with a Dobson ozone spectrophotometer be urged to make Umkehr measurements as regularly as possible (at least 10 to 15 per month). A relatively large number of observations is necessary in order to offset the rather high uncertainty of single observations using this technique. In this connexion there is a need to study the influence of dust when calculating the results of high-level Umkehr observations. Moreover, there would be great advantage in developing automatic instruments in conjunction with a computer programme to unscramble observations made when clouds are present.

As regards the location of stations making Umkehr observations, it is recommended that at a minimum the 12 or so stations already making them continue to do so and that every effort be made to initiate observations at tropical stations where any effect of CFMs on ozone should be manifested the earliest. Possible locations could be Darwin, Huancayo, Mt Kenya, Moana Loa, Samoa and Singapore. Other (extratropical) locations where Umkehr observations would be particularly valuable are Boulder, Resolute, South Pole, Toronto and Tromsö.

It is recommended that a meeting of experts on Umkehr methods be convened to clarify once and for all the accuracy and usefulness of data obtained by the Umkehr technique, and in particular whether the data are useful only in the absence of satellite measurements or can also serve to supplement satellite measurements. Careful consideration should be given to the influence of atmospheric aerosols on Umkehr measurements, specifically volcanic dust.

Finally, in view of the imprecision of the Umkehr method, it is recommended that
efforts be continued to develop alternative indirect methods, such as the dye-laser
lidar technique and the ground-based BUV technique.

Dobson Spectrophotometer total-ozone network stations. It is recommended that the
following action be taken with respect to stations on or just south of the Equator,
where the daily, annual and quasi-biennial variations in total ozone are a minimum,
and where any effect of CFMs on these trends should be most noticeable.

- The existing total-ozone station at *Huancayo* (Peru) be recalibrated as soon as
  possible (the trend at this station has been under suspicion for years, being
  upward when other tropical stations were downward).

- The newly-established total-ozone station in *Samoa* be maintained with great care
  since it must be representative of the entire equatorial Pacific.

- The total-ozone station at *Darwin* (Australia) destroyed by a tropical storm be
  reactivated as soon as possible.

- A total-ozone station be established at *Singapore* (directly on the Equator)
  and/or at *Manila* (Philippines).

- The plans for establishment of a total-ozone station on *Mt. Kenya* be carried
  forward, with the hope that some nation might contribute an inactive Dobson
  ozone spectrophotometer for this station. The proximity of Mt Kenya and
  Mahe (Seychelles) would hardly justify a Dobson instrument at both locations, and
  preference should be given to Mt Kenya since it is a background air pollution
  station.

- Every effort be made to establish a total-ozone station in *Ghana or Nigeria*.

- The total-ozone station located at *Rio de Janeiro* should be upgraded to make
  regular observations as soon as possible.

Since variations in total ozone amounts in polar regions are almost completely
dominated by atmospheric transport processes and hence reflect variations in the
atmospheric circulation, and since the most likely technique for ozone measurement
by satellite (BUV) will not provide data from these regions during the polar night,
it is recommended that moonlight observations be taken wherever possible at high-
latitude stations, and in particular that:

- every effort be made to improve the quality and quantity of the total-ozone data
  at *Amundsen-Scott* (South Pole), which for the past 10 years has provided ozone
  data only during a few summer months;

- the total-ozone station at *Halley Bay* (U.K. station in Antarctica) be reactivated
  to provide a back-up to Amundsen-Scott (Halley Bay provided excellent total-ozone
  data until 10 years ago);

- the total-ozone station at *Tromsö* be reactiviated (this station had a period of
  record in excess of 30 years, and the cessation of observations there in 1969
  was most unfortunate);

- the newly-established total-ozone station at *Point Barrow* be maintained with
  care since it must represent much of the western Arctic;

- the existing high quality of observations from *Resolute* be maintained, this
  station being especially important because of its proximity to the north
  magnetic pole (possible influence of solar proton events);

- the station at *Spitzbergen* be reactivated if feasible, considering the problems
  of personnel, supply, etc. (The station was previously in operation for only
  a few years.)

Recognizing that at present the total-ozone network in southern temperate latitudes is basically an Australian-New Zealand network and that representative hemispheric values of total ozone cannot be obtained from such a station distribution, it is recommended that:

- the total-ozone station at *South Georgia* continue in operation;

- the station at *Puerto Montt* (Chile) should be reactivated if possible; if this is not possible the Dobson instrument there (which is the property of the U.S.A.) should be recalibrated and made available for use elsewhere;

- a total-ozone station be established as soon as possible in *southern* Africa;

- consideration be given to establishing a total ozone station at *Gough Island* in the Atlantic Ocean and at *Amsterdam Island* in the Indian Ocean (members might wish to provide inactive Dobson instruments for this purpose);

- a new total-ozone station should be established if possible in South America south of about latitude 45°S;

- the total ozone station at *Buenos Aires* be continued in full operation.

Since it has not been possible to integrate total ozone data obtained using filter ozonometers into the north temperate latitude average because, though the sense of the trend is not inconsistent, its magnitude is 4-5 times greater than found in other north temperate regions (e.g. Europe, North America), it is recommended that:

- the ongoing comparison between the filter ozonometer and the U.S.A. Dobson standard at Boulder be carried out for at least a year so that some sort of trend comparison can be made, and that particular attention be given to the influence of atmospheric aerosols on the readings of the filter ozonometer;

- in view of the known differences between the types of instruments, consideration should be given to supplementing the filter ozonometer network with a small number of Dobson instruments which are not in use in other countries.

The People's Republic of China represents the largest land area in the Northern Hemisphere without any total-ozone measurements, and the inclusion of China in this type of international scientific effort is highly desirable. It is therefore recommended that:

- at least two Dobson ozone spectrophotometers be sent to China, with the suggestion that at least one of the instruments be placed in the interior of the country. Members might wish to provide inactive Dobson instruments for this purpose.

Recognizing the relatively high density of the European total-ozone network leading to a certain redundancy in data, it is suggested that European nations wishing to reduce or eliminate their Dobson ozone spectrophotometer network might be encouraged to offer their instruments through WMO for use in more remote locations.

Moreover, it is recommended that:

- the instrument at *Val Joyeux* (France) be recalibrated, since its total ozone values have been so anomalous that it has not been possible to include this station when computing the European regional average;

- the instrument at *Aarhus* (Denmark) be recalibrated and an attempt made to find out why the time trend in total ozone at this station is so different from other European stations. If this station is no longer to be considered reliable, thought should be given to reactivating the total-ozone station at *Oslo* (Norway);

- the stations of *Postdam* (German Democratic Republic), *Hradeč Kralove* (Czechoslovakia), *Lisbon* (Portugal), and *Cagliari* (Italy) attempt to upgrade their total ozone measurements;

- attention be drawn to the close proximity of the total-ozone stations at *Oxford* and *Bracknell* (United Kingdom);

- an investigation be made as to why monthly total-ozone data are often so late in arriving at the World Data Centre in Toronto (Canada) from European stations, and why no data at all are received by the Centre for some months.

The North American total-ozone network now seems to be satisfactory following some earlier problems with White Sands, nevertheless it is recommended that the proposed new total-ozone station in southwestern U.S.A. (California, Arizona or Nevada) be established as soon as possible (replacing the station at Green Bay which was too close to Toronto). In addition, consideration should be given to establishing a total-ozone station in northwestern U.S.A. (Washington, Oregon or Idaho).

The Japanese total-ozone network of 4 stations appears to be of above-average quality, although 3 of the instruments are copies of Dobson spectrophotometers. The India-Pakistan network now seems to be stable after some erratic early years. It is recommended that stations in the total-ozone networks of Japan and India-Pakistan endeavour to maintain their presently high standards.

The subtropics are of considerable interest because of the predominance of the quasi biennial oscillation in total-ozone there, and it is important to determine the longitudinal variability of this oscillation as well as its relationship to any long-term trends. Therefore it is recommended that:

- the total-ozone station at *Cairo* (Egypt) continue in full operation;
- the total-ozone station at *Casablanca* (Morocco) be recalibrated and made fully operational;
- a total-ozone station be established at *Manila* (Philippines) to replace the station in Taiwan which is no longer reporting (Members might wish to provide an inactive Dobson ozone spectrophotometer for Manila). This is a critical meridional link between Japan and Australia.
- the total-ozone station at *Moana Loa* (Hawaii) endeavour to continue its high level of performance;
- the total-ozone station at *Mexico City* be brought into full operation;
- the total-ozone station at *Kingston* (Jamaica) be reactivated; if this is not possible, the Dobson instrument there might be recalibrated and used elsewhere.

Ozonesonde network. There is now a particular need for a reliable 3-dimensional climatology of ozone which does not yet exist to the degree required for theoretical investigations. It is therefore recommended that:

- a network of ozonesonde stations be established as a special programme for a period of 3 to 4 years;
- prior to starting the programme, an ozonesonde intercomparison test be held since several different types of ozonesondes are available commercially;
- existing ozonesonde stations with substantial periods of records should be maintained and their observation programme increased to the extent possible;
- soundings be made at the tropical stations of *Cachoeiro Paulista* (Brazil),

Huancayo, Moana Loa, Nairobi and *Singapore, (or Manila)* twice a week;

- soundings be made at the polar stations of *Amundsen-Scott, Fairbanks* (Alaska), *Halley Bay, Heiss Island, Resolute,* and *Tromsö* 3 times a week in winter and spring and twice a week in summer and autumn;

- in south temperate latitudes, soundings be made twice a week at least at *Aspendale, South Georgia* and at a station in *southern Africa,* and if possible also at *Brisbane, Invercagill,* and *Macquarie Island;*

- in north temperate latitudes the present station network be augmented by soundings at least twice a week at *Boulder, Edmonton, Lindenberg, Oslo (or Stockholm), Tateno, New Delhi* and *Yakutsk;* possibly also at *Brussels, Cagliari, Fort Churchill, Goose Bay Tallahassee, Kagoshima, Sapporo, Poona, Trivandrum* and *Alma Ata;*

- a two-week period of *daily* soundings be attempted in middle and high latitudes of each hemisphere in the early spring of each year.

Measurement of Minor Atmospheric Constituents

Many of the reactive or inter-reactive atmospheric trace substances would need to be measured simultaneously through vertical and horizontal traverses with highly sensitive instruments. Instrumental capabilities are developing very rapidly in this area and attention needs to be given to intercomparison, calibration, standardization, development and information exchange.

The results of intercomparison tests should be used to identify the most reliable instrumentation for measurement of each constituent, and thereby to assemble optimum packages of instruments to measure the constituents in each reactive group. When possible, the packages for two or more groups should be combined in the same payload to obtain additional insight into the interaction between groups.

The station locations and programmes of the WMO Background Air Pollution Station Network should be reviewed periodically in the light of the potential contributions this network can make towards reaching the objectives of the World Ozone Network, and vice versa.

Recommendations for observational requirements in respect of the relevant atmospheric trace constituents are as follows:

(a) Oxides of nitrogen (NO_x) system

   N_2O - monitor at baseline stations to an accuracy of 2% twice per month;

   - measure dependence on season, latitude and altitude once per season;

   - obtain a few high-level vertical profiles.

   NO, NO_2, NH_3, HNO_3
   - make simultaneous measurements with accuracy of at least 1 ppb (10^{-9});

   - carry out intercomparisons of instruments and analysis techniques.

   N_2O_5- develop measuring techniques;

   improve laboratory data on NO_2 - N_2O_5 cycles.

   Chlorofluoromethanes, N_2O, NO, NO_2 and NH_3

   - monitor once per season in sea water within a few appropriately selected Marsden squares of the world's oceans.

(b) Oxides of hydrogen (HO_x) system

H_2O - determine climatological distribution in the stratosphere;

- continue U.S.A. balloon observations, partly as ground truth for satellites;

HO_2, H_2O_2 - develop measuring techniques;

OH - establish observational methodology and determine latitudinal and vertical distribution.

(c) Chlorofluoromethanes system

Continue U.S.A. programme of stratospheric monitoring of CFMs (at present $CFCl_3$ as well as SF_6) with an attempt being made to obtain measurements once every three months, rather than three times a year as at present. Make ground-based obser- vations at some baseline air pollution stations (at least in each hemisphere and one in the tropics) to an accuracy of 5% to monitor global amounts and detect trends (measurements are already being made at Moana Loa, Point Barrow, Samoa and the South Pole as well as at Adrigole (Ireland) but with hardly this degree of accuracy). Every effort should be made to upgrade the quality of the CFM measurement at the surface since difficulties have been experienced in this regard at all the above stations (the best estimate is that at Adrigole where "clean air" values of $CFCl_3$ increased from about 110 to about 150 ppt (10^{-12}) between 1974 and 1976). It is very important in this regard that truly "clean air" values are obtained, and at least in the case of Adrigole this requires the determination of hindcast trajectories for a period of at least a week, to ensure that the air has come across the Atlantic and not from Europe. In addition the following observations are recommended:

HCl - obtain vertical profiles through the troposphere and the stratosphere in at least three different latitudes in each hemisphere once per season;

ClO_x - continue development of stratospheric measuring techniques and analysis;

CF_2O, $CFClO$, $ClNO_3$ - attempt to identify and measure;

(Cl, Br) - initiate research efforts on bromine and on other chlorine components.

(d) Other relevant systems.

CH_4, CO, H_2 - monitor at baseline stations to an accuracy of 1%;

- obtain vertical profiles through the troposphere and stratosphere at different latitudes.

In some cases full measurements may not yet be possible. Every effort should, however, be made to obtain the vertical distribution of OH with about the same number of experiments as made for measuring the other substances.

This special effort should be based on presently developing national programmes which should be co-ordinated. As a first step, a thorough programme of inter- comparisons of different measurement techniques is needed, and further development carried out where necessary. This intercomparison programme should lead to defined experimental packages being used in a co-ordinated observational campaign during subsequent years. This campaign should consist of co-ordinated flights of multiple experiment payloads at the same location in order to obtain complementary measurements sets, as well as latitudes, including the tropics and polar regions, with several experiments per season. In the case of some semi-stable species where latitudinal cross-sections may be obtained by the use of aircraft, measurements should also be extended into the Southern Hemisphere.

As in the case of ozone soundings, not only must rapid data evaluation be carried out, but also a centralized scrutiny which will permit changes in the operational scheme to be recommended for the second year on the basis of the results obtained during the first.

The UV Monitoring Network

It is recommended that the appropriate WMO bodies (CAS and CIMO) be asked to under-take assessments of methods for monitoring ultra-violet radiation. The initiative of Australia and the U.S.A. in this field is appreciated, and high priority should be given to making an assessment of the Robertson-Berger UV meter.

In order to detect variations and trends in UV, and to correlate these with the ozone data, the following measurements are needed:

At ground level - determine UV flux as a function of ozone, haze, clouds and surface albedo, preferably with spectral resolution, in various latitude bands at least once per week.

From satellites - obtain measurements of UV flux in the band 1800-3200 $\overset{\circ}{A}$ to an accuracy of 2 to 3 percent with a 25 $\overset{\circ}{A}$ spectral band resolution.

Studies of Photochemistry

There are various fields in which research is needed in order to arrive at a fuller understanding of the stratospheric photochemical processes, of the transport mechanisms and of the role played by ozone in the radiation budget. Most of the relevant fields form part of a complex system with mutual feed-back; thus the main goal should be the investigation of all such inter-relationships, in conjunction with the development of a 3-dimensional model which incorporates full ozone photo-chemistry.

It is necessary to conduct certain experiments in the laboratory in order:

- to provide the relevant atmospheric chemical rate constants for significant reactions in the atmosphere as input to models;

- to provide photodissociation, photoionization and absorption data on the relevant molecular and atomic species occurring in the atmosphere as input data for models;

- to provide spectroscopic data on molecular and atomic species for the inter-pretation of experimental measurements;

- to carry out a critical survey and evaluation of existing experimental data on the above and recommend standard values for use in model calculations.

It is recommended that work in the following fields of atmospheric chemistry be intensified or initiated:

The nitrogen system, including the nitrogen cycle and various multidisciplinary aspects including fluxes of N_2O over different surfaces (forests, oceans, etc.). In particular, whether N_2O is predominantly from the land or the sea (to throw light on the significance of the fertilizer problem), the effect of increasing acidity of precipitation (due to anthropogenic effects) on the percentage of N_2O and N_2 released in denitrification, and the question of a tropospheric sink for N_2O. Furthermore, attention should be given to the rate constant for the reaction $O* + N_2O$, since this has recently been suggested to be much lower than values used in current models (other reactions involving $O*$ should also be examined more closely).

The hydrogen system, including the rate constants for reactions involving OH and the chlorofluoromethanes, and extending present measurements to cover pressure and temperature ranges relevant to atmospheric processes. In addition, the rate for the reaction OH + HO_2 and all other reactions involving HO_2 (especially Cl + HO_2) which have a role in the destruction of ozone should be studied, and maps of measured OH, HO_2 prepared.

The chlorine system, including the rate constants for reactions involving the higher order oxides of Cl, in particular $ClO + O_3 \longrightarrow ClO_2 + O_2$. The photo-dissociation rates for ClO_2 and ClO are also needed to estimate the amount of ClO_2 in the atmosphere. In addition more study is needed on the rate constant for Cl + CH_4, which has recently been claimed to be slower than previously thought, thus allowing more time for chlorine in the catalytic cycle, and on the ratio ClO/HCl which has recently been claimed to be higher than previously thought, suggesting a greater efficiency for the ClO catalytic chain.

The whole problem of the stability of *chlorine nitrate* and its expected effect in reducing the catalytic destruction of ozone by chlorine and nitrogen oxides.

Study of the *night-time photochemical processes*, and ways of measuring relevant trace substances at night.

Study of *heterogeneous reactions in the troposphere*, in particular the washout of NO_x, HNO_3, HCl and H_2O_2.

Improvement of *spectroscopic measurements* to permit remote sensing of several stratospheric trace constituents, for example: the absorption cross-section of HNO_3 in the 11μ band should be extended to low temperatures; infra-red spectroscopic data for N_2O_5 and HO_2; line strengths for many molecules, including HCl, in the sub-millimetre region.

Examination of the extent to which *methyl chloride* (CH_3Cl), *methyl bromide* (CH_3Br) and *carbon tetrachloride* (CCl_4) are anthropogenic or natural, thus clarifying the problem of the "natural reservoir" of chlorine in the stratosphere. Affirm that there are not large natural sources of chlorine presently in the stratosphere.

Data on the extraterrestrial solar spectrum are available, but more information on the variation of the *UV fluxes between solar maximum and minimum* would make it possible to separate long-term natural and anthropogenic variations in ozone.

Absorption by O_2 and O_3 and *multiple molecular scattering* should be included in theoretical ozone calculations.

In addition to solar fluxes, the *rate constants for photodissociation* of atmospheric gases depend upon the absorption coefficients and quantum yields at different wavelengths. While the gases of primary importance in stratospheric ozone chemistry have been extensively studied, there is a lack of information about others, e.g. H_2O_2.

A closer study of the local production of ozone in the troposphere as a result of intense industrial pollution is important in view of the importance to the global ozone budget.

Analysis and Circulation Studies

In the WMO Global Ozone Research and Monitoring Project it is recognized as essential that regular hemispheric and global analyses of the basic dynamic and thermodynamic variables (υ, P/φ, T) in the stratosphere, and of total ozone and the ozone mixing

ratio be maintained and be available to the scientific community. Such analyses
are necessary to properly identify hemispheric or global trends in thermodynamic
parameters and in total ozone to serve as the basis for comparative studies on
special experiments or special events, and also as the framework for considering
the representativeness of observations on other trace constituents.

Regular analyses up to 10 mb and analyses of total ozone distribution should be on
a daily basis; analyses for higher levels and analyses of ozone mixing ratio and
other parameters should be made at least weekly in the summer hemisphere and every
two or three days in the winter hemisphere. In general these analyses need to be
carried out where required global observational data including satellite data are
available, as well as objective analysis procedures.

Special studies of the circulation and energetics of the stratosphere are to be
encouraged. In view of the incompleteness of the observational data at these
levels, the studies should be combined with theoretical or numerical studies or
special observational programmes. In particular the magnitude and vertical extent
of the Hadley Cell and the whole question of the rapidity with which trace gases
released at the Earth's surface may reach the 30 km level needs to be investigated
(this problem is basic because it indicates how much of the ozone reduction
due to CFMs could be expected to be observed now).

The co-ordinating body for the project should ensure that such studies are undertaken in connexion with specified intensive observational and experimental periods.

In accordance with the above, every effort should be made to improve the reality
of atmospheric models including realistic values of mean meridional circulation
(not just an alteration of the eddy diffusivity values to include the effect of
mean meridional circulations).

Studies of Special Events

Since the disturbance of the "normal" ozone balance by natural or man-made events
is of major concern it follows that special events of large magnitude will provide
unique opportunities to study the underlying processes. Efforts should be made to
co-ordinate information exchange, intensify ozone measurements, and promote special
investigations of stratospheric warmings, volcanic eruptions, solar events and
unusual man-made emissions.

Ozone Modelling

Since many models already exist which simulate the atmospheric dynamical-photochemical system in varying degrees of complexity, special efforts need to be made
to co-ordinate studies using these models and to establish standard cases and
comparative model runs.

It is suggested that efforts be made to obtain the most reliable eddy exchange
coefficients for 1-D and 2-D models. A possible approach could be through studies
based on data available on the transport of relatively stable tracers.

Source Inventories

Not entirely unrelated to the topic of monitoring is the question of an inventory
of sources of CFMs, nitrogen oxides and other substances relevant to stratospheric
ozone. In order to explain trends in ozone concentration and to prepare scenarios
for the future, there is the need to keep track of the annual releases of relevant
trace gases into the atmosphere, from both man-made and natural sources. Although
the estimates must often be inferred indirectly, e.g., from manufacturers' production

records or from gasoline sales, these data are a valuable input to photochemistry models.

It is therefore recommended that UNEP be asked to obtain from the appropriate bodies annual global inventories for:

- chlorofluoromethane production and use;

- nitrogen fertilizer production and use;

- stratospheric aircraft emissions;

- volcanic eruptions that penetrate into the stratosphere;

- other factors that may be identified in the future.

Co-ordination of Atmospheric Studies

It is evident that a programme of the scope proposed would entail careful international co-ordination. It is recalled that IUGG in its Resolution XV (6 September 1975) asked WMO in particular to provide encouragement, support and co-ordination to the efforts of all nations in expanding research and monitoring efforts in order to evaluate possible global consequences of any activities which could affect the stratosphere. It was recognized that WMO had considerable experience in co-ordinating Members' activities in such international projects as the programme now proposed, and it is therefore proposed that the focal point be located at the WMO Secretariat.

The importance is stressed of putting the data obtained in the observing programme and the research results to use as quickly as possible in compiling statistics, in preparing analyses and in applying them in the various research projects. This means that efficient data management of the proposed international programme, based on the national efforts, is of very great importance. It is noted that it is planned that data obtained from the Nimbus-G satellite (U.S.A.) will be achieved and available for use within one year.

In support of the various research activities it is proposed that at an internationa level within the framework of the WMO Global Research and Monitoring Project action be taken to:

- establish an information and co-ordination centre at the WMO Secretariat which will ensure that the necessary ozone research studies and modelling experiments, standardizations and comparisons are carried out and results rapidly exchanged;

- ensure that a global stratospheric analysis programme is maintained. This should include standard mappings of wind and temperature variables, ozone mixing ratio and total ozone and regular N-S cross-sections. Arrangements should also be made for the additional requirements of special observational and experimental periods;

- encourage studies and co-ordinate the meteorological activites associated with investigating special events impacting on the ozone layer;

- organize and sponsor international conferences on subjects relevant to the objective of the Global Ozone Research and Monitoring Project, as well as informal planning meetings on specific problems arising from the implementation of the Project.

Conclusion

In conclusion, it is recommended that there should be two types of activity:

- an intensive research and monitoring effort (the Global Ozone Research and Monitoring Project) of three to four years' duration, starting during 1977, aimed at resolving current urgent problems and providing the basis for advice to Members and appropriate international organizations; and also

- the laying down of the basis for a continuing research and monitoring programme to cover at least one solar cycle (11 years) in the light of the experience and results of the project above, in order to determine trends.

The following major tasks have been identified:

- Improve the global monitoring of ozone, and monitor those quasi-permanent trace constituents which impact on the ozone system, as well as the ultra-violet flux.

- Co-ordinate national and international experiments aimed at making widely distributed atmospheric measurements, preferable simultaneous, of all pertinent trace constituents.

- Ensure efficient data management and analysis.

- Encourage and co-ordinate laboratory measurements of the behaviour and reaction rates of the chemical systems at stratospheric temperatures.

- Encourage, co-ordinate and assist in studies of the composition of, and the transport and exchange process in the atmosphere, including chemical and photo-chemical processes; also studies of the influences of cyclic and sporadic events on ozone and on other relevant trace constituents, in connexion with an assessment of the impact on global climate of any changes in the ozone content of the atmosphere.

- Support and encourage the frequent exchange of scientific information and personnel, as well as training in relevant fields, the design of better instrumentation, intercomparison tests and international network inspection.

- Encourage and facilitate investigations by competent bodies into the impact of UV variations on man and the biosphere.

Implementation of the project. Most activities will be nationally sponsored. WMO will take the lead in stimulation and co-ordination of these activities in collaboration with UNEP and ICSU.

APPENDIX A

Resolution adopted by the WMO Executive Committee

Resolution 8 (EC-XXVIII)

GLOBAL OZONE RESEARCH AND MONITORING PROJECT

THE EXECUTIVE COMMITTEE

NOTING:

(1) Resolution 13 (EC-XII),

(2) Resolution 29 (EC-XVIII),

(3) Resolution 10 (EC-XXII),

(4) Abridged report of Cg-VII, general summary, paragraph 3.2.1.3 and 3.3.5.21,

COMMENDS the president of CAS for having arranged for the preparation of the text of the WMO statement entitled "Modification of the ozone layer due to human activities and some possible geophysical consequences";

APPROVES the action taken by the Secretary-General in releasing this statment;

CONSIDERING:

(1) The urgent need to determine the extent to which human activities may contribute to the attenuation of the ozone layer due to pollution of the stratosphere, and the need to assess the geophysical consequences of any changes in the ozone content,

(2) That any such study must be conducted in such a way as to ensure that the results will command wide acceptance in the scientific community,

(3) That WMO is the appropriate body to organize such a study,

(4) That the success of the study would largely depend upon Members' own efforts in contributing to a greatly improved global monitoring system of the stratospheric environment,

(5) That there is a need to speed up the publication of data on total ozone amount

EXPRESSES APPRECIATION:

(1) To Members which have already taken steps to expand their programme of observations of ozone and other trace constituents in the stratosphere, and have undertaken related research;

(2) To the International Ozone Commission of IAMAP for its valuable collaboration

(3) To the Atmospheric Environment Service of Canada for providing the services of the World Ozone Data Centre and publishing regularly the bulletin "Ozone data for the world";

APPROVES:

The Global Ozone Research and Monitoring Project as outlined in the annex to this resolution;

URGES Members:

(1) To contribute to the implementation of the project in particular by establishing, upgrading, re-activating or maintaining ozone observing stations as required to complete the network, bearing in mind the desirability of locating stations near to rawinsonde stations and where feasible in conjunction with background air pollution stations;

(2) To co-operate in a more rapid exchange of data on ozone amounts by sending their data promptly, if possible monthly, to the World Ozone Data Centre at Toronto for publication;

(3) To make measurements of other atmospheric trace constituents relevant to ozone; and

(4) To contribute to the co-ordinated research efforts outlined in the annex to this resolution;

REQUESTS the president of CAS and other constituent bodies concerned to co-ordinate and intensify their efforts and to provide information on the accuracy of vertical profiles of temperature, humidity and wind speed in the upper atmosphere obtained from satellites;

REQUESTS the Secretary-General:

(1) To proceed, in consultation with the president of CAS, with the planning and execution of the Global Ozone Research and Monitoring Project in accordance with the outline in the annex to this resolution;

(2) To seek continued support from UNEP and other appropriate bodies in the implementation of the project;

(3) To consult with the presidents of technical commissions concerned with a view to using to the fullest extent possible existing groups of WMO technical commissions, especially CAS, in the execution of the project;

(4) To provide the necessary support for the execution of the project within the limits of available financial resources;

(5) To discuss the feasibility of further accelerating the exchange of total ozone data with the presidents of the technical commissions concerned;

(6) To bring this resolution to the attention of all concerned.

Annex to Resolution 8 (EC-XXVIII)

WMO GLOBAL OZONE RESEARCH AND MONITORING PROJECT

1. The increasing attention given to the various ways in which man may influence the chemical processes in the global atmosphere, particularly the recent concern regarding the effect on ozone of emissions from supersonic aircraft and of chorofluoromethanes, highlights the need for initiating already in 1976 an internationally co-ordinated project to improve our understanding of the complex

interactions which determine stratospheric environmental systems.

2. It is now clear that whereas ozone (O_3) is formed by the photodissociation of molecular oxygen, its catalytic destruction by nitrogen oxides (NO_x) is the predominant stratospheric removal mechanism under normal conditions. The major source of nitric oxide (NO) is oxidation in the stratosphere of nitrous oxide (N_2O) originally released by microbiological activity in the soils and oceans.

3. The possibility thus exists that additions to this NO_x source, or the addition of other interactive substances such as chlorofluoromethanes, could seriously disturb the ozone balance. The weight of expert scientific evidence, based on models and on admittedly scant atmospheric measurements, is that future large aviation fleets flying at higher levels may perturb the NO_x levels sufficiently to require controls of their emissions and that the continued release into the atmosphere of man-made chlorofluoromethanes poses an even more serious threat.

4. The distribution of ozone is influenced to a major extent by dynamical processes and existing data are not sufficient to allow a full understanding of the extent and nature of this interdependence.

5. The foregoing considerations clearly point to the need for a co-ordinated joint international project to find scientifically sound answers to questions concerning the role of anthropogenic pollutants in reducing the quantity of ozone in the stratosphere, with particular emphasis on chlorofluoromethanes and nitrogen oxides, and on the possible impact of changes in the stratospheric ozone content on climate.

6. It is recognized that WMO must play a leading role in the organization, execution and evaluation of this project, and it is expected that its outcome will be of great significance to Members and to some intergovernmental organizations such as UNEP, since the results of the project may justify calls for measures to limit the production of particular substances, with inherent economic implications.

Objectives

7. The Global Zone Monitoring and Research Project should enable WMO to provide advice to Members and to the United Nations and other appropriate international organizations concerning:

(a) The extent to which man-made pollutants might be responsible for reducing the quantity of ozone in the stratosphere, with particular emphasis on the role played by chlorofluoromethanes and nitrogen oxides;

(b) The possible impact of changes in the stratospheric content of ozone on climatic trends and on solar ultra-violet radiation at the Earth's surface;

(c) The establishment of the basis and identification of the needs for strengthening the long-term monitoring programme of the ozone system for determination of trends and of future threats to the ozone shield.

Strategy

8. The means by which the objectives may be achieved are as follows:

(a) To gather and evaluate existing knowledge on atmospheric ozone;

(b) To provide information on the atmospheric concentrations of substances of relevance to the N_2O, NO_x, HO_x, ClO_x, BrO_x cycles leading to a better appreciation of their currently known important sources and sinks and their impact on the ozone budget;

(c)   To extend and improve analyses and circulation studies of the stratosphere based on actual data, with a view to clarifying and predicting the exchange processes between the troposphere and the stratosphere;

(d)   To assess the impact on the atmospheric environment of possible ozone depletion, recognizing that this involves a variety of interdisciplinary interactions requiring input from the fields of meteorology, oceanography, agriculture and microbiology;

(e)   To arrange for an exchange of knowledge through reports, newsletters, scientific meetings and exchange of personnel.

Programme of Work

9.   There should be two types of activity:

(a)   Intensive research and monitoring effort (the Global Ozone Research and Monitoring Project) of three to four years' duration, starting during 1976, aimed at resolving current urgent problems and providing the basis for advice to Members and appropriate international organizations;

(b)   The laying of the basis for a continuing research and monitoring programme to cover at least one solar cycle (11 years) in the light of the experience and results of the project above, in order to determine trends.

10.   The following major tasks have been identified:

(a)   Improve the global monitoring of ozone, and monitor those quasi-permanent trace constituents which impact on the ozone system, as well as the ultra-violet flux;

(b)   Co-ordinate national and international experiments aimed at making widely distributed atmospheric measurements, preferably simultaneous, of all pertinent trace constituents;

(c)   Ensure efficient data management and analysis;

(d)   Encourage and co-ordinate laboratory measurements of the behaviour and reaction rates of the chemical systems at stratospheric temperatures;

(e)   Arrange, with other responsible agencies as necessary, for an inventory to be maintained of the annual emissions of chlorofluoromethanes, nitrogen oxides and other important species, as well as of national production rates;

(f)   Encourage, co-ordinate and assist in studies of the composition of, and the transport and exchange process in the atmosphere, including chemical and photochemical processes; also studies of the influences of cyclic and sporadic events on ozone and on other relevant trace constituents, in connexion with an assessment of the impact on global climate of any changes in the ozone content of the atmosphere;

(g)   Support and encourage the frequent exchange of scientific information and personnel, as well as training in relevant fields, the design of better instrumentation, intercomparison tests and international network inspection;

(h)   Encourage and facilitate investigations by competent bodies into the impact of UV variations on man and the biosphere.

Implementation of the Project

11.   Most activities will be nationally sponsored. WMO will take the lead in co-ordinating these activities in collaboration with UNEP and ICSU.

12.   As regards the scientific direction and arrangement of the project, this should be carried out by:

(a)   *A CAS Working Group on Stratospheric Ozone*, whose role is to consider and recommend scientific goals and plans for the project. With the help of additional expertise where necessary, the group provides the Executive Committee and the Secretary-General with advice on details of the objectives of the project and how these could be achieved in principle;

(b)   *A project co-ordinator*, assisted by consultants as necessary, working at the WMO Secretariat, using the available experience and support of the Secretariat and other relevant working groups. He will co-ordinate the implementation of the project in line with guidance given by the CAS working group and the Secretary-General.

13.   The WMO Executive Committee should be informed annually of the development of the project and should ensure its implementation to the maximum extent possible. The Secretary-General should arrange for information on the implementation of the project to be made available to Members as well as to UNEP and ICSU.

APPENDIX B

ASSESSMENT OF THE EFFECT OF NITROGEN OXIDES ON OZONE

This appendix gives a short review of the present state of knowledge regarding the effect of nitrogen oxides on stratospheric ozone, including the effect of a number of human activities. The importance of nitrogen oxides for stratospheric ozone follows from the fact that NO and NO_2 destroy ozone via the set of reactions

$$(NO + O_3 \longrightarrow NO_2 + O_2) + (NO_2 + O \longrightarrow NO + O_2) + (O_3 + h\nu \longrightarrow O + O_2) =$$

$$(2\ O_3 \longrightarrow 3\ O_2).$$

Natural Sources and Sinks of NO and NO_2 in the Stratosphere

Oxidation of N_2O. The reaction

$$N_2O + O\ (^1D) \longrightarrow 2\ NO \tag{1}$$

is now considered to be the main source of nitrogen oxides in the stratosphere, but estimates of the yearly global production of NO following from this reaction varies from approximately 2×10^{33} to 2×10^{34} molecules. The differences between these estimates can be explained mainly by different assumptions regarding the vertical diffusion of N_2O in the stratosphere. Besides reaction (1) there are other processes leading to the loss of N_2O, namely

$$N_2O + O\ (^1D) \longrightarrow N_2 + O_2 \tag{2}$$

and especially

$$N_2O + h\nu \longrightarrow N_2 + O\ (^1D) \hspace{5cm} (3)$$

Electronically excited $O\ (^1D)$ atoms are produced in the stratosphere mainly by the photolysis of O_3

$$O_3 + h\nu \longrightarrow O\ (^1D) + O_2 \quad \lambda \leqslant 310\ nm \hspace{3cm} (4)$$

Most of the $O\ (^1D)$ atoms are deactivated by collisions with O_2 and N_2

$$O\ (^1D) + (N_2, O_2) \longrightarrow O\ (^3P) + (N_2, O_2) \hspace{3cm} (5)$$

Besides different assumptions about the diffusion of N_2O in the stratosphere, the above-mentioned difference in global NO production rates can be explained by different assumptions regarding the quantum yields of $O\ (^1D)$ production in reaction (4), the deactivation rate coefficient of $O\ (^1D)$, and the photolysis rates of N_2O, especially the contribution from solar radiation below 100 nm penetrating in the Schumann-Runge band of molecular oxygen.

Recent developments have shown that further revisions of the estimates of NO production rates are necessary because of temperature dependence of the $O\ (^1D)$ production quantum yield in reaction (4), and of the reaction rates of reactions (1), (2) and (5). Furthermore, there is considerable doubt about the quality of the N_2O absorption cross-section at wavelengths larger than about 250 nm.

Galactic cosmic rays. A continuous flow of energetic protons reaches the Earth mainly at high geomagnetic latitudes (>60°). These protons cause ionization and production of NO mainly in the lower stratosphere. There is an interesting variation in the production of nitric oxide following solar activity with a period of about 11 years. The estimated global yearly production of NO varies between about 1.2×10^{33} (at solar maximum) and 1.8×10^{33} molecules (at solar minimum). It appears that these estimates of the production of NO due to the action of galactic cosmic rays is rather well known. It has been proposed that the afore-mentioned solar-cycle variations in NO production would be the cause of variations in total ozone, which have been remarkably large at middle- and high-latitude stations (e.g. ± 5% at Tromsö, 70°N and ± 2% at Arosa, 45°N). Recent, independent, detailed calculations with two-dimensional models (Brasseur and Crutzen) cast considerable doubt on the proposed link between variable input of NO over the solar cycle and the observed variations in total ozone. A search for a different mechanism seems to be justified.

Solar proton events. Sporadic bombardment of the Earth's upper atmosphere at high geomagnetic latitudes ($\ell > 60°$) by energetic solar particles (mainly protons) produce large amounts of NO in the same way as galactic cosmic rays. The average energy of solar protons being lower than that of galactic cosmic rays, the production of NO from solar protons occurs generally above 30 km. A recent analysis of three solar proton events gave the following approximate global productions of NO in the stratosphere: 2×10^{33} (Nov. 1960), 6×10^{32} (Sept. 1966), 6×10^{33} (Aug. 1972). According to these calculations the amount of NO formed in the stratosphere during these three solar proton events was comparable to, or larger than, the total annual production of NO from the action of galactic cosmic rays and its variations over the solar cycle.

The solar proton event of August 1972 especially constituted a major perturbation in the chemical conditions in the stratosphere and studies of stratospheric ozone and temperature records are very important to increase our knowledge on the response of the stratosphere to major disturbances and provide a means for validating and improving model simulation of the stratosphere.

Other possible sources for stratospheric NO. Large quantities of NO are produced above 100 km due to the action of short-wave ultra-violet radiation in auroral activity. Fortunately, most of the high altitude NO does not reach the lower portions of the atmosphere due to the photodissociation process

$$NO + h\nu \longrightarrow N + O \quad \lambda \leqslant 192 \text{ nm} \tag{6}$$

followed by the reaction

$$N + NO \longrightarrow N_2 + O \tag{7}$$

However, there is the possibility of some leakage of upper-level NO in the winter hemisphere and this possible source of stratospheric NO should be investigated.

Nitric oxide may be produced in the troposphere by biological processes, by lightning and by conversion of some NH_3 to NO by gas phase reactions. Little is known of the NO_x budget of the troposphere and it may well be that most NO_x in the troposphere is of industrial origin. It is normally assumed that transport of tropospheric NO_x to the stratosphere is not possible due to the efficiency of rainout and washout processes in the troposphere. Although this is very likely the case for sunlight conditions during which there is a conversion from NO_2 to very soluble and reactive HNO_3 by the reactions

$$NO_2 + OH \ (+M) \longrightarrow HNO_3 \ (+M) \tag{8}$$

there is some doubt whether removal of NO_x is efficient enough during winter conditions to prevent significant transfer of NO_x from ground level to the stratosphere. This is an area of research which deserves particular attention.

Anthropogenic Sources of NO in the Stratosphere

A number of human activities may lead to increases in the stratospheric NO content and may consequently lower the atmospheric ozone abundance. Among these activities should be mentioned especially:

— changes in the conditions of the soil;
— increased industrial injection of NO at ground level;
— injections of NO_x in the lower stratosphere by stratosphere-flying aircraft (both subsonic and supersonic).

Special attention should be given to the possibility that human activities may cause an increase in the input of N_2O into the atmosphere by intensified agricultural activity and by affecting the physical and chemical conditions of the soil — especially by increasing soil acidity (see section above on the Nitrogen System).

Nothing quantitative can be said presently of the possible size of the input of industrial gound-level NO_x into the stratosphere and it remains to be seen whether this may in fact constitute a significant problem.

Some consequences of stratospheric flights. Extensive research activities have recently been devoted to the assessment of the impact of stratosphere-flying aircraft on the ozone layer. Research documents and conclusions have been published in the U.S.A., and recently results have become available from programmes in some other countries.

The principal results of the CIAP Report of Findings are summarized in Table 6.3, those of the U.S. National Academy of Sciences report on Environmental Impact of Stratospheric Flight in Table 6.4, and both of these tables refer to the Northern

TABLE 6.3   Estimated Percent Ozone Reduction per 100 Aircraft

Aircraft Type	Fuel Burned per Year* (kg/yr)	Altitude km (kft)	NO_x Emission Index (EI) Without Controls (g per kg fuel)	% Ozone Reduction in Northern Hemisphere		
				Without Controls	EI Controls	
					1/6 Today	1/60 Today
Subsonic**						
707/DC-8	1×10^9	11 (36)	6	0.0034	0.00070	0.000070
DC-10/L-1011	1.5×10^9	11 (36)	15	0.010	0.0020	0.00020
747	2.1×10^9	11 (36)	15	0.014	0.0025	0.00025
747-SP	2.0×10^9	13.5 (44)	15	0.079	0.014	0.0014
Supersonic						
Concorde/TU-144	4×10^8 / 3×10^9	13.5 (44) / 16.5 (54)	18	0.39	0.068	0.0068
Advanced SST	3×10^8 / 6×10^9	16.5 (54) / 19.5 (64)	18	1.74	0.32	0.032

* Subsonics assumed to operate at high altitude, 5.4 hours per day, 365 days per year. Supersonics assumed to operate at high altitude, 4.4 hours per day, 364 days per year.

** The present subsonic fleet consists of 1,217 707/DC-8s, 232 DC-10/L-1011s, and 232 747s flying at a mean altitude of 11 km (36 kft) and is estimated to cause a 0.1% ozone reduction.

TABLE 6.4   Percentage Reduction in Stratospheric Ozone per
Hundred Aircraft Operating in the Stratosphere

Aircraft Type	% per 100 Aircraft	Uncertainty Factor
Present Subsonic	0.02	10
Projected Subsonic	0.2	10
Present Supersonic	0.7	3
Formerly Projected Supersonic	3	2

Hemisphere where the major stratospheric operations are likely to take place. They
are separate assessments of virtually the same set of model results, the higher
estimates of the ozone reductions in the MAS Report being mainly due to the introduction of different vertical eddy diffusion coefficients into the model results as
derived from a later study of Carbon 14 distributions in the stratosphere. The
uncertainties in the results are stated to be twice to one-tenth of the given percentage ozone depletions in the CIAP Table 6.3. The NAS results have been stated
to have uncertainties of a factor of two for flights at 20 km, a factor of three at
17 km and as much as ten for flights at lower altitudes.

Whereas a percentage depletion of 0.39 is suggested in the CIAP report (Table 6.3)
and 0.7 in the NAS report (Table 6.4), COMESA (United Kingdom) estimates on the
basis of injection of 10^{11} g of NO_x per year at an altitude of 16-18 km by 100
Concorde/TU-144 aircraft flying 7 hours per day, that the percentage ozone depletion
would be only 0.11%. It should be noted that different assumptions as to the
injection rate of NO_x were used by CIAP and NAS.

It is clear that according to these reports there may well be significant reductions
in the ozone amounts in the statosphere if large fleets of SSTs are operated at
20 km or above. On the other hand the effect will be smaller for small aircraft
fleets operating at lower altitudes and is likely to be insignificant and undetec-
table for subsonic fleets now flying and the presently planned number of SSTs
(given as about 30 in the CIAP report) likely to operate at 17 km.

APPENDIX C

THE CHLOROFLUOROMETHANES AND THE OZONE DEPLETION PROBLEM

Stratospheric Chlorine Chemistry

The two chlorofluoromethane gases, CF_2Cl_2 (Fluorocarbon-12) and $CFCl_3$ (Fluorocarbon-
11), are now being produced at rates of approximately 500,000 and 300,000 tons per
annum respectively, in the world, being used almost entirely in techological applications involving eventual release to the Earth's atmosphere. The only important
sink known at present for these molecules is mid-stratospheric photolysis by solar
ultra-violet radiation in the 190-220 nm band. Such photolysis results in the
immediate release of one chlorine atom per molecule, and a second chlorine atom is
rapidly released by the reaction of the remaining radical with O_2. The third atom
of Cl from CCl_3F is usually released later by the photolysis of the residual $CClFO$.

The UV radiation which causes chlorofluoromethane photodissociation is also absorbed
both by molecular O_2 and by O_3 in the Earth's atmosphere, and has been removed from
solar radiation which penetrates to the lower atmosphere. However, as the chlorofluoromethane molecules rise into the stratosphere above most of the O_2 and O_3 due
to the atmospheric motions, they encounter this short wave-length solar UV radiation
and are photodissociated. The median altitude for relase of these Cl atoms is

about 30 km.

Free chlorine atoms have long been known from laboratory studies to react very
rapidly with ozone by reaction (1),

$$Cl + O_3 \longrightarrow ClO + O_2 \tag{1}$$

and ClO is itself very reactive with two stratospheric components, O and NO, by
reactions (2) and (3),

$$ClO + O \longrightarrow Cl + O_2 \tag{2}$$

$$ClO + NO \longrightarrow Cl + NO_2 \tag{3}$$

since the concentration of O atoms increases rapidly with altitude, the combination
of reactions (1) and (2) is most important about 30 km, and together comprise the
ClO_x catalytic chains which result in the removal of two equivalents of ozone by
(4)

$$O + O_3 \longrightarrow 2 O_2 \tag{4}$$

while returning the Cl to its original form for another cycle.

The repetition of cycles (1) + (2) or (1) + (3) is interrupted by the occasional
reaction of Cl with CH_4, H_2, or other H-containing substance to form HCl, as in (5).
The catalytic chain can then be restarted by attack of OH on HCl, as in (6)

$$Cl + CH_4 \longrightarrow HCl + CH_3 \tag{5}$$

$$OH + HCL \longrightarrow Cl + H_2O \tag{6}$$

At any particular altitude in the stratosphere, the photochemical distribution of
chlorine among HCl, ClO and Cl can be calculated in equilibrium if (a) the concentrations of the other reactants (O_3, O, NO, CH_4, H_2, OH) are known; and (b) the
rate constants of the reactions (1), (2), (3), (5) and (6) are known by laboratory
measurements at stratospheric temperatures. Concentrated research effort during
the past year has provided both additional stratospheric measurements and evaluated
laboratory rate constants, and this partition of chlorine among HCl, ClO, Cl can
now be calculated with reasonable accuracy. At all altitudes, HCl is the most
abundant species, but ClO reaches the 10-30% range between 25 and 40 km. Chlorine
atoms are always present in lower concentration because of the rapidity of their
reaction with ozone.

The current experimental evidence in the stratosphere indicates that this description is essentially correct, and suggests that no undiscovered additional stratospheric chlorine chemistry will materially alter the overall description. A few
high altitude samples taken by balloon and analysed in ground laboratories have
shown that the mixing ration of $CFCl_3$ decreases rapidly with altitude above 15 km
and is essentially absent above 35 km, as expected on photochemical grounds.
CF_2Cl_2 rises higher in the stratosphere because of its lesser photochemical sensitivity, but also has been observed to decrease with altitude. These measurements
confirm that both molecules are being decomposed in the mid-stratosphere.

Airbone filters carried into the stratosphere either by aircraft or balloon have
been analyzed after return for chlorine content. Neutral filters do not retain any
appreciable chlorine when used below 25 km, although they are quite efficient in
retaining the particles of sulfuric acid found throughout the lower stratosphere.
When coated with a chemical base, however, these filters do retain chlorine in the
form of chloride (HCl). The analyses have shown that the amount of chloride
retained increases with altitude to about 25-30 km, and agrees with the amounts

expected from stratospheric decomposition of $CFCl_3$, CF_2Cl_2, CCl_4, and CH_3Cl. Although the present agreement is much closer, the combined accuracies of the various measurements indicate only that the amount of HCl is within a factor of 2 of that expected. The observed increase of HCl mixing ratio with altitude is inconsistent with tropospheric injection of chloride, either as NaCl from sea salt or as HCl from industrial sources or volcanoes, but is consistent with mid-stratospheric release of chlorine atoms from these organo-chlorine molecules. Since HCl, ClO and Cl are all interconnected by the chemical reactions given above, any unknown reaction of Cl of ClO would also reduce the concentration of HCl. Further verification that HCl is present in approximately the amounts expected would indicate that the major chlorine reactions have probably been included.

Direct measurements of Cl in the stratosphere will be very difficult because of its low concentrations. Direct measurement of ClO has not been made (although instruments for this purpose are being constructed and tested) so that it has not yet been possible to obtain further confirmation in this way. The neutral filter measurements of chloride have, however, consistently shown positive measurements about 25 km, indicating the presence of a new chlorine species. The amounts are consistent with those expected for ClO, but laboratory tests are needed to identify the species involved.

Ozone Depletion of ClO_x

The relative rates of removal of ozone can be estimated from direct stratospheric measurements of the concentrations of the important species. In the natural atmosphere, the dominant process involves the NO_x-catalyzed chain, proceeding largely through reactions such as (7)

$$NO_2 + O \longrightarrow NO + O_2 \tag{7}$$

Lesser amounts of ozone are also removed by the direct reaction of O with O_3 as in (4), and the various HO_x-catalyzed chains. A minor contribution is also made by ClO_x-catalyzed chains of natural origin (e.g. Cl from CH_3Cl).

The present contributions to ozone removal from anthropogenic ClO_x are small, but that part caused by CF_2Cl_2 and CF_3Cl can be expected to increase with time because of the long lifetime of these molecules in the atmosphere. The lifetimes for $CFCl_3$ for a stratospheric photochemical sink has been calculated using one-dimensional models to be about 50-70 years, and comparison of actual atmospheric concentrations with known world production indicates an observed lifetime of more than 30 years, with no other appreciable sinks yet discovered. With continued release to the atmosphere it can be expected that the concentrations of both CF_2Cl_2 and $CFCl_3$ will increase, the amounts of ClO in the stratosphere will increase correspondingly, and the amounts of ozone removed by reactions (1) + (2) will then also increase. At present rates of world use, the amounts can be expected to increase to at least 10 times the present concentrations. Accelerated use (world use of $CFCl_3$ and CF_2Cl_2 has increased by about 13% per year from the 1950s to 1973) would result in even higher concentrations. At these higher concentrations, direct comparison or removal rates of O by reactions with NO_2, O_3, and ClO (as in (7), (4) and (2)) indicate that anthropogenic ClO would be an important additional process for removal of ozone, and the average amount of ozone in the Earth's atmosphere would be appreciably decreased.

Several quantitative evaluations, based on calculations using one-dimensional models with transport represented by vertical eddy diffusion, have now been made of progressive effects of anthropogenic ClO_x on the average amount of ozone. These calculations involve average concentrations of the important atmospheric species (taken from the limited observations so far available) and the best current estimations of reaction rates (as measured in laboratories) for about one hundred chemical reactions. On an average world-wide basis, without allowance for variation

with latitude and longitude, the calculations indicate that the present anthropogenic depletion of ozone by CF_2Cl_2 and $CFCl_3$ is in the range of 0.5 to 2.0% and the
long-term steady-state effect of continued release at the 1972 world rate of
production would be about 10% average ozone depletion with the factor of uncertainty
of about two. Preliminary calculations, using models which include latitude
variations, indicate ozone depletion of about the same value.

Thus whilst a fuller understanding of the chemistry involving chlorine in the
stratosphere will require further measurements and calculations, including the
naturally-occurring chlorocarbons (e.g. methyl chloride — CH_3Cl, and possibly
others), the present evidence strongly supports the view that a continued release
of chlorofuoromethanes into the atmosphere may lead to a significant reduction in
stratospheric ozone.

Other Chlorine Compounds

Chlorine atoms can also be released in the stratosphere by reactions of other
chlorine-containing molecules such as CCl_4, CH_3Cl, CH_3CCl_3, etc. The atmospheric
chemistry of CCl_4 is similar to that of CF_3Cl and CF_2Cl_2, with the solar ultraviolet sink probably of major importance. However, the amount of CCl_4 in technological application involving release to the atmosphere appears currently to be
small compared to that of the chlorofluoromethanes.

Chlorocarbon molecules containing C-H bonds also have a major tropospheric sink
through reaction (8), illustrated with CH_3Cl, while chlorocarbon molecules

$$OH + CH_3Cl \longrightarrow H_2O + CH_2Cl \qquad\qquad (8)$$

containing C = C bonds, e.g. $CCl_2 = CCl_2$, are also removed for the most part by
reactions in the troposphere. Nevertheless, a minor fraction of each species will
penetrate into the stratosphere and release Cl atoms there. It is therefore
important to monitor these additional species continually, and to evaluate the
relative importance of tropospheric and stratospheric sinks for each.

The concentration of CH_3Cl in the atmosphere is much too large for the input to be
entirely from ordinary industrial sources, and it probably arises chiefly from
natural sources. However, the possibility that some of it may be anthropogenic
origin through agricultural practices, e.g. burning of vegetation, requires
thorough investigation. Similarly, the anthropogenic and/or natural sources should
be determined for the other halocarbons found to be widely distributed in the
atmosphere, e.g. CCl_4, $CHCl_3$.

Chlorine Nitrate

The chlorine and nitrogen radicals may interact through

$$ClO + NO_2 \longrightarrow ClNO_3 \qquad\qquad (9)$$

forming chlorine nitrate. This could be an important reaction from the point of
view of ozone photochemistry because it removed radicals from both the chlorine and
nitrogen systems, and it is these radicals that may play such a large role in the
catalytic destruction of ozone. The above reaction was recognized only recently as
being of potential importance in ozone photochemistry, and partly because of it,
the NAS report was delayed a few months. On the basis of the above reaction, the
NAS reduced the ultimate ozone depletion (assuming continued fluorocarbon release
at the 1973 rate) by a factor of 1.85, or from 14% to 7.5%. Recent calculations
and observations suggest that the chlorine nitrate concentration in the stratosphere
is less than 2 ppb, or considerably less than thought possible a few months ago.
This may be because the hydroxyl radicals combine more easily with nitrogen dioxide

than does chlorine monoxide, i.e.

$$ClO + NO_2 + M \longrightarrow ClONO_2 + M \tag{10}$$

is much slower than the reaction

$$OH + NO_2 + M \longrightarrow HNO_3 + M \tag{11}$$

This in turn implies that the NAS may have overestimated the factor by which the ozone reduction should be decreased. One recent opinion is that the peak ozone reduction should be more nearly 10% than 7.5%.

APPENDIX D

THE NITROGEN CYCLE

The bulk of the Earth's volatile nitrogen, approximately 4×10^{15} tonnes, resides in the atmosphere as molecular nitrogen (N_2). The remainder, approximately 6×10^{14} tonnes, is distributed between sediments, the ocean, the soil and the biosphere, in concentrations as shown in Table 6.5 below. The biological processes play a major role in regulating the manner in which nitrogen is exchanged between the various reservoirs.

TABLE 6.5   Nitrogen Content of Various Terrestrial Reservoirs

Reservoir	Nitrogen Content (tonnes)	Reference*
Atmosphere	4×10^{15}	(1)
Land Biomass	1×10^{10}	(2)
Humus	6×10^{10}	(3)
Soil Inorganic	1×10^{11}	(4)
Ocean Biomass	8×10^{8}	(5)
Ocean Organic	2×10^{11}	(5)
Ocean Nitrate	6×10^{11}	(6)
Ocean Ammonia & Nitrate	1×10^{10}	(6)
Ocean N_2	2×10^{13}	(6)
Sediments	6×10^{14}	(1)

* (1) H.D.Holland, private communication, 1975.
  (2) McElroy, assuming a C to N ration of 30 (the C to N ration of trees is typically in the range 20-40).
  (3) McElroy, assuming a C to N ratio of 15   (I. Chet, private communication, 1975).
  (4) C.C.Delwiche, *The Biosphere*, W.H.Freeman & Co., San Francisco, 1970.
  (5) McElroy, assuming a C to N ratio of 12   (R. Mitchell, private communication, 1975).
  (6) R.F.Vaccaro, *Chemical Oceanography*, Vol. I, Academic Press, London, New York, 1965.

Nitrogen is an essential nutrient and yet in the form in which it may be utilized by the biosphere it is in remarkably limited supply. Before atmospheric nitrogen can be incorporated into the tissue of living things it must be fixed, i.e. it must be transformed from the relatively abundant though chemically inert form of N_2 into more useful compounds such as ammonia (NH_3) and NO_3^-. Such transformation requires a significant input of energy; 226 kilocalories (946 kJ) per mole is needed to break the N-N bond. This is accomplished either as a byproduct of combustion — in automobiles and stationary power sources, or as a result of the application of fertilizers in agriculture. In 1974, combustion accounted for fixation of approximately 4×10^7 tonnes of nitrogen, and about the same amount was produced from nitrogeneous fertilizers. In the same year natural processes, due mainly to symbiotic associations between certain bacteria (Rhizobium) and plants of the family *Leguminosae*, accounted for a total source of fixed nitrogen estimated at 1.8×10^8 tonnes.

Fixation of atmospheric nitrogen must be balanced on a geological time scale by the reverse process, denitrification, through which N is returned to atmosphere in its most stable forms N_2 and N_2O (nitrous oxide). If denitrification did not occur, in about 10^7 years the supply of N_2 would be exhausted. It appears that most of the biospheric nitrogen is returned to the atmosphere as N_2; the global yield of N_2O relative to N_2, according to current estimates, is about 8%. Most of the N_2O released into the atmosphere decomposes harmlessly to form N_2, but a small fraction (about 5%) is decomposed chemically to yield nitric oxide (NO) and it is this source of NO which is believed to represent the major contribution to the stratospheric budget of NO_x. In this way nature, through complex biotic reactions, has the means of exerting a long-term control over ozone, and the biosphere has a capability of influencing the radiation environment at the surface of the planet. Major components of the nitrogen cycle, with estimates for the rates at which the element is transferred between various reservoirs, are illustrated in Figure 6.13.

Man can influence the nitrogen cycle in several ways. He can do so by agriculture; e.g. fixation of nitrogen in chemical fertilizer plants has grown by a factor of 30 over the past 25 years and will grow by at least a factor of 5 over the next 25 years. This could lead to a very significant increase in the rate of global fixation of N by the turn of the century, and the increase may be even greater if recent advances in genetics should permit the development of new crop species capable of symbiotic fixation in the manner which applies now to the legumes. With the increased production of fixed nitrogen, it is to be expected that there would be a corresponding rise in the rate of denitrification, and thus in the formation of N_2O. Since ozone is largely controlled by the N_2O flux into the stratosphere it is clear that a large increase in the rate of production of N_2O would lead to a significant reduction in ozone.

The rise in the amount of N_2O would be even larger if one considers the possible short-term shifts between aerobic and anaerobic processes, and also the possible changes which may take place in the acidity of rain. A decrease in the pH of rain (i.e. an increase in its acidity) may leads to an increase in the production of N_2O relative to N_2. There are indications that various forms of industrial activity, for example, the release of sulphur due to burning of fossil fuels and the smelting of metallic sulphides, are already affecting shifts of this nature. This trend must be watched extremely carefully in order to forestall possible harmful effects on the global atmospheric and biospheric environments.

It seems clear that the matters raised here require further attention. They raise problems of enormous complexity, which affect agriculture, soil science, microbiology, public health and oceanography as well as meteorology.

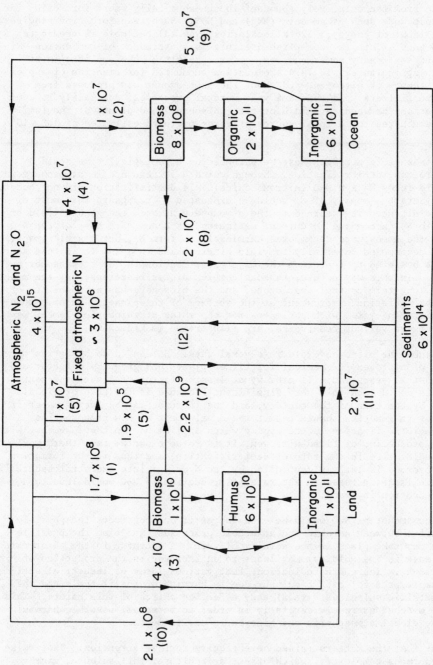

Fig. 6.13. Representation of the nitrogen cycle. Numbers in rectangular boxes give the abundance of N measured in metric tons. Transfer rates are given in tons per year, and the numbers in parentheses define the transfer reactions as follows: (1) land biological fixation, (2) marine biological fixation, (3) fixation in chemical fertilizer, (4) fixation due to combustion, (5) fixation due to lightning, (6) transfer due to volatilization of NH_3, (7) transfer due to rain over land, (8) transfer due to rain over sea, (9) marine denitrification, (10) land denitrification, (11) transfer from land to sea in river runoff, (12) transfer due to raising of sediments, with a comparable amount due to deposition.

PAPER 7

*Ozone Layer and Development**

Presented by the Department of Economic and Social Affairs

INTRODUCTION

The United Nations Department of Economic and Social Affairs, which reports to the Economic and Social Council, is particularly concerned with the economic and social progress of developing countries, through the preparation of the Third Development Decade (1981-1990) and of the New International Economic Order.

At first sight, some of the problems related to the causes and effects of a reduction of the ozone layer seem remote from the specific concerns of developing countries. For instance, the problem of propelling a hair cosmetic seems to be irrelevant. On the other hand, the consequences of a depletion of the ozone layer concerns both developing and developed countries.

However, a more careful analysis leads to the opinion that some aspects of this problem area, are of particular importance to developing countries. We will review briefly hereunder the economic and social implications (a) of a reduction of the ozone layer; and (b) of possible regulatory measures, from the particular point of view of development.

ECONOMIC AND SOCIAL IMPLICATIONS OF A REDUCTION OF THE OZONE LAYER

A reduction of the ozone layer may have an impact on health, climate and the ecosystems.

Health

A reduction of the ozone layer may produce an increase in ultra-violet radiation, which may cause skin cancer, sunburn or eye damage. Developing countries are mainly located in tropical areas. The geographical distribution of the ozone layer differs with the latitude: in temperate climates, where most developed countries are located, ozone distribution is higher, 300 to 400 m-atm-cm (Dobson units); in tropical countries, the ozone layer varies from 240 to 300 **.

---

*   Prepared by Bertrand H. Chatel, Chief of the Technology Application Office for Science and Technology, Department of Economic and Social Affairs.

** World Meterological Organization - "Atmospheric Ozone" - UNEP/WG. 7/5 of January 1, 1977, p. 13, Fig. 1.

Therefore when the ozone layer is decreased by 10% through human activities, the ozone layer above northern countries will be 270 to 360; while that above tropical countries may go down to 216 Dobson units. The resulting ultra-violet radiation will be more acute in tropical countries, which are already less protected by ozone layer. It is the responsibility of the WHO to investigate the impact of ozone layer reduction on health.

Climate

Any significant change in climate would have an impact on world agricultural production, a factor of great importance to developing countries, where famine and malnutrition still exist.

Ecosystems

Similarly, a change in aquatic or terrestrial ecosystems - particularly the phyto-plankton - would have an effect on the agricultural production of developing countries.

ECONOMIC AND SOCIAL IMPLICATIONS OF POSSIBLE REGULATORY MEASURES

Chlorofluoromethane (CFM)

The economic and social consequences of a possible ban on the use of chlorofluoro-methane (CFM) has been the subject of a particular study by the United Nations Economic and Social Affairs Department, and a report will be presented by N. Michae Betts, Consultant on this topic.

It seems that research should be able to re-orient technologies so that they are no harmful to the ozone shield.

The refrigeration industry is of great importance to developing countries, notably for the conservation of food, and for the air conditioning of offices and homes in tropical areas, to improve comfort and stimulate intellectual creativity.

Aircraft

Large fleets of aircraft flying at altitudes above 17 km may have a significant effect on the ozone layer. It appears that a limitation in the number of such air-craft and a reduction in their altitudes of flight would have little impact on the process of development. This is the concern of the ICAO.

Fertilizers

The influence of nitrogen on ozone depletion seems to be one of the most important topics for the attention of the United Nations system in this area, particularly for FAO.

Agricultural production, at present, mainly relies on the use of nitrogen fertil-izers. More research would focus on the impact of these fertilizers on the ozone layer.

If an appropriate technological assessment would demonstrate that limits should be set to the use of such fertilizers to protect the ozone layer, this matter would be of great concern to Governments of developing countries.

Fuel Burning

The control of smoke emission of industrial and domestic burning of fuel for the purpose of protecting the ozone layer will be a subject of study for UNIDO and UNEP.

Nuclear Explosions

Nuclear explosions, for peaceful purposes, which fall within the terms of reference of IAEA, have been used for civil engineering work. An assessment of this technique from the point of view of its effect on the ozone barrier would be useful.

CONCLUSIONS

The General Assembly has decided that a United Nations Conference on Science and Technology shall be convened in 1979. Governments may wish to consider whether the problems related to the ozone layer might be submitted for consideration by this conference.

PAPER 8

Protection of the Ozone Layer - Some Economic and Social Implications of a Possible Ban on the Use of Fluorocarbons*

Presented by the Department of Economic and Social Affairs

INTRODUCTION

In 1974, two papers, reporting independent work by two sets of researchers, indicated that stratospheric ozone could be catalytically reduced by a chain reaction
involving chlorine atoms [1]. Molina and Rowland further hypothesised that an increasingly significant source of this chlorine was chlorine atoms released by photodissociation of chlorofluoromethanes - CFMs [2]. CFMs, it was further argued,
accumulate in the atmosphere largely due to the activities of man, their having been
used increasingly as spray propellants, refrigerants, foam-blowing agents and solvents in many countries since the 1930s.

These, and subsequent papers, generated a great deal of scientific activity,
especially in North America. A Joint Report prepared, with remarkable urgency, a
review of all of the scientific information then to hand [3]. The United States
Environmental Protection Agency commissioned two studies, the one with a view to
assessing the more direct commercial impacts of regulating CFM production and
release [4], and the other a broader cost/benefit evaluation of environmentally-
motivated regulation of CFM production [5]. In an attempt to assess both the
technical and economic alternatives to CFM use in its four major outlets, USEPA's
Office of Toxic Substances produced its own report [6]. Inevitably, with the growing number of reports some duplication has occurred, but some of these performed a
valuable function, either by up-dating earlier reports, amplifying them, or giving
a slightly different insight into the problem area. One such study was produced
by the US Bureau of Domestic Commerce [7].

It will, by now, be realised that by far the strongest concern for the possible
effects of anthropogenic CFMs and their potential degradation in the stratosphere
has been manifested in the USA. 'Position Papers' have been issued by the governmental agencies concerned in other countries, for example the United Kingdom [8].
At an international level, OECD, with the collaboration of UNEP, has a work programme underway (under the guidance of the Environment Directorate, and largely
being implemented by the United States and Canada) to 'update the data base on
fluorocarbon production, use and estimated release into the atmosphere' [9].

*Prepared by the staff of Environmental Resources Ltd., under the direction of
Michael P. Belts, senior consultant. We should also like to express our sincere
thanks for the extensive assistance rendered by John Butlin of the Department of
Agricultural Economics, University of Manchester, U.K.

Concurrently, the major chlorofluorocarbon producers of the world have been funding a substantial research programme in this area, the programme being supervised and monitored through the Manufacturing Chemists Association [10]. A recent summary of expenditures on different aspects of the MCA programme is shown in Table 8.1.

TABLE 8.1 Fluorocarbon Research Programme – Administered by Manufacturing Chemists Association – (Manufacturing Chemists Association [10]).

	Completed Projects	Active Projects	Total
Tropospheric Measurement	$ 169,671	$ 340,979	$ 510,650
Stratospheric Measurement	210,306	973,408	1,183,714
Analytical Methods	225,692	106,661	332,353
Chemical Reactions and Kinetics	225,420	296,719	552,139
Modelling	148,144	272,217	420,361
Consulting	22,297	23,039	45,336
TOTAL	$1,001,530	$2,013,023	$3,014,553

This flurry of activity by both national and international organisations was motivated by the potential effects that a possible depletion of stratospheric ozone could produce. Amongst these are the following:

- the reduction in stratospheric ozone would probably increase the quantity of ultraviolet rays in the 'b' range of the ultraviolet spectrum incident on the earth's surface, thereby increasing the evidence of various forms of skin-cancer

- a reduction of stratospheric ozone may produce a global climate change, most likely in the surface temperature, thereby affecting, *inter alia,* agricultural production, especially at the current extensive margins of production, and, possibly having an adverse effect on world food supplies;

- rates of materials degradation may be speeded up by the increased incidence of UV-B;

- the balance of major ecological systems may be disturbed by the increased ultra-violet irradation, and the transition towards a new ecological equilibrium may produce several manifestations. Amongst the more important of these are the possible adverse effects on marine phytoplankton, with the subsequent reduction in fish stocks.

The purpose of this paper is three-fold. The first is to explain, *a priori,* why we should expect, with growing global population and growing (albeit unevenly distrib-uted) *per capita* income, abuse of specific environments such as the atmosphere. The second purpose is to examine the costs and benefits of continuing current fluorocarbon emissions (in the light of the scientific evidence to hand) and of banning fluorocarbon use in part or in whole. (A slightly unconventional cost/benefit approach will be advocated due to the global and, particularly long-term nature of the problem). The third part of the paper examines the merits of alternative forms of controlling CFM production and/or use, and the paper concludes with a discussion of the case for and against regulatory action.

GLOBAL COMMON PROPERTY RESOURCES [11]

The problems of overuse of natural resources to which no party or parties have exclusive title has been studied in detail during the post-war period. Without recourse to the more theoretical economic discussions which have been used in much of this work, the conclusions can be simply stated: if the resource is non-renewable, it will be depleted more rapidly than it would be if access and depletion could be controlled by a single owner; if the resource is renewable, its level of productivity will be lower than that under private ownership. With a certain amount of licence, this second approach can be used when discussing problems of disruption of the stratospheric environment. It provides, globally, services in the form of control of the level of ultraviolet irradiation, temperature control, and so on. It is also available, potentially, as a sump into which global society's air-borne wastes may, in part, be deposited. The absence of private ownership of, or national sovereignty over, the stratosphere means that we should expect overuse of its services in at least some direction. Technology has yet to advance to the stage at which we can influence the flow of ultraviolet radiation - controlling services rendered by the stratosphere, and so we might expect, *a priori*, that its overuse, or lowered productivity, would occur via its potential to store air-borne wastes originating from industrial production and consumption.

There are, conceptually, several ways in which such use may be limited. They are, basically, of two types: either controls, which set definite technical limits on the amount of waste released from a production process and require the firm or industry to bear the cost of these controls; or they may be fiscal incentives, which either require the firm or industry to bear the societal costs of wastes disposal into the environment thereby creating an incentive to reduce the level of emissions (taxes), or encourage the firm to use less pollution intensive techniques (subsidies). All this assumes that there exists some body or institution capable of formulating the various categories of controls, and of implementing and enforcing them. The problem with stratospheric pollution is that, as a global common-property resource, there is no global body with supreme authority for its management. In the absence of such an institution, many of the above options for regulation are foreclosed. No individual nation (unless it is a major source of the pollutants under discussion) would countenance unilateral regulatory or fiscal measures. It will rarely make sense for a country, in the name of environmental improvement, to put itself at an economic disadvantage for a marginal environmental improvement from which its citizens may benefit only slightly, and perceive even less. (at the same time, citizens of other countries would receive similar slight benefits, without the associated economic disadvantage).

This, then, is the situation. Before even looking at the evidence to hand, we must realise that by the very nature of the resource (and by the nature of the institutional framework within which industrial man operates) the stratosphere can potentially be overutilised as a sump for industrial processing and consumption wastes. Furthermore, because it is global in extent, and because the potential problems can arise from many nations and affect many nations, the regulatory options that exist are both limited in extent and particularly difficult to implement.

CFM RELEASE AND STRATOSPHERIC OZONE DEPLETION: THE CURRENT EVIDENCE

Whilst the class of chlorofluoromethanes is quite large, we shall concentrate our attention on three which are by far the most important in the four above-mentioned uses. The three are: trichlorofluoromethane (F-11), dichlorodifluoromethane

(F-12), and chlorodifluoromethane (F-22). Table 8.2 indicates the proportion of
total world production of CFMs represented by F11 and F12 for the first five years
of this decade.

TABLE 8.2 Global F11 and F12 Production as a Proportion of Global
CFM Production, %, 1970-74. (OECD [9])

	1970	1971	1972	1973	1974
F11	32	31	33	33	32
F12	51	52	53	52	51
F11 & 12	83	83	86	85	83

Thus, without adding in F22 production (for which there are currently no global
production estimates) it can be seen that we are concentrating on approximately 85
percent (probably more than 90 percent when F22 *is* added in) of the global output
of CFMs which constitute the source of emissions into the atmosphere and stratosphere.

The National Academy of Sciences report, *"Halocarbons: Effects on Stratospheric
Ozone"* contains the following conclusions:

> "All the evidence we examined indicates that the long-term release of F11 and
> F12 at present rates will cause an appreciable reduction in stratospheric ozone
> In more specific terms it appears that their continued release at the 1973
> production rates would cause the ozone to decrease steadily until a probable
> reduction of about 6 to 7.5 per cent is reached, with an uncertainty range of
> at least 2 to 20%, using what are believed to be roughly 95 percent confidence
> limits. The time required for the reduction to attain half of this steady-
> state value (3 to 3.75 per cent) would be 40 to 50 years." [12].

It is not our place here to enter the scientific discussion, and, indeed, it would
be presumptuous to do so. There have, however, been criticisms of previous work in
this area on the grounds that the data base might be inadequate, that the model of
the atmosphere used may be inadequate to give satisfactory results, that CFMs may
be removed from the stratosphere without ozone depletion, that there may be larger
natural sources of stratospheric chlorine. The NAS report does attempt, however,
to take the additional uncertainty that these problems raise into account in its
2 to 20 percent estimated reduction of stratospheric ozone, with a probability of
95 percent, at 1973 release rates. It also raises the question of other sources of
stratospheric pollution, mainly the US Space Shuttle, the release of oxides of
nitrogen from nitrogenous fertilizers, and as a result of supersonic flight in the
stratosphere [13]. NAS emphasises the need to monitor the effects of all sources
of stratospheric pollution. It is to be regarded as a limitation of this paper
that it tackles only the CFM aspect of the problem (although this is by far the major
source of concern at the moment).

The final category of information provided by this NAS report is the sensitivity
analysis of its findings to alternative postulated release rates. Whilst noting
that releases of CFMs had decreased during 1975/6, the report continues:

> "Resumption of exponential growth in production and use of CFMs could well
> occur and lead to a doubling of their release rate in 10 years. Even if the

release rates became constant at that point they would cause a doubling in
the expected ozone reduction, to a value of 12 to 15 percent, with a range
of 4 to 25 percent, once a steady state was reached." [14, 15].

This far we have established only that a certain set of chemical reactions is likely
to occur in the stratosphere. What are the likely effects of these?

In a second report, *"Halocarbons: Environmental Effects of Chlorofluoromethane
Release"* (op. cit.), the NAS Committee on the Impact of Stratospheric Change
addresses this problem. It will again be sufficient for us to concentrate on their
summary findings. The environmental impacts of ozone reduction are listed as
follows [16].

- "The major effects of DUV (direct ultraviolet radiation [17] increase due to
  ozone reduction could involve:

  increased incidence of malignant melanoma, a serious form of skin cancer frequently causing death, and thus an increase in mortality from this cause;

  increased incidence of basal and squamous cell carcinomas, less serious
  but much more prevalent forms of skin cancer, rarely causing death but causing
  much more expense and, occasionally, more or less serious disfigurement;

  effects on plants and animals of unknown magnitude".

- "If the increased infrared absorption and emission due to the presence of CFMs
  in the atmosphere were to alter our climate by small amounts, the most important
  effects would be on agriculture, particularly through the boundaries of the
  regions in which particular crops can be grown effectively." [18].

From the scientific evidence, then, we find that there is a *potential* environmental
hazard in the release into the atmosphere, and accumulation in the stratosphere, of
CFMs. Our path from here is clear. We must first identify the possible effects of
global emissions of CFMs to date. We must then try to identify the effects of an
outright global ban on CFMs (laying aside the problems of implementation of such
a ban as discussed above). Obviously, there would be associated costs, mainly
due to economic and social disruption caused in the industries concerned. Finally,
we will attempt to identify alternative approaches which may achieve the same aim
as a global moratorium on CFM production and use, with a greater net benefit to
global society [19].

THE COSTS AND BENEFITS OF BANNING CFM PRODUCTION AND USE

The exercise which we are about to perform is, in part, an application of cost/
benefit analysis. Without entering into a rigorous discussion of the theoretical
basis for cost/benefit analysis, and the problems in implementing, we shall assert
the following: any action taken by an individual in a particular society may have
an effect (of which the instigator may or may not be aware) upon other members of
that society. The effect may be beneficial or detrimental. It may be experienced
by the recipient either through his expenditure, or more indirectly, by making his
life experience in some way more or less pleasant. The general aim of cost/benefit
analysis is to evaluate the sum of the costs and benefits of a particular proposal,
cancelling out the effects which simply represent a gain to one person equivalent
to a loss to another ('transfer payments') and, using some criterion, evaluating
the remaining costs and benefits and seeing if the difference is positive or
negative. A slightly more elaborate approach is needed when the analysis is used
to choose between two or more competing projects. Problems which often arise in
using the technique include:

- identifying *all* relevant costs and benefits;

- identifying transfer payments as such to avoid 'double counting';

- pricing benefits and costs not evaluated by the market mechanism;

- determining whether existing prices determine the true (social) value to society of using those resources in that way.

An example may help to clarify matters here. Assume a family moves to a house in the country, and that, with their given income they are able to enjoy a particular standard of living. A decision is then made to build an airport, say, three miles from the house. It is reasonable to assume that this family's standard of living will fall. Extra expenditure may have to be incurred on sound proofing the house, family relations may become strained because of excessive noise, the amenity value of the countryside will have been at least partly destroyed, and so forth. This loss in living standard by the family concerned would be counted as a cost (approximated as nearly as possible by the extra costs incurred by the family *as a direct result of* the airport's construction and operation) in a cost/benefit analysis. A private entrepreneur would have no incentive to consider such costs if he was deciding whether or not to build a commercial airport. At the same time, the airport would probably confer benefits that the entrepreneur would have no incentive to consider either. Amongst these would be the time savings realised by travellers who would otherwise have had to fly from more distant airports. Thus, cost/benefit analysis attempts to incorporate all society's costs and benefits and not simply those borne by the instigators of a particular project.

There are one or two other problems associated with cost/benefit analysis, especially as applied to natural resource and environmental problems. The first of these is that, while the costs of implementing the projects may impact soon after the project is implemented, the benefits may not be felt for a long time into the future. Thus, if any attempt is made to represent costs and benefits in current value terms, the implicit weighting system automatically acts against the interest of those who are likely to experience some of the benefits accruing to the project. Such problems are referred to as 'paternalistic bias', or 'intergenerational equity' problems. The second problem arises because depletion of natural resources, or the use of an environment as a waste disposal sump, may bring about changes which within in a given time-scale (sometimes in terms of several centuries, sometimes in terms of millenia) are irreversible. It is argued by some [20] that a simple cost/benefit analysis does not take into account the costs of foreclosing of options to future generations that such a policy represents [21].

An attempt to undertake a global cost/benefit analysis of the effects of a ban on CFM production, and use would be, in part (and for technical reasons), a misuse of the technique. This would require a massive amount of information which is simply not available, and, to be of policy significance, would need to be disaggregated, at least continent by continent, and preferably on a smaller regional basis. This is not possible, however. The exercise we are about to indulge in is, rather a hybrid economic/environmental/social impact assessment, using the rules of cost/benefit analysis to provide an orderly basis and to ensure that the assessment is reasonably comprehensive.

THE EFFECTS OF CONTINUING CFM RELEASE

The data to hand are not easy to assess in terms of the possible effects of unregulated CFM use, as there is some difficulty in interpreting data and projecting future production and use patterns in an unregulated environment. Recent history indicates that US production increased at an average rate of 10 percent per annum for the two decades up to 1974, but then fell by 15 percent in 1975 and 1976. This is shown in Table 8.3.

TABLE 8.3 Production and Release of F-11 ($CFCl_3$ and F-12 (CF_2Cl_2) to date[a] (10^3 metric tons/yr). (NAS - Committee on the Impacts of Stratospheric Change [3])

Year	F-11 United States	F-11 World[b]	F-12 United States	F-12 World[b]
To 1958		172.5		570.0
1958	23	29.7	60	74.0
1959	27	35.8	71	88.5
1960	33	50.0	75	100.6
1961	41	60.8	79	110.2
1962	56	78.7	94	130.6
1963	64	94.2	99	149.9
1964	67	112.0	104	175.0
1965	77	124.5	123	196.4
1966	77	141.9	130	227.1
1967	83	163.6	141	257.8
1968	93	187.1	148	277.1
1969	109	223.2	167	311.5
1970	111	245.9	170	335.4
1971	117	274.6	177	355.9
1972	136	317.5	200	398.6
1973	148	367.8	221	441.0
1974	158	400.2	231	437.6
1975	121	357.3	178	416.3
TOTAL		3437.3[b]		5089.5[b]
N. Hemisphere		3326.8		4891.1
S. Hemisphere		110.5		198.4
Total Released		2934.1		4414.1
Percentage Released		85.36		86.73

[a]MCA Survey, 1975 and 1976

[b]This includes Eastern Bloc production, which is estimated to be 155 and 209 thousand metric tons for F-11 and F-12, respectively, through 1975.

Department of Economic and Social Affairs

TABLE 8.4 Estimated World and US Production and Atmospheric Emissions of F-11, F-12 and F-22 - 1973. (A.D. Little [4])

Halocarbons	World Production, (thousands of metric tons)	U.S. Production, (thousands of metric tons)	World Emissions, (thousands of metric tons)	U.S. Emissions, (thousands of metric tons)	Lifetime in the Atmosphere
Fluorocarbons[a]					
F-11	) 930	150	) 700	140	)greater than 10)years
F-12	)	220	)	170	
F-22	120	60	60	28	1 to 10 years
Total	1,050	430	760	340	

[a]Production of F-11, F-12, and F-22 accounts for approximately 90 percent of total fluorocarbon production.

Whether the 1975/6 reduction is due to increased environmental consciousness, or to the global recession at that time, is difficult to verify. Effects of unregulated hydrocarbon release will be taken for production and emission at 1973 levels, unless otherwise indicated.

1973: The Status Quo

The situation with respect to the production and emissions of fluorcarbons has been summarised by an Arthur D. Little study (op.cit.). This information is summarised in Table 8.4.

One important feature of this table is the proportion of the CFMs produced in the USA. In 1973, the United States accounted for 45 percent of world F11 and F12 production, and 55 percent of world F22 production. If we are to consider forms of control other than a total moratorium, information on emission sources is also necessary. This is not available on a global basis but Table 8.5 presents estimates for the United States.

THE EFFECTS OF REGULATING CFM PRODUCTION AND RELEASE

Any attempt to regulate CFM emissions is going to have at least a short-run impact upon the chemical manufacturing industries of many countries in the Western industrialised world. Table 8.6 gives a listing of the twenty major companies involved in the production of CFMs throughout the non-Communist world.

The A.D. Little Study (op.cit.) identifies ten sectors of industry in the United States which would be affected if attempts were made to regulate the production and emission of fluorocarbons in the United States. The information is reproduced in Table 8.7. Whilst the sectors affected in other countries might differ because of the nature of each country's industry, this is a useful and comprehensive indication of the sectors that may be affected by any attempt to limit fluorocarbon production in any country.

It is obvious that a complete moratorium on CFM production and use would have an immediate and severe impact on the chemical industries concerned in the short-run. In a longer run perspective, the effect of such a moratorium would depend crucially upon the extent to which substitutes for CFMs could be found and the lead-time involved in producing and marketing these on a commerical scale. Again, the A.D. Little Study's estimates of these magnitudes are the most comprehensive. These estimates are reproduced in Table 8.8. [22]

An alternative way of looking at this problem in terms of direct employment impacts is provided by the US Bureau of Domestic Commerce Study (op.cit.). The essential data are given in Table 8.9.

Apart from conventional considerations of the effects of CFM regulation on employment in the chemical industries, and industries using them, packaging them, and distributing them, there are other considerations which, in a fully comprehensive study, need to be considered. The major concerns here are [23]:

- the special uses of fluorocarbons;

- effects upon energy flows;

- health effects;

- effects upon trade flows.

Of the special/unique role of CFMs, the US Bureau of Domestic Commerce identifies the following uses as being important [24].

TABLE 8.5 Estimated US Atmospheric Emissions of F-11, F-12 and F-22 (thousands of metric tons) - 1973. (A.D. Little [4])

Chemical	Production	Production, Storage and Transport Losses	Estimated Annual Emission from Use and Disposal					Total
			Propellant	Refrigerant	Solvent	Blowing Agent	Use as Intermediate Chemical	
F-11	150	1.5	107	5	Small	13	-	125
F-12	220	1.2	113	59	-	3.4	-	180
F-22	60	0.6	Small	27	-	-	Small	28
Total	430	3.3	220	91	Small	16.4	Small	333

TABLE 8.6 Major Companies Involved in the Production and Distribution
of CFMs throughout the Non-Communist World.   (OECD [9])

1. Akzo Chemie N.V. (Holland)

2. Allied Chemical Corporation (USA)

3. Asahi Glass Co., Ltd. (Japan)

4. Australian Fluorine Chemical Pty., Ltd. (Australia)

5. Daikin Kogyo Co., Ltd. (Japan)

6. E.I. du Pont de Nemours & Company, Inc. (USA)

    (a) Du Pont de Nemours (Nederland) N.V.
    (b) Ducilo S.A. (Argentina)
    (c) Du Pont do Brazil S.A. (Mexico)
    (d) Halocarburos S.A. (Mexico)

7. Du Pont of Canada Limited (Canada)

8. Farbwerke Hoechst AG (West Germany)

    (a) Hoechst Iberica (Spain)
    (b) Hoechst do Brasil Quimica e Farmaceutica S.A.

9. Imperial Chemical Industries Limited (England)

    (a) African Explosive & Chemical Industries Ltd.

10. I.S.C. Chemicals Ltd. (England)

11. Kaiser Aluminum & Chemical Corporation (USA)

12. Kali-Chemie Aktiengesellschaft (West Germany)

    (a) Kali-Chemie Iberia SA (Spain)

13. Mitsui Fluorochemicals Co., Ltd. (Japan)

14. Montedison S.P.A. (Italy)

15. Pennwalt Corporation (USA)

16. Racon Incorporated (USA)

17. Rhone-Poulenc Industrie (France)

18. ShowaDenko, K.K. (Japan)

19. Ugine Kuhlmann, Produits Chimiques (France)

    (a) Pacific Chemical Industries Pty. Ltd. (Australia)
    (b) Ugimica S.A. (Spain)

20. Union Carbide Corporation (USA)

TABLE 8.7 Estimated Employment and Production Value Related to Fluorocarbon (and Chlorocarbon) Production and Consumption – 1973. (A.D. Little [4])

	Total Industry Employment	Directly[c] Related Employment	Total Industry Production Value 1973, ($ millions)	Directly Related Production Value 1973 ($ millions)
Raw Materials:				
chlorine	10,300	1,350	490	60
hydrofluoric acid	800	330	140	55
Basic Chemical Production:				
fluorocarbons	2,700	2,700	240	240
chlorocarbons	6,600	1,320	600	120
Aerosols:				
containers[a]	68,200	2,500	4,900	190
valves, caps & related materials	NA[b]	2,000	NA[b]	60
concentrate ingredients	NA	NA	NA	NA
aerosol fillers[d]	14,000	7,000	250	130
aerosol marketers[d] (total)	15,000	7,500	2,000	1,000
production	6,000	3,000	–	–
support (R & D, marketing, etc.)	9,000	4,500	–	–
Refrigeration: (1972)				
refrigeration equipment	120,000	120,000	5,600	5,600
household refrigerators & freezers	32,000	32,000	1,600	1,600
Blowing agent applications:				
foam and derived products	45,000	30,000	1,000	600
raw materials	10,000	5,000	460	230
Solvent Applications	NA	NA	NA	NA

TABLE 8.7 contd.

	Total Industry Employment	Directly[c] Related Employment	Total Industry Production Value 1973, ($ millions)	Directly Related Production Value 1973 ($ millions)
Chemical Intermediate	NA	3,000	NA	25

a. Includes only cans, which represent 95 percent of total aerosol containers.

b. NA – not available.

c. Related employment refers to production of F-11, F-12, F-22, carbon tetrachloride and methyl chloroform.

d. Directly related employment and production of aerosol fillers and marketers was estimated as 50 percent of total industry employment and production.

- sterilisation of equipment and material in hospitals;

- sterilisation of surgical instruments, gauzes, bandages, and related textile materials;

- propellants for a selected number of essential pharmaceutical applications;

- industrial cleaning or degreasing of electronic equipment and semi-conductor products;

- special situations for fire extinguishing where low toxicity products leaving no residue are required, e.g. fire control in aircraft engines and computer mainframe units.

Whilst some of these uses involve immediate discharge, others involve relatively slow release of the particular CFMs involved. It is estimated that these essential uses represent 5 per cent of total US use of CFMs.

The contribution of chlorofluorocarbons to energy conservation in advanced industrial economics is both direct and indirect in character. The rigid polyurethane foams that are produced using CFMs as a blowing agent are claimed to have 'about twice the thermal insulating value of such foams without fluorocarbon' (US Bureau of Domestic Commerce, op.cit., p. 101). The indirect contribution to energy conservation lies in their use in refrigerators and air-conditioners. If CFMs were banned, and the next-most-efficient technology adopted, the energy efficiency of refrigerators and air-conditioners would fall substantially.

The contribution of CFMs to health has already, in part, been touched upon, in the paragraph concerned with essential uses of fluorocarbons. Hence, we want to focus on the use of CFMs as refrigerants. The provision of refrigeration based on CFMs has contributed significantly to the control of diseases caused by decaying food. Thus any policy aimed at reducing CFM output and use must take into account not only the increased energy costs of refrigeration, but also the possibility of the reduced use of refrigeration, and consequent increase in health problems associated with food storage [25].

The effects of a ban on CFM production and use on trade flows is difficult to esti-mate, and must, in any case be used with caution. The difficulty in estimation arises from there not being adequate data on trade flows, not only in the final products, that is aerosols, refrigerators and freezers, foams and solvents, but also in raw materials, both for the manufacture of CFMs, and the products in the manu-facture of which CFMs are used. This would apply especially to the materials demand of the domestic appliance industry in several countries, if a ban on the use of CFMs as refrigerants was implemented. The other problem of taking international trade effects separately is that it involves a danger of double-counting, inasmuch as some effects of trade changes are reflected in price changes already accounted for. With this proviso in mind, however, separate consideration of trade flows can help to highlight any balance of payments problems that may arise. For several countries such problems are of immediate economic and political importance. Tables 8.10 and 8.11 show trade flows for the US in raw fluorocarbons, (not including fluorocarbons contained in products, such as refrigerators, aerosols, air-conditioning units in automobiles, and so forth).

These, then, are the incipient benefits of fluorocarbon production. The societal costs of unregulated production, whilst being much more difficult to catalogue and quantify, also need to be taken into account. The problems in assessing these costs arise because the costs are much more indirect, and have to be assessed, if possible, by indirect methods. The costs which are most likely to be of concern are:

TABLE 8.8 Consuming Industry Response Times to Chemical Use Restrictions. (A.D. Little [4])

Consuming Industry	Emission Reduction by Equipment Up-grading (years)	Primary Response to Ban of Chemical Use[a] (years)		Conversion to Substitute Chemicals[b] (years)
		to absorption	to F-22	
Intermediate Chemical Applications	Not an option	2-7		5-10
Propellant Applications	Not an option	1-2		4-10
Refrigerant Applications[c]				
appliances	2-3	4-6	3-4	5-11
mobile air conditioners	3-4	Indefinite	Indefinite	5-11
home air conditioners	2-3	3-5	zero	5-11
commercial refrigeration	2-3	4-6	3-4	5-11
commercial air conditioners				
– reciprocating	2-3	3-5	2-3	5-11
– centrifugal	2-3	2-3	1-3	5-11
Blowing Agent Applications				
flexible foams	2-3	six months		six months
rigid foams	2-3	3		4-9
Solvent Applications	1	1-2		4-9

a The primary response times are the elapsed times required for the consuming industries to introduce substitute products to meet the demand now satisfied by the controlled chemicals or products using the chemicals.

b The conversion to substitute chemicals times are those required to develop new chemicals with properties similar to the banned compounds and to modify the products using the banned chemicals.

c In the event of a ban of F-11 and F-12, most refrigerant applications could be converted to F-22. If F-22 is also banned, other refrigerants could be used or some products could be converted to absorption systems.

TABLE 8.9 Comparison of US F/C* Employment Dependence with F/C* Emissions and Production – 1974. (US Bureau of Domestic Commerce [7])

Production	F/C* Dependent Employment (1000's)	Percentage Distribution of US F/C*		1000's of F/C Dependent Employees per Percentage of US F/C*	
		Emissions	Production	Emissions	Production
Aerosol Propellants	31.3	62	49	0.5	0.6
Refrigerants	496.6	26	28	19.1	17.7
Foamed Plastics	57.3	5	7	11.5	8.2
Fluoroplastic Materials	11.1	0	4	0.0	2.8
Other	N/A	7	12	N/A	N/A
Total (T) or Mean (M)	(T)596.3	(T)100	(T)100	(M)6.0	(M)6.0

*F/C - Fluorocarbon

TABLE 8.10 Fluorocarbon Exports, Shipment by Destination. (US Bureau of Domestic Commerce [7])

Country of Destination	Shipments - 12 Months (1000 pounds)	
	1970	1974
Canada	1160.4	3553.4
Mexico	568.7	2385.8
Guatemala	59.6	142.8
Salvador	75.4	232.3
Honduras	62.0	90.9
Nicaragua	210.9	1472.4
Costa Rica	121.2	314.3
Panama	278.6	534.6
Bermuda	28.6	52.2
Bahamas	123.3	137.9
Jamaica	237.8	396.8
Dominican Republic	188.1	387.0
Leeward and Windward Islands	-	31.2
Barbados	34.0	69.2
Trinidad	239.4	272.2
Netherlands Antilles	155.7	292.3
Columbia	651.9	2457.1
Venezuela	1093.4	2242.2
Guyana	29.0	101.1
Surinam	-	51.8
Ecuador	135.9	327.6
Peru	107.5	356.6
Chile	443.2	186.7
Brazil	160.2	4411.9
Uruguay	30.9	33.7
Argentina	596.1	3067.5
Denmark	-	1.8
United Kingdom	70.3	281.2
Netherlands	4652.0	6644.9
Belgium	2299.2	367.1
France	171.6	95.7
West Germany	26.7	69.7
Switzerland	138.1	110.6
Spain	45.1	-
Poland	-	1.5
Italy	7.4	87.4
Greece	90.8	68.0
Lebanon	-	91.9
Iran	222.3	21.0
Israel	-	59.1
Kuwait	235.4	105.9
Saudi Arabia	111.7	17.7
Qatar	-	15.3
Arabia	48.1	-
Arabia Emirates	-	37.2
India	65.7	76.7
Pakistan	181.1	150.7
Thailand	376.3	611.1
South Vietnam	103.8	316.2
Malaysia	105.2	1021.0

TABLE 8.10 contd.

Country of Destination	Shipments - 12 Months (1000 pounds)	
	1970	1974
Singapore	546.8	1929.2
Indonesia	153.4	641.3
Philippines Republic	855.6	1611.4
Korea Republic	47.6	257.6
Hong Kong	732.6	694.9
China (Taiwan)	241.0	966.9
Japan	1438.3	3482.9
Australia	1020.0	2403.4
New Zealand	506.1	1709.8
Libya	36.5	-
Egypt	32.3	18.8
Cameroon	-	34.6
Ivory Coast	36.8	22.3
Ghana	-	19.5
Liberia	38.5	-
Nigeria	-	38.2
Angola	-	78.4
Kenya	-	27.0
Congo	34.6	-
Republic of South Africa	1278.9	1976.3
Other Countries	296.1	307.3
Total Exports (1000 pounds)	23,100.8	50,073.1
Total Value of Exports ($000)	8,618.6	22,121.5

TABLE 8.11 Fluorocarbon Imports, Shipment by Source Country. (US Bureau
of Domestic Commerce [7])

Country of Destination	Shipments (thousand pounds)	
	1970	1974
Canada	190.8	4,527.1
Sweden	2.7	-
United Kingdom	1,610.8	4,680.7
Belgium	-	44.1
Netherlands	4.4	131.3
France	73.9	469.8
W. Germany	818.2	813.1
Switzerland	3.3	1.4
Italy	0.8	47.2
Israel	60.7	166.4
Japan	820.6	318.9
Mexico	-	179.0
Other Countries	8.1	-
Total Imports (thousand pounds)	3,594.3	11,378.8
Total Value of Imports ($000)	1,929	6,597

TABLE 8.12 Qualitative Costs and Benefits of a Global Ban on Production and Distribution of CFMs

	Benefits		Costs
(i)	Reduced incidence of melanoma and non-melanomous cancers, and associated personal costs.	(i)	Costs of increased unemployment insurance for employees of industries formerly producing chlorofluorocarbons.
(ii)	Reduced health costs of treating skincancers induced by UV radiation.	(ii)	Increased energy costs due to less efficient refrigerants and less efficient thermal insulation.
(iii)	Increased standard of living due to lower food costs.	(iii)	Increased health costs due to increase in disease related to inefficient food storage.
(iv)	Reduced damage to natural ecological systems caused by UV radiation.	(iv)	Costs of financing any balance of payments distortions brought about by the policy.
(v)	Increased industrial output due to a reduction in workdays lost caused by skin-cancer.	(v)	Opportunity cost of capital tied up in operating plant and equipment which was not obsolescent at the time the ban was implemented.

TABLE 8.13 Summary of Estimated Economic Impact Assessment Resulting from a Restriction on U.S. Use of F-11, F-12 and Carbon Tetrachloride. (A.D. Little [4])

Impact Sectors

Regulatory Options	Emission# percentage	Propellant Applications: Aerosol Industry					Air	
		Basic Chemical Producers	Independent Fillers	Car Manufacturers	Valve Manufacturers	Aerosol Marketers	Conditioning Manufacturers	Plastic Foam Producers
(1)	92	2-3 #	1	3	1	2	1	2
(2)	82	3	1	3	1	2	4	3
(3)	70	3	1	3	1	2	5	4
(4)	80	3	1	3	1	2	4	3-4

6 months

Regulatory Options

(1) Ban all but replacement uses of controlled chemicals after six months.
(2) Regulation of non-propellant uses and ban of propellant uses after six months.
(3) No regulation of non-propellant uses and ban of propellant uses after six months.
(4) Government control of total chemical production after six months.

Emission reductions represent percent reduction in U.S. F-11, F-12 and CC14 emissions to the atmosphere over a 20 year period beginning in 1976. A 5 percent growth in the absence of controls is assumed: no restriction on critical propellant uses (5 percent of total), a 5 year half-life of refrigerants and 50 percent control of emissions from plastic foams are assumed.

Impact Code: 1 - severe
               2 - substantial
               3 - moderate
               4 - limited

- increased morbidity from basal-cell and squamous-cell skin concerns, and increased morbidity and mortality from a melanoma (a form of skin cancer which mo: frequently is fatal); the costs of caring for the increase in disease, the loss in output to the economy, and other associated costs;

- costs of any change in climate upon agricultural output and, thereby, food price

- damage to other ecological systems, especially (from a food-availability aspect, the effect upon the marine environment.

The expected magnitude of the effects of ozone depletion upon melanoma mortality i difficult to quantify. The NAS Committee on Impacts of Stratospheric Change (op.cit., pp. 8-16 to 8-17) concludes:

> "We find an increase in melanoma deaths is likely, but not certain, to occur as a consequence of a continuing increase in the rate at which DUV received at the ground accumulates. Such a melanoma increase, if it occurred, would be delayed, beyond any delay in the DUV increase, while the accumulated dose builds up in the individuals. A 7 percent ultimate reduction in ozone, with a consequent 14 percent ultimate increase in DUV accumulation rate, might be expected, if most melanoma deaths are solar UV radiation-related, to produce a somewhat smaller percentage increase (less than 15 percent) in melanoma deaths. Thus, a few hundred deaths per year would be expected after all dela have taken place. (With a constant rate of CFM release, the increase in DUV arriving at the ground would first reach two-thirds of its ultimate level after 80 years, so that more than a century of continued release at a constant level would be required for three fourths of such an effect to be manifest)."

This NAS study, however, does not give information on several aspects of this problem. For example, no estimate of the increase in non-melanoma cancers (not usually fatal) is given. The dose/response relationship is not well understood. Hence, we do not know the possible effects of the level of ozone depletion being at the upper 95 percent confidence limit (20 percent), rather than at the 7 percent mean. If the response relationship is non-linear, then it may be that, at constant release rates, the incidence of melanoma deaths might increase dramatically. Anoth aspect of the problem is that the incidence of ultraviolet radiation differs with latitude. Thus, any attempt to assess the effects of ozone depletion on melanoma would need to take this into account, as well as the difference in effects noted between different races.

It can be seen that the problems associated with estimating the costs resulting fro CFM release are difficult. In a slightly different context, d'Arge attempted to assess the global costs of the effects of supersonic aircraft on the stratosphere [26]. Working on a far superior data base, d'Arge still encountered horrendous problems, and the estimates presented must be regarded as being very broad approximations.

We have decided to take a rather less ambitious approach. We shall first list the costs and benefits of an immediate ban on production and distribution of CFMs. We shall then consider if there is another means of achieving essentially the same end as that to which a complete moratorium is directed.

Table 8.12 is a listing of the costs and benefits of a policy of banning CFM production and use globally. The table is general rather than detailed. In the absence of detailed data there appears to be little advantage in specifying the ramifications of the broad categories of costs and benefits in greater detail. The table may, in fact, best be considered as the broad specification of a research project on the costs and benefits of a ban on CFM production and release if the relevant data were available. In considering Table 8.12 several points ought to be borne in mind:

- An immediate ban on CFM release will not produce an immediate cessation of ozone depletion. The life of CFMs in the stratosphere is variable, but mostly in excess of ten years. It would take decades for the stratosphere to adjust to a steady state ozone level in response to CFM emissions that have occurred over the past several decades.

- CFMs are not the only source of stratospheric ozone depletion. Other sources have been indicated above. A global ban CFMs would, however, reduce what is currently considered to be, by far, the most important anthropogenic source of ozone depletion in the stratosphere.

- If conventional cost/benefit criteria were used, then normal time-discounting of both benefits and costs would bias the analysis in favour of the contemporary costs and against the more distant benefits. We shall address this in more detail below.

In general, a complete ban on CFM production and release (setting aside the problems of implementing such a global ban) is likely to be an inefficient way of handling the CFM release problem. If the CFM problem is considered to require regulatory control, then a control on emissions by source would seem to be more efficent (in the sense of achieving the same ends with less industrial and social disruption). For the use of CFMs as aerosol propellants in non-essential uses (mostly domestic) a ban would appear most appropriate. There might be some loss of consumer conveniences, in having to revert to more primitive technologies for personal hygiene, insect repelling, furniture polishing, and so forth, but the gain is likely to be greater than this loss. In the case of refrigeration and air-conditioning equipment, a significant reduction in emissions could be achieved by reduced leakage from systems, and by the mandatory reclamation of CFMs from discarded equipment. Alternatively, a charge on CFMs refundable on the return of a disused refrigerator, air-conditioner or car to a centre where CFMs could be drained without loss, might be considered [27].

For the use of CFMs in foam blowing and as solvent and degreaser it may be possible to require that these activities are undertaken in centres designed to prevent the release of CFMs to the atmosphere. Alternatively, it may be desirable to accept the loss in insulating efficiency and associated costs that would be associated with a ban on CFM use for these purposes.

A.D. Little (op.cit.) have studied the direct economic impact of several regulatory alternatives for fluorocarbons. Their findings are, in part, summarised in Table 8.13. In essence, regulatory option 1 is equivalent to a total ban, and regulatory option 2 represents the alternative considered above. For the US, 92 percent and 82 percent emission reductions are achieved respectively. The sectors most severely effected by the complete ban are those aspects of aerosol production concerned with can-filling and valve production (representing a relatively small proportion of the labour force) and the manufacture of refrigeration and air-conditioning equipment, (representing a much larger proportion of the labour force). The second option reduces the most severe impact of the total ban, but at the cost of a 10 percent loss in emission reduction.

We now need to face two difficult issues. The first relates to the problem of the remoteness, in time, of any benefits resulting from CFM regulation, as compared to the proximity in time of the associated costs. The second is the practical problem of implementing a global control of any sort. We shall address the latter problem first.

The problems of regulating the use of a global common-property resources are numerous. For some (e.g. fisheries) the allocation or assumption of property rights by nations is proving to be effective. This approach would not, however, be appropriate for the stratosphere. The two main alternatives open seem to be

voluntary unilateral and multi-lateral action, and international control by an
agency in which all nations are represented. For non-essential aerosols the US
appears to be taking the former approach. As the major world producer of CFMs,
this action will have a significant effect upon global emissions of CFMs in the
short-run. In the absence of similar action by other countries, however, it is
difficult to see how the US could sustain this action in the long-run especially if
the rest of the world's production and consumption of CFMs continues to increase.

The precedents for effective control by an international agency are, similarly, not
encouraging. The evidence for international common property problems, when the
numbers of recipient and donor nations are large (and not mutually exclusive),
suggests that there may indeed be no solution [28]. In the absence of a global
recognition that there is a global problem that can have global impacts possibly of
severe magnitude, it is difficult to see a basis for the formation of an international regulatory agency.

The problem of how to weight costs and benefits is equally intractable. It would
seem more likely that the interests of future generations would be better looked
after by an international agency rather than by the independent actions of individual nations. One alternative, the evaluation of costs and benefits of such a
programme without time - discounting, does cope with this problem by weighting each
generation's interest equally. There are, however, problems associated with this
approach.

Associated with the issue of intergenerational equity is the problem of possibly
foreclosing options to future generations by acting in a way that irreversibly
deteriorates the environment at the moment. This problem was discussed above.
Much of the discussion on the depletion of stratospheric ozone assumes a constant
rate of global emission. An increase in emission rates over time would bring about
a level of stratospheric ozone deterioration which, even if a decision was made in
ten year's time to ban production and emission of CFMs, would impose on that (and
successive) generations a further deterioration in stratospheric quality. It is
partly because of this problem of irreversibility (within a time scale measured in
centuries) that a conventional cost/benefit analysis of this problem is not
entirely appropriate.

SOME CONCLUDING COMMENTS

In this paper, we have had to rely entirely on indirect sources of evidence relating
to the possible reduction in stratospheric ozone due to anthropogenic release of
CFMs, and the consequences of this. In the light of our evaluation of the evidence,
we offer the following closing comments:

- It appears on available evidence that CFM emission is likely to have adverse
  effects upon the quality of life on a global scale, the particular effects having
  been listed above. Estimates of the steady-state levels of ozone depletion, and
  the results of this, are available. Should the steady-state reduction approach
  the upper confidence limit (20 percent) the adverse effects may be more than
  proportional to the associated extent of ozone depletion.

- Further information is needed, especially on response to the increased UV
  radiation that would result from a 20 percent as opposed to a 7 percent reduction
  in stratospheric ozone. Also, general equilibrium (multi-dimensional) rather
  than partial equilibrium (one dimensional) models of the atmosphere need to be
  developed in order to better predict the effects of anthropogenic activities on
  the stratosphere [29]. Such information would enable a better assessment of the
  potential benefits to society resulting from regulation of CFM production and
  use, or alternatively, would facilitate a more satisfactory estimate of the
  social costs that would be incurred if CFMs were not regulated.

- Any regulation to control CFM use *would* adversely affect industries producing and using CFMs, though the severity of the ensuing disruption would depend very much on the *form* of regulation adopted. Our analysis of the available evidence would seem to suggest that regulatory action should not take the form of a complete moratorium on production and distribution of CFMs. Rather that, if and when regulations are introduced, selective banning of CFMs in non-essential uses (particularly aerosols, other than medical and pharmaceutical applications) should be used. Refrigeration and air-conditioning systems should be required to be leak-proof, and the CFMs contained in them reclaimed (either through regulation or by using financial incentives) upon becoming obsolete. Use of CFMs as foam-blowing agents and solvents should be subject either to stringent control or, in some cases, a ban.

- If the approach outlined in the above comment were to be applied, then we believe that the ensuing economic and social disruption would be substantially less than if a complete across-the-board ban on CFM production and use were adopted. In any case, from available evidence, we suspect that the extent of these economic and social effects may have been overstated.

- The unilateral or multilateral action of certain major CFM producing countries to control their production and distribution, while going some way towards alleviating the potential depletion of stratospheric ozone by CFMs, would as a long-term strategy result in only a marginal impact on the problem of ozone depletion. The problem is, above all, global in nature, requiring global agreement and action.

- In the longer-term interests of stratospheric protection, it would seem highly desirable for an international regulatory body to be formed to monitor and control the whole gamut of man's activities of consequence for the stratosphere. This need is founded not only on the potentially severe effects of a reduction in stratospheric ozone, but also on the fact that longer-term protection of the stratosphere does not begin and end with the problem of CFMs.

- At the present time, the NAS Committee on the Impacts of Stratospheric Change has suggested that no regulations should be considered until a further two years' scientific investigation has been completed. On the other hand, it can be argued that the problems of irreversibility, associated with the potential depletion of stratospheric ozone by anthropogenic CFMs, suggest a need for more immediate action, and indeed the history of depletion of other common property resources (e.g. fisheries [30]) is replete with delays resulting in damage and loss, whilst more conclusive scientific evidence was obtained. In the final analysis, the decision on timing of regulatory action must be based on an implicit weighing-up of the potential consequences of delaying action compared with the more severe impact on industries which would occur if immediate controls were to be instituted.

REFERENCES AND NOTES

[1]   Stolarski, R.S. and R.J. Cicerone: Stratospheric Chlorine: a Possible Sink for Ozone, Canadian Journal of Chemistry 52, 1974, pp. 1610-1615 and Molina, M.J. and F.S. Rowland: Stratospheric Sink for Chlorofluoromethanes: Chlorine Atom Catalysed Destruction of Ozone, Nature, 249, 1974, pp. 810-812.

[2]   Crudely-speaking this implies the splitting up of the CFM molecules by the sun's rays acting on them in the stratosphere.

[3]   This information appears in two reports. They are: Halocarbons: Effects

on Stratospheric Ozone and Halocarbons: Environmental Effects of
Chlorofluoromethane Release. Preliminary copies of both were published
by the National Academy of Sciences, Washington, DC, September, 1976.

[4]   Preliminary Economic Impact Assessment of Possible Regulatory Action to
        Control Atmospheric Emissions of Selected Halocarbons, A.D. Little, Inc
        (under contract to USEPA) Cambridge, Massachusetts, September 1975.

[5]   Benefit/Cost Analyses for Regulating Stratospheric Emissions of Fluoro-
        carbons 11 and 12. R.C. d'Arge, University of Wyoming, Laramie,
        Wyoming (for USEPA) draft, February, 1976.

[6]   Chemical Technology and Economics in Environmental Perspectives: Task 1 -
        Technical Alternatives to Selected Chlorofluorocarbon Uses, USEPA,
        Washington, DC, February 1976.

[7]   Economic Significance of Fluorocarbons, US Bureau of Domestic Commerce Staff
        Study, December, 1975.

[8]   Chlorofluorocarbons and their Effect on Stratospheric Ozone, Central Unit
        on Environmental Pollution, Department of the Environment, HMSO, London
        February, 1976.

[9]   Fluorocarbons: an Assessment of Worldwide Production, Use and Environmental
        Issues (First Interim Report). Environment Directorate, OECD, Paris,
        1976, p. 4.

[10]   Summary: Research Programme on Effect of Fluorocarbons on the Atmosphere,
        Revision No. 4. Manufacturing Chemists Association, Washington, DC,
        September 1976.

[11]   For exposition, this section draws from Walter, I. (ed.): Studies in Inter-
        national Environmental Economics, Wiley, New York, 1976; Part Three:
        The Transnational Dimension, and especially the contributions to that
        section by Scott, d'Arge and Butlin.

[12]   Op.cit. p. 1-4.

[13]   Op.cit. pp. 1-21 to 1-24.

[14]   Op.cit. p. 1-24.

[15]   Presumably the time taken to achieve half the steady-state value would also
        increase, although the report does not mention this.

[16]   Op.cit. pp. 1-5 to 1-7.

[17]   Our expansion in parentheses.

[18]   Loc.cit.

[19]   It may occur to some people that this paper duplicates the work of others,
        especially that done by d'Arge for EPA. We should emphasise that, at
        the time of writing, we have not seen this paper, or résumés of it, nor
        have we had the chance to meet and talk with anybody familiar with this
        work.

[20]   See, for example, Fisher, A.C., J.V. Krutilla and C.V. Cichetti; The

Economics of Environmental Preservation: a Theoretical and Empirical
Analysis, American Economic Review 62, September 1972, pp. 605-619.

[21]   The option that is foreclosed is, of course, that of enjoying the original,
        undeveloped environment.

[22]   The extent to which substitutes for the propellant can include those using
        different dispersal technologies, (for example, roll-on or stick, as
        compared to spray, deodorant).

[23]   We do not include here uses of CFMs for refrigeration and air-conditioning
        uses. There are those, including the US Bureau of Domestic Commerce,
        who (because of the lack of adequate substitutes) regard CFMs for these
        purposes as an essential use. We discuss this below under 'Health
        Effects'.

[24]   US Bureau of Domestic Commerce, op.cit. pp. 91-93.

[25]   To the extent that other techniques (for example, freeze-drying) are
        substituted for refrigeration in the event of a ban on fluorocarbons,
        this problem will be limited.

[26]   d'Arge, R.C.: Transfrontier Pollution: Some Issues on Regulation, Chapter
        12 in Studies in International Environmental Economics (I. Walter, ed.)
        Wiley, New York, 1976.

[27]   Indeed, if the charge was associated with the durable good problem, the
        system could be so designed as to prevent littering by irresponsible
        disposal of these goods, as well as to reclaim the chlorofluorocarbons
        used in them.

[28]   See Butlin, J.A.: Problems of Intergovernmental Negotiations and Pollution
        Control, pp. 349-352 in Walter (ed.) op.cit.

[29]   After the initial draft of this paper was completed, a paper was received
        suggesting that the photodissociation of CFMs may increase the extent of
        the reduction in stratospheric ozone due to 'molecular scatter'. See:
        Sunde, J. and I.S.A. Isaksen: Stratospheric Chlorine Nitrate: Reduced
        Effect on the Catalytical Destruction of Ozone due to Molecular
        Scattering and Reflection of Photon Fluxes in the Atmosphere, Report
        No. 21, Institute Report Series, Geophysical Institute, University of
        Oslo, November, 1976.

[30]   A.C. Burd, The North East Atlantic Herring and The Failure of an Industry,
        UK Ministry of Agriculture, Fisheries and Food, Lowestoft, 1972.

PAPER 9

A Review of the Industry-sponsored Research into the Effect of Chlorofluorocarbons on the Concentration of Atmospheric Ozone*

Presented by the International Chamber of Commerce on behalf of the Manufacturing Chemists Association

This review covers the research program funded by the six U.S. and 14 foreign manufacturers of chlorofluorobarbons that is administered by the Manufacturing Chemists Association, and also the research undertaken internally by several of the companies.

Although there has been little new scientific information that would change the conclusions of the National Academy of Sciences reports [1,2] issued in September, there is continued high interest in the chlorofluorobarbon-ozone issue on the part of government agencies, industry, and certain public interest groups. It is therefore worthwhile to review the present gaps in our scientific knowledge, and the steps that are being taken to reduce them in industry-sponsored research programs. This is best done using the framework of the NAS Panel report [1]. It will be recalled that the prediction of 2-20% steady state ozone depletion, which would not be achieved for ca. 100 years, was calculated from known data and particular assumptions for (1) release rates of FC-11 and FC-12, (2) atmospheric transport of these molecules, and (3) stratospheric chemistry. The existence of as-yet-unknown factors that could substantially alter this ozone depletion estimate was clearly recognized by the NAS Panel. Indeed, their first recommendation for future work reads:

> "In general terms, the greatest need is for verification of stratospheric ozone chemistry through measurements of trace atmospheric constituents, both stable and reactive. The dearth of such measurements is a serious limitation to establishing the reliability of calculations and predictions of stratospheric effects." [3]

Dr Tukey, in his statement before the Subcommittee on the Upper Atmosphere of the U.S. Senate, December 15, reemphasized the two kinds of uncertainties in the NAS reports: (1) uncertainties associated with imprecise knowledge of numerical data, which necessitated giving the predicted ozone depletion as a range of possible values from 2-20%, and (2) presently unknown factors that could drastically alter the predicted depletion, possibly taking it below the 2-20% range. As stated in the NAS Committee report, "This range does not allow for possible inadequacies in the bases of the calculation," including "essential chemical reactions not so far recognized as such" and "the possibility of unexpected effects of tropospheric sinks" [4].

---

*Prepared by J.R. Soulen.

Stratospheric chemistry related to the chlorofluorocarbon-ozone question was felt to be particularly uncertain by the NAS Panel. Earlier this year the molecule $ClONO_2$, which was dismissed in the earliest considerations of the fluorocarbon-ozone question, was again brought up. The NAS Panel determined that inclusion of $ClONO_2$ chemistry would reduce ozone depletion estimates by a factor of 1.85. $ClONO_2$ is just one example of how new chemistry could affect the calculated amount of ozone depletion strongly. As stated in the NAS Committee report, "The probable importance of $ClONO_2$ was a surprise. Whether or not there is another such surprise in store remains to be seen" [5].

The NAS Panel listed six specific objectives to reduce the uncertainty in ozone depletion estimates due both to factors already known and to the possibility of finding subsequently an important unidentified factor. The most important studies expected to require one to two years, are as follows:

1. *"Establish the role of $ClONO_2$ in the stratosphere* – This might be accomplished by comparing infrared measurements of its concentration with calculated results. However, particularly if disparities are revealed by such comparisons, there will need to be further study of the stratospheric processes that might convert the $ClONO_2$ back into catalytically active species." [6]

The chlorofluorocarbon industry is supporting the work of Dr D.G. Murcray of the University of Denver in his efforts to determine the stratospheric concentration of $ClONO_2$. This support is for balloon flights to make measurements by infrared absorption spectroscopy, laboratory studies to obtain the necessary calibration spectra for $ClONO_2$, and assistance in the preparative work to obtain samples for the laboratory studies. Several more months will be required before definitive information will be available from flights carried out in 1976. On the basis of indications to date, significant amounts of $ClONO_2$ in the stratosphere are neither proved nor ruled out.

A major problem in positive identification of molecules using the infrared technique and in determining their concentration is possible interferences from other molecules that absorb in the same spectral region. Greater resolution can minimize this problem and the chlorofluorocarbon manufacturers are supporting Dr H.J. Buijs of Bomem, Inc., to construct a new spectrometer with an anticipated resolution of 0.01 cm^{-1}. This instrument will be useful in the wavelength region of interest for $ClONO_2$ and should be completed in early 1977.

For long-term monitoring of stratospheric molecules, continuous ground-based observations will be much more economical than in situ measurements for balloons, airplanes, or rockets. Scientists from the Du Pont Company and the National Bureau of Standards have collaborated to determine the microwave spectrum of $ClONO_2$ [7], which will be useful for both air and ground-based measurements by millimeter wavelength techniques.

The rates of formation and destruction of $ClONO_2$ are important to the understanding of its stratospheric chemistry, as is knowledge of its mechanism of decomposition (e.g., photolysis, reaction) and the products of its decomposition. The chlorofluorocarbon industry is sponsoring research at the University of Illinois under Dr J.W. Birks, who determined the rate of formation of $ClONO_2$ via the reaction $ClO + NO_2 + M \rightarrow ClONO_2 + M$ (where M is a molecule such as N_2 or He). Dr Birks is also studying the rate of destruction of $ClONO_2$ by both homogeneous and heterogeneous reactions that could result in removal of reactive species from the catalytic ozone destruction cycle.

The photochemistry of $ClONO_2$ that is relevant to its stratospheric behavior can also be studied in the laboratory. The chlorofluorocarbon manufacturers are supporting work by Dr J. Wiesenfeld at Cornell University and Dr R.B. Timmons of Catholic University to determine photoabsorption characteristics and photodecomposition products of $ClONO_2$ as a function of wavelength and temperature.

2. *"Verify the reliability of predicted ozone reductions* - By this we mean a more direct means than has been available so far for estimating and narrowing the *total* uncertainty of the ozone reductions predicted for the CFMs, including the possibility of as yet unidentified processes. Observations of ClO and Cl profiles in the stratosphere and their comparison with calculated values seem to be the most likely means of accomplishing this objective. However, the analysis of any disparities probably will be difficult, and requires comparisons with simultaneous measurements of other species such as HCl and O_3. Preferably, the observations of ClO and Cl should be by two independent methods, such as resonance fluorescence and microwave emission in the case of ClO." [8]

An approach to verify predicted ozone reductions not emphasized in the shorter-term objectives of the NAS Panel is that based on direct observations of stratospheric ozone itself. Dr William J. Hill and his colleagues at Allied Chemical Corporation have applied the technique of time series analysis to data from nine long-term ozone stations well-spaced around the world. Their conclusions [9] are of considerable importance:

- No detectable abnormal decline in ozone has occurred during the six-year period 1970-75.

- The predicted cumulative ozone depletions due to FC-11 and FC-12, subsonic and supersonic aircraft, bromine compounds, nitrogenous fertilizers, and a number of halocarbons, when treated cumulatively appear to be too large.

- The calculated upper limit of 20% steady state depletion (ca. 100 years hence) due to FC-11 and FC-12 appears very unlikely.

- The time series analysis of data from nine globally representative stations could detect a 1.5% abnormal ozone change. Appropriate action following this early warning could limit the maximum reversible change to only 2.3%. Use of data from a larger number of ground-based or satellite stations could increase the detectability and decrease the impact on the ozone layer.

The chlorofluorocarbon manufacturers are supporting Drs E. Parzen and M. Pagano of the State University of New York at Buffalo to carry out further research on time series analysis of ozone data. Work of this type clearly can have great importance to both the stratospheric ozone problem and the research and measurement programs addressing it by:

- Providing a direct test of ozone depletion estimates from model calculations.

- Maximizing the value and usefulness of the extensive ozone measurements, past, present, and future.

- Providing a global early warning system for abnormal ozone changes, i.e., man-made or natural trends that are different from past variations.

- Allowing practical limits to be set on the scope of the biological and climatic effects research programs in accordance with the maximum ozone change allowed by the early warning system plus appropriate action to limit the change.

To date the principal method used to test the ozone depletion theory has been indirect: measure the concentrations of certain species the amounts of which are

quite small and compare the measurements with theoretical predictions. This has
required the development of new data, techniques, and instruments, and the chlorofluorocarbon manufacturers have since 1974 had this as an important part of their
research program. Among a number of projects relating to the analysis of ClO and
other chlorine-containing species, one under the direction of Dr D.H. Stedman at
the University of Michigan measured successfully both ClO and Cl by titration with
NO and detection of Cl by resonance fluorescence [10]. NASA provided support for
Dr J.G. Anderson to develop a flight instrument based on this technique and make
measurements of ClO and Cl this year. The values from the several sets of measurements vary several-fold from each other, and all are higher than theory. It will
therefore be important to follow up this work in accordance with the NAS recommendations for:

- measurement of a greater number of different species simultaneously to provide
  a more rigorous test of the theory; and

- measurement of the most important species simultaneously by two different
  techniques to check on the accuracy of the analytical methods.

Work supported by the chlorofluorocarbon manufacturers to help accomplish these
tasks includes laboratory investigations of ultraviolet spectra of ClO by Dr R.W.
Nicholls of York University [11] and Dr D.D. Davis of Georgia Tech and of the laser
magnetic resonance technique for detection of ClO by Dr C.J. Howard at the NOAA
Laboratory at Boulder. Support is being provided for stratospheric measurements of
ClO by Dr D.G. Murcray using a balloon-borne ultraviolet spectrophotometer and to
Dr P.M. Solomon of SUNY Stony Brook for ground-based microwave measurements. In
the latter program the 3 mm maser detector that will be employed will be the most
sensitive receiver in the world for this type of observation.

To help resolve the serious questions that have been raised by the finding of larger
than predicted amounts of chlorine species in the stratosphere, the chlorofluorocarbon manufacturers are supporting a project under Dr A.E.J. Eggleton at AERE
Harwell to develop methods for determining total chlorine and total fluorine. They
are to be employed during balloon flights to compare their sum with the total
amounts of these elements found with the sum of individual Cl- and F-containing molecules measured by other investigators.

3. *"Develop an accurate materials balance"* - Inasmuch as a materials balance can
detect, in principle, otherwise unidentified sinks for the CFMs, a detailed analysis
should be made of how best to obtain a materials balance accurate enough to be
meaningful. The study should define a minimum set of observations (including its
time span) sufficient to establish a true global burden, and develop standards of
sensitivity, calibration and accuracy necessary for the results to be useful. If
such a study supports its feasibility, there should then be a coordinated observational program and data analysis for both F-11 and F-12." [12]

In spite of assertions to the contrary from elsewhere in the scientific community,
the possibility of an unidentified sink for the chlorofluorocarbons indicated above
by the NAS Panel has been underscored by the NAS Committee:

> "Although the materials balance can be interpreted as consistent with
> little or no active removal of the CFMs, the uncertainty limits range
> from zero inactive removal to a rate sixfold faster than that of the
> stratospheric photolysis." [13]

Modeling work sponsored by the chlorofluorocarbon manufacturers led to the realization that chlorofluorocarbon measurements to date did not rule out the possibility of sinks for these molecules [14]. The most recent work indicates the most probable tropospheric lifetime for FC-11 is between 15-20 years [15]. The ozone depletion calculations in the NAS reports are based on an infinite tropospheric lifetime and would be reduced as much as five-fold by this shorter tropospheric lifetime.

In its recommendation above the NAS Panel calls for development of both strategy and techniques to determine if the materials balance approach can be successful. The chlorofluorocarbon manufacturers have encouraged one of their modeling contractors to examine measurement problems and strategy. Several reports were prepared including one that was presented to the NASA Stratospheric Research Advisory Committee in December 1975 and has appeared subsequently as a publication [16].

Three contractors in the industry program are now studying problems of calibration and accuracy of chlorofluorocarbon measurements, Drs J.E. Lovelock, R.A. Rasmussen, and D.H. Stedman. Two are investigating new methods of absolute calibration, and one will conduct an interlaboratory comparison program.

4. *"Reduce errors in rate constants* - Considerable improvement in the rate constants for the $Cl + CH_4$ and $HO + HO_2$ reactions should be sought and may be expected within the next year or two. Further work on several of the other reactions would serve to reduce their lesser contributions to the uncertainties. A strong effort should be made to reduce the uncertainty factor of the $HO + HO_2$ reaction to 1.5 and those of the other reactions to 1.3, or less, including the effects of temperature dependences." [17]

The uncertainty in the rate of the reaction $HO + HO_2$, of all the reactions in the model, accounts for the largest uncertainty in calculated ozone depletion. Because this is a reaction between two radicals, it is especially difficult to measure. The chlorofluorocarbon manufacturers are supporting two investigators to determine the rate of this and other reactions of the HO_2 radical, Drs B.A. Thrush of the University of Cambridge and C.J. Howard of the NOAA-Boulder laboratory. Dr J.W. Birks has measured the rate of reaction $ClO + NO_2$, which is listed as another of the important reactions in the NAS Panel report.

5. *"Detailed evaluation of identified sinks* - As a complement to the materials balance approach, inactive removal processes of CFMs, once identified, should be characterized in quantitative terms. In particular, oceanic removal of F-11 should be investigated more thoroughly by additional measurements of its concentration gradient in surface waters, by observations of its transport across the air-water interface, by redetermination of its solubility in sea water, and by a search for mechanisms that might contribute to its degradation in the surface waters. The same types of studies should be made for F-12. Also, efforts should be made to place narrower limits on the removal of F-11 and F-12 by photolysis and ion-molecule reactions in the troposphere." [18]

A number of potential sinks for the chorofluorocarbons have been investigated in the industry-sponsored program including photochemical changes by Dr C. Sandorfy of the University of Montreal [19], reactions with plants and soil and with OH and the $O(^1D)$ atom [20] by Dr J.N. Pitts of the University of California at Riverside; reactions involving charged species by Dr M.J. Campbell of Washington State University [21], Dr V.A. Mohnen of SUNY Albany, and Dr R.G. Hirsch of Du Pont [22]; and

trapping in Antarctic ice by Dr R.A. Rasmussen. These projects have not identified
with certainty a tropospheric sink for chlorofluorocarbons; studies by Dr J.E.
Lovelock are continuing.

6. *"Improve other aspects of atmospheric chemistry* – Photochemical processes and
concentrations of natural species (NO_x and HO_x) contribute appreciable uncertaintie
to the prediction of ozone reduction by the CFMs. These aspects should be studied
further, to place better limits upon their importance and to seek the most produc-
tive ways of reducing their contributions." [23]

OH is among the HO_x species contributing to the uncertainties in the prediction of
ozone depletion by the chlorofluorocarbons. The industry is supporting development
of an instrument by Dr D.D. Davis for measuring stratospheric OH concentrations by
laser-induced fluorescence. A miniturized version of a unit flown successfully
earlier on an airplane is being prepared for balloon flights, which will allow
observations over extended periods of time, and of diurnal changes.

In accordance with the NAS Panel recommendation that important species be measured
simultaneously by two different methods, Dr D.G. Murcray is also being supported to
measure the concentration of stratospheric OH by solar ultraviolet absorption.

REFERENCES

[1]   Halocarbons: Effects on Stratospheric Ozone, Panel on Atmospheric Chemistr
        National Academy of Sciences, September 13, 1976.

[2]   Halocarbons: Environmental Effects of Chlorofluoromethane Release, Committ
        on Impacts of Stratospheric Change, National Academy of Sciences,
        September 13, 1976.

[3]   Ref. 1, p. F-1a.

[4]   Ref. 2, pp. 1-3, 4.

[5]   Ref. 2, p. 5-12

[6]   Ref. 1, p. F-2.

[7]   R.D. Suenram, D.R. Johnson, L.C. Glasgow and P.Z. Meakin, Geophys. Res.
        Lett. 3, 611-14, October, 1976.

[8]   Ref. 1, p. F-2.

[9]   W.J. Hill, P.N. Sheldon and J.J. Tiede, Analyzing Worldwide Total Ozone for
        Trends, paper accepted for publication in Geophys. Res. Lett.

[10]   D.H. Stedman, Res./Dev. 27, 22-4, 26, January, 1976.

[11]   M. Mandelman and R.W. Nicholls, The Absorption Cross-Sections and f-Values
        for the v"=O_2 Progression of Bands and Associated Cortinuum for the
        ClO ($A^2\Pi_i \leftarrow X^2\Pi_i$) System, accepted for publication in J. Quant. Spectros
        Radiat. Transfer.

[12]   Ref. 1, p. F-2, 3.

[13]   Ref. 2, p. 5-10.

[14]   N.D. Sze and M.F. Wu, Measurements of Fluorocarbons 11 and 12 and Model
       Validation: An Assessment, 27 pp., 15 December 1975.

[15]   J.P. Jesson, P. Meakin and L.G. Glasgow, The Fluorocarbon-Ozone Theory II:
       Tropospheric Lifetimes: An Experimental Estimate of the Tropospheric
       Lifetime of CCl_3F, accepted for publication in Atmos. Environ.

[16]   R.G. Prinn, F.N. Alyea and D.M. Cunnold, Bull. Am. Meteorol. Soc. 57,
       686-94, June, 1976.

[17]   Ref. 1, p. F-3.

[18]   Ref. 1, p. F-3.

[19]   C. Sandorfy, Atmos. Environ. 10, 343-51, May, 1976.

[20]   R. Atkinson, G.M. Brewer, J.N. Pitts, Jr and H.L. Sandoval, Tropospheric
       and Stratospheric Sinks for Halocarbons, accepted for publication in
       J. Geophys. Res.

[21]   M.J. Campbell, Geophys. Res. Lett. 3, 661-4, November, 1976.

[22]   R.G. Hirsch, Atmos. Environ. 10, 703-5. September, 1976.

[23]   Ref. 1, pp. F-3, 4.

PAPER 10

*A Summary of the Research Program on the Effect of Fluorocarbons on the Atmosphere**

Sponsored by the Fluorocarbon Industry and Administered by the
Manufacturing Chemists Association

This summary describes work supported by the manufacturers of chlorofluorocarbons
(commony called fluorocarbons) in an attempt to assess the possible impact of these
chemicals on the environment and, in particular, on the stratospheric ozone layer.

THE INDUSTRY-SPONSORED PROGRAM

In July 1972, E.I. du Pont de Nemours & Company issued to fluorocarbon manufacturers
worldwide an invitation to a "Seminar on the Ecology of Fluorocarbons". Its purpose
was to establish a technical program because, as stated in the invitation,

> "Fluorocarbons are intentionally or accidentally vented to
> the atmosphere worldwide at a rate approaching one billion
> pounds per year. These compounds may be either accumulating
> in the atmosphere or returning to the surface, land or sea,
> in the pure form or as decomposition products. Under any
> of these alternatives, it is prudent that we investigate any
> effects which the compounds may produce on plants or animals
> now or in the future."

Representatives of 15 companies attended the meeting, agreed that such a program
was important, and established and funded a fluorocarbon research program under the
administration of the Manufacturing Chemists Association (MCA). Thus, in 1972, with
no evidence that fluorocarbons could harm the environment, the producers of these
chemicals agreed unanimously that there was a need for more information and
proceeded to act.

The 20 fluorocarbon producers supporting this program (see Table 10.1) represent
almost the total free world production of fluorocarbons. The research is directed
by an MCA Technical Panel on Fluorocarbon Research with one member from each
supporting company. This Technical Panel meets for two full days each month to
review progress on current research, evaluate new proposals, and exchange data with
contractors and with government agencies and researchers. Three projects were
initiated in 1973:

— Dr James Lovelock, U. of Reading, "Fluorocarbons in the Environment".

— Dr Camile Sandorfy, U. of Montreal, "Spectroscopy and Photochemical Changes in
  Fluorocarbons".

---

*Originally issued: September 16, 1975. Revision No. 5: December 31, 1976.

TABLE 10.1 Fluorocarbon Manufacturers Represented on the
MCA Technical Panel on Fluorocarbon Research

Akzo Chemie nv (Holland)
Allied Chemical Corporation (U.S.)
Asahi Glass Co. Ltd. (Japan)
Australian Fluorine Chemical Pty. Ltd. (Australia)
Daikin Kogyo Co. Ltd. (Japan)
E.I. de Pont de Nemours & Company, Inc. (U.S.)
Du Pont of Canada Limited (Canada)
Hoechst AG (West Germany)
Imperial Chemical Industries Limited (England)
I.S.C. Chemicals Ltd. (England)
Kaiser Aluminum & Chemical Corporation (U.S.)
Kali-Chemi Aktiengesellschaft (West Germany)
Mitsui Fluorochemicals Co. Ltd. (Japan)
Montedison S.p.A. (Italy)
Pennwalt Corporation (U.S.)
Racon Incorporated (U.S.)
Rhône-Poulenc Industrie (France)
Showa Denko K.K. (Japan)
Ugine Kuhlmann, Produits Chimiques (France)
Union Carbide Corporation (U.S.)

— Dr O.C. Taylor, U. of California (Riverside), "Monitoring and Atmospheric
  Reactions of Fluorocarbons".

Publication of the Rowland-Molina hypothesis in 1974 identified a potentially
serious problem, so the MCA research program was expanded considerably. The fluorocarbon-ozone relationship attracted the attention of many scientists in academic
and government laboratories, legislative and regulatory bodies, and the press.
MCA's program is concentrating on research most likely to answer the critical
question: To what extent will fluorocarbons affect the stratospheric ozone layer?

To strengthen the overall effort to find the answer, MCA has attempted to coordinate
its efforts with other working on the same or related problems such as the SST and
the space shuttle. All of these concern the federal government, and interactions
with a numbers of agencies have been especially helpful in:

— taking advantage of the knowledge and experience gained in the Climatic Impact
  Assessment Program;

— coordinating funding of programs addressing the halogen-ozone problem;

— planning joint experiments with government research groups; and

— helping to set priorities for industry-sponsored research.

About 165 research proposals have been reviewed to date, and projects totaling over
$3,200,000 have been funded (see Table 10.2). Calendar 1977 commitments are expec-
ted to exceed $2,000,000, with total expenditures through 1977 expected to exceed
$5,000,000. Approximately $650,000 has been committed to studies concerning the
measurement of reactive chlorine species in the stratosphere, a difficult and
important area. A promising method was developed at the University of Michigan
with MCA support, and funding for the follow-on program for development of equipment
and stratospheric measurements has been provided by the National Aeronautics and
Space Administration (NASA). This illustrates an important contribution by

TABLE 10.2 Fluorocarbon Research Program (administered by
the MCA) - Financial Summary

	Completed Projects	Active Projects	Total
Chemical Reactions and Kinetics	$ 242,050	$ 351,600	$ 593,650
Analytical Methods	259,312	105,418	364,730
Tropospheric Measurement	169,107	339,846	508,953
Stratospheric Measurement	211,187	1,036,530	1,247,717
Modeling	147,652	285,005	432,657
Consulting	31,475	22,646	54,121
TOTAL	$1,060,783	$2,141,045	$3,201,828

industry: prompt funding of promising work that, if successful, can be taken over
by a government agency for incorporation into its more comprehensive program.

MCA Technical Panel members have visited knowledgeable scientists throughout North
America and Europe, including many at government laboratories. The Panel held its
September 1975 meeting in Montreal where, in addition to hearing presentations by
Canadian contractors and potential contractors, it was briefed on the stratospheric
measurement program of the Canadian government. The February 1976 Panel meeting
was held near Milan, Italy, and the October 1976 meeting in Erbach, Germany. The
Panel was briefed on pertinent European activities and heard presentations by
European atmospheric scientists at these meetings.

Discussions in early 1975 with the Energy Research and Development Administration's
(ERDA) Sandia Laboratory at Albuquerque and the Army's Atmospheric Sciences Laboratory at White Sands Missile Range, New Mexico, indicated that industry could
contribute significantly to the stratosphere composition (STRATCOM) program.
Cooperation began immediately with the Panel proving funding for two additional
STRATCOM experiments and an additional balloon for stratospheric measurements in the
fall of 1975. Similarly, the Panel provided funding for several experiments that
were flown on balloons included in the May 1976 Army COSMEP flights in Alaska and
the STRATCOM flights in September 1976. In the 1976 flights there was also
cooperation between MCA contractors and the Atmospheric Environment Service of
Canada on Fourier transform spectrometer measurements. Through these types of
collaboration it is hoped to reduce substantially the time required to obtain critical research results.

GAPS IN SCIENTIFIC KNOWLEDGE

Two groups were charged during 1975 with looking exclusively at the scientific
aspects of the halocarbon-ozone problem and making recommendations for further work.
In May 1975 the government's Interdepartmental Committee for Atmospheric Sciences
(ICAS) made 12 recommendations for research and monitoring programs to assess the
possible impact of fluorocarbons and halogens on ozone. Six of these were singled
out as of major importance. In July 1975 the Panel on Atmospheric Chemistry of the

National Academy of Sciences (NAS) identified a number of areas in which relevant
data are nonexistent, fragmentary, or insufficient and listed 19 high priority
areas requiring study. The final report of the NAS Panel, which was issued in
September 1976, contains six recommendations for pertinent studies that are expected to be completed in the period of up to two years. Its parent committee, the
NAS Committee on Impacts of Stratospheric Change, recommends that this period
should be allowed before a decision is made on the necessity for restrictive action
The Panel also made fifteen recommendations for longer-range studies, many of which
are relevant to the possible impact of halogens on ozone.

Table 10.3 shows the relationship between the latest NAS recommendations and the
research program sponsored by the fluorocarbon manufacturers and with the preliminary recommendations of NAS and those of ICAS. There is generally good agreement
between the ICAS and NAS recommendations. In addition to other small differences,
the former include some aspects of monitoring not in the latter, and the latter
include photolysis studies and bromine chemistry not in the former. The second
column of Table 10.3 lists the MCA-administered industry program. Its emphasis is
overwhelmingly in the major areas recommended by the NAS Panel and ICAS. The
industry-sponsored program, therefore, aims to fill in the most important gaps in
existing scientific knowledge. The only major area not in the MCA program is
acceleration of ozone monitoring, which can be accomplished more appropriately by
governmental and international agencies. In addition, the MCA program includes
work to correlate UV radiation reaching the ground with ozone measurements and to
improve the statistical treatment of ozone data so that very small abnormal changes
can be detected in a much shorter period of time than was previously believed
possible.

More detail on the MCA program is given in Tables 10.4, 10.5A and 10.5B. Table
10.4 lists summaries of the projects by type of research activity. Table 10.5A
lists completed projects, and Table 10.5B lists active projects in chronological
order of funding. Table 10.6 lists publications resulting from industry-sponsored
work. These important contributions have not always been recognized as industry-
funded activity.

In addition to the work supported by the industry as a whole there are also studies
under way in the laboratories of individual member companies who have staff
scientists able to make significant contributions to the resolution of the problem.
Examples are the modeling, reaction, and spectroscopic studies at Du Pont and the
statistical analysis of ozone trends at Allied.

SUMMARY AND RECOMMENDATIONS

In 1972 the fluorocarbon manufacturers began supporting a program to investigate the effects of fluorocarbons on the environment. This program has been expanded greatly to help determine the extent, if any, to which these compounds may affect the stratospheric ozone layer. Industry and government-sponsored scientists working on the halogen-ozone problem have cooperated effectively. Continuation of this cooperation is essential, with special attention to providing periodically updated summaries of research priorities, programs, and results, together with critical analysis of the reliability and significance of the data. Such a program will produce most rapidly the information required for rational decision-making.

TABLE 10.3 Fluorocarbon Ozone Problem — Recommended Studies and MCA Program

NAS Recommendations, September 1976[a,b]	MCA Supported Work	
	Completed	In Progress
I.1 Establish role of $ClONO_2$ in the stratosphere.		Birks — U. of Illinois Murray — U. of Denver Timmons — Catholic U. Wiesenfeld — Cornell U.
I.2 Verify the reliability of predicted ozone reductions. (a) via ClO and Cl profiles simultaneously with HCl, O_3, etc. (b) via decrease in total ozone column (c) via changes in ozone profile	Davis — U. of Maryland[f] Ekstrom — Battelle Northwest Stedman — U. of Michigan[f] Young — Xonics, Inc.[f]	Davis — U. of Maryland/ Georgia Inst. Technol.[f] Howard — NOAA-Boulder[f] Murcray — U. of Denver Nicholls — York U.[f] Parzen & Pagano — Frontier Science & Technology Research Foundation, inc. Solomon — State U. of N.Y.-Stony Brook
I.3 Develop an accurate materials balance for F-11 and F-12. (a) Establish requirements for sensitivity, calibration, and accuracy. (b) Define minimum observational program.	Lovelock — U. of Reading Manufacturers via MCA Rasmussen — Washington State U. Taylor — U. of Calif.-Riverside	Eggleton — AERE Harwell[f] Lovelock — Private Murcray — U. of Denver[f] Manufacturers via MCA Rasmussen — Washington State U. Stedman — U. of Michigan[f]
I.4 Reduce errors in rate constants with emphasis on $Cl + CH_4$, $OH + HO_2$, $ClO + NO$, $ClO + NO_2$, $OH + HCl$, $Cl + O_3$		Birks — U. of Illinois Howard — NOAA-Boulder Thrush — U. of Cambridge Timmons — Catholic U. Wiesenfeld — Cornell U.
I.5 Make a detailed evaluation of sinks once they are identified. (a) Oceanic removal of F-11 and F-12. (b) Photolysis of F-11 and F-12. (c) Ion-molecule reactions of F-11 and F-12 in the troposphere.	Campbell — Washington State U. Mohnen — State U. of N.Y.-Albany Pitts — U. of Calif.-Riverside	Birks — U. of Illinois Kaufman — Emory U. Lovelock — U. of Reading Martin — Aerospace Corp. Rasmussen — Rasmussen Associates
I.6 Improve other aspects of atmospheric chemistry. (a) Photochemical processes of natural species (NO_x, HO_x). (b) Concentrations of natural species.		Buijs — Bomem, Inc. Davis — U. of Maryland/ Georgia Inst. Technol. Girard — ONERA (France)[g] Murcray — U. of Denver
II.1 Monitor all major releases of volatile halocarbons and other potential stratospheric pollutants on a global scale.	Lovelock — U. of Reading Rasmussen — Washington State U. Taylor — U. of Calif.-Riverside	Lovelock — Private Murcray — U. of Denver Rasmussen — Washington State U. Ridley — York U.
II.2 Determine sinks for every substance identified as a potential or actual hazard to the stratosphere.		fc. I.5
II.3 Determine for each pollutant the complete sequence of events from injection to removal of the final products from the atmosphere. (a) Fate of halocarbon radicals remaining after photodissociation. (b) Effect of HF and HCl on rain and snow.		
II.4 Improve reliability of 1-D calculations. (a) More extensive and more accurate data for CH_4, N_2O, HF and O_3 for "fine-tuning" models. (b) Treatment of diurnal variations.		Cunnold, Alyea, Prinn — CAP Associates Sze — ERT, Inc.
II.5 Expedite development of 2- and 3-D models.		Cunnold, Alyea, Prinn — CAP Associates

ICAS Recommendations, May 1975[c,d]	NAS Recommendations, July 1975[e]
1. Measure Cl and ClO concentrations in the stratosphere. 5. Accelerate program for monitoring of ozone on a global basis. Establish trends in atmospheric ozone.	1. Measure Cl and ClO upper atmosphere concentration profiles.
3. Make systematic search for other chlorine-bearing gases in the stratosphere. A2. Expand monitoring of F-11, F-12, and CCl_4 in the troposphere and develop standards for their measurement.	14. Determine total chlorine in the atmosphere and compare with sum of all identified compounds.
6. Confirm and improve certain reaction rate measurements crucial to ozone destruction.	6. Measure rate constants for ClO + NO, ClO + O, Cl + HO_2, Cl + CH_4. 7. Measure rate constants for OH + H_2O; H, O, NO, O_3 + HO_2; H_2O, CH_4, N_2, O_2 + $O(^1D)$.
4. Conduct promising research to determine sinks for chlorine in the stratosphere.	16. Seek out and characterize other possible sinks for removal of chlorofluorocarbons. 17. Determine source and further reactions of chlorine in stratospheric particulates.
2. Measure [F-11, F112, CCl_4,]OH, H_2O, and other gases at 25-40 km.	2. Measure OH, HCl, HF, NO, etc., upper atmosphere concentration profiles. 12. Sample at low altitude to find additional possible natural and anthropogenic sources of halogen-containing compounds. Establish concentrations.
2. Measure F-11, F-12, [CCl_4, CH, H_2O,] and other gases at 25-40 km. A1. Obtain a full record of past and present halocarbon production and release.	4. Measure F-11, F-12, CCl_4, $CHCl_3$, and N_2O concentration profiles 0-40 km. Establish standards. 13. Obtain accurate data on production and uses of each halocarbon manufactured worldwide.
Cf. I.5 A3. Study the lifetimes of halocarbons, HCl, and phosgene in the troposphere.	Cf. I.5
	15. Measure atmospheric abundance of decomposition products such as COF_2, $COCl_2$, COFCl, and determine whether COFCl releases additional Cl.
A4. Expand the stratospheric modeling studies of ozone formation and destruction. Improve chemistry and transport.	3. Measure global distribution of certain trace gases to improve models. 5. Develop better atmospheric models.
	Cf. II.4

TABLE 10.3 (cont'd.)

NAS Recommendations, September 1976[a,b]	MCA Supported Work	
	Completed	In Progress
II.6 Determine accurate rate constants for a large number of chemical reactions involving Cl_x and atmospheric species, HO_x chemistry, and $O(^1D)$ quenching and reaction rate constants.		cf. I.4
II.7 Determine reaction rate parameters over the tropospheric and stratospheric temperature range for the reaction of OH with all important chloro- and fluorocarbons.		
II.8 Determine kinetic data for all important reactions of Br_x and photolytic properties of the various bromine-containing species.		
II.9 Measure solar flux and its variability, especially at wavelengths greater than 175 nm, and stratospheric scattering parameters.		
II.10 Study photochemistry of halocarbons and Cl_x further. (a) Photolysis parameters of F-11, F-12, OClO, ClOO, $ClNO_2$, and $ClONO_2$. (b) Photolysis parameters of ether halocarbons such as $CHFCl_2$, CCl_4, CH_3CCl_3, CH_2Cl_2, and CH_3Cl.	Sandorfy — U. of Montreal	
II.11 Define and establish a global network for atmospheric measurements.	Ekstrom — Battelle Northwest	Solomon — State U. of N.Y.-Stony Brook
II.12 Measure concentration profiles in upper atmosphere for species related to the chemistry of chlorine, including OH, HCl, HF, and NO.		Buijs — Bomem, Inc. Davis — U. of Maryland/ Georgia Inst. Technol. Girard — ONERA (France)[g] Murcray — U. of Denver
II.13 Make global monitoring of ozone sufficiently sensitive, accurate, and complete to detect small changes due to man's activities.		Parzen & Pagano — Frontier Science & Technology Research Foundation, Inc.
II.14 Give further attention to the problem of water vapor feedback.		
II.15 Study the reduction in stratospheric ozone associated with the use of nitrogen fertilizers including the effects of the interaction between the NO_x and ClO_x cycles.		
	Nicholls — York U.	Davis — U. of Maryland/ Georgia Inst. Technol.

[a] "Halocarbons: Effects on Stratospheric Ozone", Panel on Atmospheric Chemistry, Committee on Impacts on Stratospheric Change, Assembly of Mathematical and Physical Sciences, National Research Council, NAS, September, 1976, Appendix F.
[b] Group I items are short-term (1-2 year), Group II longer range.
[c] "The Possible Impact of Fluorocarbons and Halocarbons on Ozone", ICAS 18a-FY, May 1975, pp. 65-66.
[d] 1-6 indicate major recommendations, A1-A6 other.
[e] "Interim Report of the Panel on Atmospheric Chemistry", Climatic Impact Committee, Assembly of Mathematical and Physical Sciences, National Research Council, NAS, July, 1975, pp. 3-8.
[f] Laboratory study to develop data, analytical method, or instrumentation required for stratospheric measurements.
[g] Office National d'Etudes et de Recherches Aérospatiales.

ICAS Recommendations, May 1975[c,d]	NAS Recommendations, July 1975[e]
Cf. I.4	Cf. I.4
Cf. I.4	8. Measure rate constants for OH + CH_2FCl, CHF_2Cl, CH_3Cl, CH_3CCl_3, CH_2Cl_2, and all other potentially usable halocarbons with C-H bonds.
	19. Obtain kinetic data for all important reactions of bromine, Br + O_3, HO_2; BrO + O, NO, O_3; CH_3Br + OH.
	9. Measure solar flux and scattering parameters for the stratosphere.
	10. Measure photolysis parameters for chlorohydro-carbons and chlorofluorocarbons.
A2. Expand monitoring of F-11, F-12, and CCl_4 in the troposphere and develop standards for their measurement.	18. Accurate baseline monitoring of ground level composition and absorption spectrum of "clean air" including measurements of O_3, HCL, HF, N_2O, NO, NO_2; monitoring column densities of HCl, HF, ClO, OH, NO_2, O_3.
F. I.6 and II.1	Cf. I.6 and II.11
Cf. I.2	Cf. I.2
A5. Expand the general circulation modeling of the effects of reduced stratospheric ozone in climate.	
A6. Continue to study the effects of releasing halo-carbons on humans and the biosphere.	
	11. Make laboratory studies of the photolysis of ClO.

TABLE 10.4   Fluorocarbon Research Program — Types of
Research Activities — Summaries

A.   REACTION RATE CONSTANT MEASUREMENTS

Dr. J.W. Birks — University of Illinois — 75-1, 76-117A. Measurement of
Reaction Rates Relevant to the Fluorocarbon-Ozone Problem.

Reaction rates are measured at various temperatures by a
discharge flow technique with a quadrupole mass spectrometer
for detection of products. Reactions studied include
$ClO + O_3 + M$ (yields $OClO$, unimportant in daytime),
$ClO + NO_2 + M$ (important in lower stratosphere, $k = 1.5 \times 10^{-13}$
cm^6 molecule^{-2}s^{-1} at 297°K), $ClO + SO_2$ (no appreciable
reaction observed), $NO + O_3$ (in agreement with previous
studies), $O + ClONO_2$ (preliminary results in agreement with
Davis), and $BrO + NO_2$ (in progress). The products of the
four-center reactions $Cl_2O + Cl$, O, N, or ClO have been
investigated by molecular beam mass spectrometry and indicate
that O, N, and ClO all react with the O in $ClOCl$. Future
work will include $ClO + H_2O_2$ and reactions of HO_2.

Dr. C.J. Howard — NOAA, Boulder — 76-100. Laser Magnetic Resonance Study of
H_2O Chemistry.

The object of this study, which is just getting under way, is
to use the laser magnetic resonance technique to measure rate
constants for stratosperically important H_2O reactions.

Dr. J.N. Pitts — University of California at Riverside — 74-2. Atmospheric
Reactions of Fluorocarbons.

Reaction rate constants have been measured for the reactions
of $O(^1D)$ with fluorocarbons 11, 12, 22, 113, and 114, and of
OH with fluorocarbons 11, 12, and 22. The results indicate
that in the stratosphere the reaction of $O(^1D)$ atoms with
fluorocarbons 11 and 12 is secondary to photolysis, whereas
the reaction of OH with fluorocarbon 22 is much more important
than photolysis. The photooxidation products of 12, 13, and
22 at 184.9 nm, i.e. $COCl_2$, $COFCl$, and COF_2 as appropriate,
are also observed to be the products for reaction with $O(^1D)$.

Dr. B.A. Thrush — University of Cambridge — 75-58, 75-58II. Reactions of the
HO_2 Radical Studied by Laser Magnetic Resonance.

The rate coefficient of the reaction $O + HO_2$ has been
measured for the first time. The value found, 4.3×10^{-11}
cm^3s^{-1}, improves the fit of the calculated OH profile with
Anderson's recent measurements. Work on the $HO + HO_2$
reaction is in progress.

Dr. R.B. Timmons — Catholic University — 76-129. Photochemical and Chemical
Kinetics Measurements of Stratospheric Importance with Respect to the Fluorocarbon Issue.

It is planned to measure accurately the UV absorption cross-
section of $HOCl$, study its photochemistry, and perform $ClONO_2$
photochemical studies to complement Wiesenfeld's proposal (q.v.,
below). $HOCl$ is a possible stratospheric sink the magnitude
of which would depend on the absorption cross-section.
Earlier spectral measurements were inaccurate.

TABLE 10.4 (cont'd.)

Dr. J. Wiesenfeld — Cornell — 76-128. Photochemistry of Small Chlorinated Molecules.

> The photochemistry of chlorine nitrate will be studied. While not yet thoroughly treated in the models, chlorine nitrate may be an important stratospheric sink. A flash photolysis-resonance absorption system is used for photolysis and analysis of products and incorporates either a conventional flash lamp with filters or a tunable laser. The wavelength and temperature dependencies of the quantum yield and product formation will be determined.

B. SOURCE AND SINK STUDIES

Dr. J.W. Birks — University of Illinois — 76-117B. Studies of Heterogeneous Reactions.

> Potentially important heterogeneous reactions, which will be chosen with the advice of ERT, are to be studied, including the heterogeneous reaction of $ClONO_2$ and H_2O. Equipment is being assembled for measurements in early 1977.

Dr. M.J. Campbell — Washington State University — 75-53. Chlorofluoromethane Destruction by Natural Ionization.

> Laboratory measurements under conditions approximating the real atmosphere in all respects except ion age show large rate constants for removal of CCl_4 and fluorocarbon 11, apparently by atmospheric ions. The rate constant for fluorocarbon 12 is much smaller. If the same rate constants were to hold for older ions, this mechanism might constitute a sink that could decrease the stratospheric concentration of fluorocarbon 11 by a factor of two.

Dr. M. Kaufman — Emory University — 76-126. Studies of Compounds of Sulfur, Oxygen, and Chlorine.

> Reactions between SO_x and Cl_x species of possible stratospheric importance are being investigated. They may yield compounds with long lifetimes or lead to incorporation of Cl into aerosols. Currently, the rate constant of the reaction of Cl with SO_2 is being measured.

Dr. J.E. Lovelock — University of Reading — 75-67. Unidentified Factors in the Fluorocarbon-Ozone Problem.

> The absorption and reactivity characteristics of the atmospheric dust plume from the Sahara for atmospheric halocarbon vapors (fluorocarbon 11 and CCl_4) are being evaluated. Investigations are in progress on the relationship between photochemically produced atmospheric peroxy compounds (e.g. peroxyacetyl nitrate) and the incidence of skin carcinoma).

TABLE 10.4 (cont'd.)

Dr. L.R. Martin — Aerospace Corp. — 75-81. Laboratory Investigation of the
Heterogeneous Interaction of Cl and ClO with H_2SO_4.

> A flowing afterglow apparatus is being used to measure the
> rate of the heterogeneous reactions of Cl and ClO with
> sulfuric acid, simulating the stratospheric aerosol. Pre-
> liminary data suggest that the reaction rate of Cl is too
> slow for its reaction to constitute a significant sink,
> although rates were markedly increased by the presence of
> certain metal salts in the sulfuric acid. Measurements of
> the ClO rates, to which the ozone depletion cycle is more
> sensitive, are planned. The nature of the surface inter-
> action is also to be characterized.

Dr. V.A. Mohnen — SUNY, Albany — 75-64. Ion Molecule Reactions Involving
Fluorocarbons.

> A drift tube in which ions are formed and their mobility
> determined and an electric quadrupole mass spectrometer were
> used to study the role of ion-molecules in the reactions
> between fluorocarbon 12 and positive and negative ion
> clusters that form in air-like gas mixtures. There was no
> clustering of fluorocarbon 12 and no evidence for charge
> transfer or cluster ion reactions with positive ions of the
> form $H^+ \cdot (H_2O)_n (NH_3)m$. Reaction with negative ions of the
> form $O^- \cdot X$ and $CO_4^- \cdot X$ results in the formation of the ion-
> cluster series $Cl^- \cdot (H_2O)_x$.

Dr. J.N. Pitts — University of California at Riverside — 75-12. Monitoring and
Atmospheric Reactions of Fluorocarbons.

> Long-term photolysis studies in simulated sunlight revealed
> that fluorocarbons 11 and 12 are photochemically stable, even
> when irradiated for several weeks. An attempt to study
> fluorocarbon-NO_x-air mixtures under simulated stratospheric
> conditions could not be completed because the background
> level of hydrocarbons in the environmental chamber could not
> be made sufficiently small.

> Plant tissues did not absorb measurable quantities of fluoro-
> carbons 11, 12, or 22, and no adverse effects could be
> measured. Fluorocarbons penetrate into the soil atmosphere,
> and concentrations change in direct relationship with changes
> in concentration in the atmosphere above ground.

Dr. R.A. Rasmussen — Washington State University — 75-71. Measurement of
Fluorocarbon Content of "Antique" Air Samples.

> A sensitive method for the determination of low parts per
> trillion analysis of fluorocarbons 11 and 12 in small-volume
> air samples in containers was developed and applied to a
> wide variety of vessels believed to contain antique air.
> Both of these fluorocarbons were found in all samples
> analyzed, suggesting either the presence of natural fluoro-
> carbons before 1930 or contamination during storage or
> handling.

TABLE 10.4   (cont'd.)

Dr. R.A. Rasmussen — Washington State University — 75-84. Collection and
Analysis of Antarctic Ice. Cores.

> The concentration of halocarbons in air obtained from
> Antarctic snow shows no enrichment in samples obtained from
> the Ross Ice shelf (mainly -30°F), whereas there is enrich-
> ment in samples obtained from the South Pole (-50 to -60°F).

Dr. R.A. Rasmussen — Private — 76-140. Lower Stratospheric Measurement of
Non-methane Hydrocarbons.

> This study is now beginning. It is important because non-
> methane hydrocarbons, if present in the stratosphere, will
> act as a sink for Cl in the same way that methane acts and
> thus reduce the ozone depletion attributable to chlorine-
> containing species.

Dr. C. Sandorfy — University of Montreal — 73-2. Spectroscopy and Photochemical
Changes of Fluorocarbons.

> The vacuum ultraviolet and photoelectron spectra of all
> industrially important chlorofluorocarbons were measured
> and published. The photochemical vulnerability of these
> molecules was predicted from their spectra.

LABORATORY STUDIES RELATED TO POTENTIAL ATMOSPHERIC MEASUREMENTS

Dr. H.J. Buijs — Bomem, Inc. — 75-90. Contruction of a Fourier Transform
Spectrometer.

> A spectrometer with a projected resolution of 0.01 cm^{-1} is
> being completed for use in the simultaneous determination of
> $ClONO_2$ and either HCl or HF, or of HCl and HF.

Dr. D.D. Davis — University of Maryland — 74-10. Laboratory Determination of
the Sensitivity of Laser-Induced Fluorescence for the Detection of ClO Under
Atmospheric Conditions.

> Ground-state stationary ClO concentrations of about $10^{12} cm^{-3}$
> were scanned at several electronic absorption wavelength
> regions with a tunable UV laser. Laser-induced fluorescence
> proved to be unusable for measuring ClO because (1) at 300°K
> the highest populated K level, 12, contains only 5% of the
> total ClO and (2) the majority of the ClO undergoes pre-
> dissociation to Cl and O.

Dr. D.D. Davis — University of Maryland/Georgia Institute of Technology — 75-73.
Laboratory Measurement of Spectroscopic Absorption Cross-Sections of ClO.

> Work is being completed on the measurement of the absorption
> cross-sections for ClO, data required for determining concen-
> trations from UV absorption measurements of this molecule.

TABLE 10.4   (cont'd.)

Dr. D.D. Davis — University of Maryland/Georgia Institute of Technology — 75-87.
Development of Instrument for Stratospheric OH Measurement by Laser-Induced
Fluorescence.

> This program is to measure stratospheric OH concentrations
> over extended periods of time, including some covering
> diurnal changes. A miniaturized dye laser module for balloon
> flights has been built and tested. An engineering flight is
> planned after a high-altitude, pressurized chamber for housing
> the dye laser and Nd-Yag driver has been built and required
> data storage and control facilities are ready.

Dr. A.E.J. Eggleton — AERE, Harwell — 76-116. Total Chlorine Measurements in
the Troposphere and Stratosphere.

> The feasibility of measuring total chlorine and fluorine in
> the atmosphere by neutron activation and γ photon activation,
> respectively, after collection of reactive species and
> particulate material on a filter and collection of gaseous
> compounds on activated charcoal is being studied.

Dr. C.J. Howard — NOAA, Boulder — 75-47. Laboratory Determination of the
Feasibility of Laser Magnetic Resonance for ClO Detection and Reaction Studies.

> This study is to detect ClO using laser magnetic resonance
> and to utilize this method for measuring rate constants for
> stratospherically significant ClO reactions. It has been
> demonstrated that ClO can be detected by this method, and
> the sensitivity is under evaluation.

Dr. J.E. Lovelock — Private — 76-120. The Electron Capture Detector as a
Reference Standard for Analysis of Atmospheric Halocarbons.

> The electron capture gas chromatograph is being investigated
> as a reference standard for analysis of atmospheric halocarbons.

Dr. D.G. Murcray — University of Denver — 75-92. Laboratory Measurement of
High Resolution Infrared Spectra of Chlorine-Containing Molecules of Stratospheric Interest.

> Laboratory measurements of the high resolution infrared
> spectra of chlorine-containing molecules are to be made.
> Statistical-band-model analyses and integrated intensity
> measurements for the 10.8 μm band of fluorocarbon 12 and
> 11.8 μm band of fluorocarbon 11 have been published. Addi-
> tional spectra for methane, methyl chloride, and carbonyl
> fluoride have been run, and work is in progress on chlorine
> nitrate.

Dr. R.W. Nicholls — York University — 75-11 and 75-11-II. Experimental and
Theoretical Studies on the UV Spectrum of ClO With Stratospheric Applications.

> Absolute absorption coefficients and cross-sections have been
> measured for all bands and the photodissociation continuum
> of the $V'' = 0$ progression for ClO. The very complicated
> emission spectrum that has been excited over the wavelength
> range 2500-4500 Å in discharges through ClO_2 and Cl_2O is
> currently undergoing measurement, identification, and

TABLE 10.4 (cont'd.)

analysis. Computer-based synthetic spectra of various ClO
bands have been calculated. Emphasis is now being placed
on the high resolution spectrum of ClO in the spectral
region above 3000 Å, which should lay the groundwork for
the detection and measurement of stratospheric ClO from the
ground and from balloon-borne equipment. Possible interferences by O_3 are being evaluated.

Dr. R.W. Nicholls — York University — 75-30b. Laboratory Studies of the Infrared Vibration-Rotation Spectrum of ClO.

Work in this area was suspended to allow greater effort in
the UV measurements (75-11).

Dr. R.A. Rasmussen — Private — 76-142. Interlaboratory Comparison of Fluorocarbon Measurements.

A second round of identical samples of rural air will be
circulated blind to participating laboratories for analysis
for fluorocarbons 11 and 12, chloroform, methyl chloroform,
carbon tetrachloride, and N_2O. The results will be analyzed
statistically.

Dr. D.H. Stedman — University of Michigan — 74-7. Atmospheric Determination of
ClO Concentration: A Feasibility Study.

Laboratory studies have demonstrated the feasibility for
detecting stratospheric ClO by chemical conversion to Cl (by
reaction with NO) accompanied by vacuum ultraviolet resonance
fluorescence. In-flight use of this technique is being
supported by NASA.

Dr. D.H. Stedman — University of Michigan — 76-132. Absolute Calibration of
Fluorocarbon Measurements.

The final design of a feed-back flow system for the calibration
of fluorocarbon samples is being built.

D. TROPOSPHERIC AND STRATOSPHERIC MEASUREMENTS

Dr. H.L. Buijs — Bomem, Inc. — 75-98. Measurement of HCl and HF in the Stratosphere by Fourier Transform Spectroscopy.

Data have been successfully collected from balloom flights
in Alaska (May 1976) and New Mexico (September 1976) and are
being analyzed for HCl and HF concentrations.

Dr. P.A. Ekstrom — Battelle Memorial Institute, Pacific Northwest Laboratories
— 75-27. Ground-Based Millimeter-Wavelength Observations of Stratospheric ClO.

About 500,000 data points were obtained in the microwave
spectra near 93 GHz with the Kitt Peak radiotelescope.
Excessive noise made interpretation difficult, but base-line
corrected spectra suggested an upper limit on ClO of one
hundred times model predictions.

TABLE 10.4   (cont'd.)

Dr. A. Girard — ONERA-France — 75-88.   Measurement of HCl, HF, ClO, etc., in
the Stratosphere by High Resolution Infrared Spectroscopy.

> Balloon-borne slights and supporting laboratory work are
> planned to measure fluorocarbons, HCl, HF, ClO, ClNO$_3$, and
> other compounds by high resolution infrared spectroscopy.
> This program is part of a cooperative program including
> aircraft flights with the Centre National d'Etudes Spatiales.
> A flight in 1976 provided an HCl concentration profile
> between 25 and 30 km. Equipment failure occurred in a
> September 1976 flight. A flight from Brazil in December 1976
> and additional flights in the spring of 1977 are planned.

Dr. J.E. Lovelock — University of Reading — 73-1, 74-3, 75-67.   Fluorocarbons
in the Environment.

> The electron-capture gas chromatograph (ECGC) has been
> developed and applied to the measurement of chlorocarbons
> and chlorofluorocarbons in the troposphere and lower strato-
> sphere, particularly over Europe and the Atlantic Ocean.
> Current fluorocarbon 11 levels are about 130 ppt (U.K.) and
> 80 ppt (Southern Hemisphere). ECGC has been adapted to
> measure CH$_3$Cl at atmospheric concentrations of about 10^{-9} v/v,
> with sources for CH$_3$Cl identified as the ocean and smouldering
> vegetation.

Dr. D.G. Murcray — University of Denver — 75-13, 76-101, 76-135.   The Measurement of the Stratospheric Distribution of Fluorocarbons and Other Constituents
of Interest in the Possible Effect of Chlorine Pollutants on the Ozone Layer.

> Measurement of stratospheric distribution by balloon-borne,
> high-resolution, infrared absorption measurement at large
> solar zenith angles has been achieved using a specially-
> constructed grating spectrometer. The distributions for
> fluorocarbon 11, fluorocarbon 12, and CCl$_4$ obtained show a
> concentration increase of about 2.5, with a rather wide
> range of uncertainty, from 1968 to 1975 for fluorocarbon 11
> and fluorocarbon 12. Subsequent flights also yielded HCl
> and HNO$_3$ profiles. Profiles or upper limit measurements
> on several other species, including ClONO$_2$, are potentially
> accessible.

> Balloon-borne experiments will be continued to measure
> stratospheric ClO (300-500 nm) and OH (306.4 nm) by solar
> UV absorption.

Dr. R.A. Rasmussen — Washington State University — 75-2, 75-59.   Fluorocarbon
Research.

> An attempt was made to obtain halocarbon concentration
> measurements as far into the stratosphere as could be reached
> by an available commercial aircraft. A small portable gas
> chromatograph was used for on-board measurements, and cannister
> samples were collected for subsequent detailed halocarbon
> analysis on the ground. One phase of the study consisted of
> samples collected over a wide area of the Pacific Northwest,
> a second of samples collected frequently to the maximum
> attainable altitude over Alaska. The halocarbon concentrations

TABLE 10.4   (cont'd.)

are either constant or decrease very slowly with altitude
in the troposphere, decrease rapidly in the tropopause from
the tropospheric concentration to an average value identi-
fiable with the stratosphere, and do not show a clear pattern
of concentration gradients above the tropopause.

A trans-Pacific flight from 80°N to 60°S has been completed,
and the air samples collected have been analyzed for halocarbons. Most of the samples were collected at 39,000 to
43,000 ft. Fluorocarbon 12 concentrations are about 10%
higher in the north than in the south at ground level, and
the difference is apparently greater for fluorocarbon 11.

Dr. B.A. Ridley — York University — 76-102A, 76-102B. Measurement of Fluorocarbons and Related Chlorocarbons in the Stratosphere by Collection and
Analysis.

Balloon-borne grab sample devices have been developed and
flown to obtain fluorocarbon 11, fluorocarbon 12, and N_2O
concentrations at 10-40 km altitude. The data show a relatively consistent drop-off with height above the tropopause,
whereas their consistency in drop-off is much worse as a
function of height above the surface. The spread in data
is larger than the accuracy of the measurements, which
indicates variability in stratospheric concentrations due
to horizontal motions.

Dr. P.M. Solomon — SUNY, Stony Book — 76.130. Millimeter Wave Observations of
Chlorofluoromethane Byproducts in the Stratosphere.

This study, which is directed toward the development of a
gound-based method for continuous determination of ClO, is
just getting under way. One of the millimeter wave
observing systems will be based on a 3 mm maser, a unique
instrument that is the most sensitive detector in the world
in the 83-94 GHz range.

Dr. O.C.Taylor — University of California at Riverside — 73-3, 74-2. Monitoring
and Atmospheric Reactions of Fluorocarbons.

An electron capture gas chromatograph was used to measure
the concentrations of fluorocarbons 11 and 12 in the troposphere over southern California and in the lower stratosphere
over New Mexico and Colorado. The tropospheric concentrations
were found to vary from day to day as climatic conditions
affected dispersion and dilution. Concentration decreased
with increasing altitude in the lower stratosphere.

Dr. R.A.Young — Xonics, Inc. — 75-50, 75-86. Development of an Instrument to
Measure O, ClO, O_3, and Total Cl in the Stratosphere.

The preliminary experiment to measure stratospheric ClO and
total chlorine by resonance fluorescence flown on the September 1975 STRATCOM balloon was not completely successful;
however, several key components appeared to have functioned
satisfactorily. Follow-up work has been undertaken in a
stratospheric simulator to provide design information for
further development of stratospheric instrumentation.

TABLE 10.4   (cont'd.)

E. MODELING

Drs. D.M. Cunnold, F.N. Alyea and R.G. Prinn — CAP Associates — 75-24, 76-122, 76-122S. Meteorological and Multi-Dimensional Modeling Considerations Relating to Atmospheric Effects of Halocarbons.

> Studies to assess the accuracy and shortcomings of the 1-D model used to estimate ozone depletion indicated that a tropospheric lifetime for fluorocarbons 11 and 12 as short as 10 years was not inconsistent with atmospheric measurements. Thus, ozone depletion estimates might be considerably less than present estimates. The great variability of stratospheric measurements indicates that simultaneous measurements for many important stratospheric species are needed and that seasonal dependence of species must be considered.

> The absence of circulation in the radiative models used to calculate warming (greenhouse effect) limits the value of calculated effects.

> Because it has been shown that meteorological conditions can cause variations in both tropospheric and lower stratospheric fluorocarbon measurements in Alaska, meteorological analysis will be performed for selected fluorocarbon measurements, and temporal and latitudinal variability of fluorocarbon concentration will be determined using 2-D and 3-D models.

Dr. N.D. Sze — Environmental Research and Technology, Inc. — 75-32. 76-115. Model Analysis of the Fluorocarbon Problem.

> A one-dimensional model has been used to evaluate the role of stratospheric water in the NO_x and ClX cycles, the relationship of eddy diffusion coefficient and fluorocarbon lifetime, the use of fluorocarbon measuremets to calculate lifetime, and the effect of chlorine nitrate. The importance of OH concentration on calculated ozone depletion was shown, and key reactions were identified. Analyses showed 10-20 year tropospheric lifetimes were not inconsistent with measurements and helped to define quantitatively the undertainties associated with ozone depletion calculations. Current work is to continue a sensitivity analysis of the 1-D model, develop a diurnal 1-D model, and include multiple scattering and heterogeneous (aerosol) reactions in the model.

F. OTHER

Dr. D. Berger — Temple University — 75-62. Gound-Level Monitoring of Ultraviolet Solar Radiation.

> The monitoring of solar ultraviolet radiation, initiated by the Climatic Impact Assessment Program (CIAP), is being continued. The number of instruments and stations is being increased. Calibration of operating instruments is maintained, and comparisons between different types of monitoring equipment are being made.

TABLE 10.4 (cont'd.)

Drs. E. Parzen and M. Pagano — Frontier Science and Technology Research
Foundation, Inc. — 76.106. Total World Ozone Level: Statistical Analysis.

> Ozone column measurements from at least 20 stations are
> being evaluated statistically to detect trends in recorded
> ozone concentrations and to establish the limits of detection
> for such trends. Time series analysis has been shown to be
> substantially more sensitive in detecting non-random ozone
> changes than the estimates of such sensitivity made by the
> Federal Task Force on Inadvertent Modification of the Strato-
> sphere (IMOS) in 1975. The absence of detectable trends
> provides an upper limit for actial depletion and a test of
> model predictions.
>
> If sufficient sensitivity is achieved, this technique will
> enable an effective early warning system for ozone depletion
> to be established.

G. CONSULTANTS

Dr. J.G. Anderson	University of Michigan	Stratospheric Measurements
Dr. A.W. Castleman, Jr.	University of Colorado	Heterogeneous Chemistry
Dr. F.C. Fehsenfeld Dr. R.R. Ferguson }	NOAA Environmental Research Laboratories	Reactions of Charged Species
Dr. D.R. Herschbach Dr. W. Klemperer }	Harvard University	Homogeneous Chemistry, Kinetics, & Stectroscopy
Dr. L.E. Snyder	University of Illinois	Millimeter Wavelength Spectroscopy

TABLE 10.5(a)   Research Funded by the Fluorocarbon Industry
                and Administered by the Manufacturing Chemists
                Association — Work Completed

Program

Measurement of fluorocarbons in the atmosphere*

Monitoring of fluorocarbons in the atmosphere and simulation of atmospheric
reactions of fluorocarbons

Investigation of spectroscopy of and photochemical changes in fluorocarbons

Continuation of 73-1*

Laboratory determination of sensitivity of laser-induced fluorescence for the
detection of ClO under atmospheric conditions

Laboratory investigation of the feasibility of measuring ClO in the atmosphere
by the chemical conversion-resonance fluorescence detection method

Continuation of 73-3

Measurement of fluorocarbons and related chlorocarbons in the stratosphere and
upper troposphere*

Laboratory and theoretical studies of the ultraviolet and visible electronic
spectra of ClO*

Continuation of 74-2

Ground-based millimeter wavelength observations of stratospheric ClO

Investigation of the destruction of chlorofluoromethanes by naturally occurring
ions

Development of an instrument to measure O, ClO, O_3, and total Cl in the
stratosphere

Investigation of ion-molecule reactions involving chlorofluorocarbons

Modeling of the fluorocarbon-ozone system*

Critique of models used to estimate chlorofluorocarbon effects on ozone*

Studies of reactions of HO_2 by laser magnetic resonance*

Continuation of 75-50

---

*Work is continuing in a follow-on contract.

Investigator	Organization	Proposal Number	Completion Date
Lovelock	U. of Reading	73-1	10/27/74
Taylor	U. of Calif., Riverside	73-3	10/16/74
Sandorfy	U. of Montreal	73-2	10/11/74
Lovelock	U. of Reading	74-3	12/31/75
Davis	U. of Maryland	74-10	5/31/75
Stedman	U. of Michigan	74-7	2/28/75
Pitts	U. of Calif., Riverside	74-2	12/31/75
Rasmussen	Washington State U.	75-2	4/15/76
Nicholls	York U.	75-11	6/14/76
Pitts	U. of Calif., Riverside	75-12	4/15/76
Ekstrom	Battelle Northwest	75-27	5/24/76
Campbell	Washington State U.	75-53	4/23/76
Young	Xonics, Inc.	75-50	4/7/76
Mohnen	State U. of N.Y., Albany	75-64	4/1/76
Sze	ERT, Inc.	75-32	8/10/76
Cunnold, Alyea, Prinn	CAP Associates	75-24	9/10/76
Thrush	U. of Cambridge	75-58	11/8/76
Young	Xonics, Inc.	75-86	11/15/76

TABLE 10.5(b)   Research Funded by the Fluorocarbon Industry
                and Administered by the Manufacturing Chemists
                Association — Work in Progress

Program

Measurement of reaction rates relevant to the fluorocarbon–ozone problem

Measurement of stratospheric distribution of fluorocarbons and related species by
infrared absorption spectroscopy

Laboratory studies of the infrared vibration-rotation spectrum of ClO

Laboratory determination of the feasibility of laser magnetic resonance for ClO
detection and reaction studies

Collection and analysis of Antarctic ice cores

Measurement of fluorocarbon content of "antique" air samples

Exploration for unidentified factors in the fluorocarbon–ozone problem

Laboratory measurement of spectroscopic absorption cross sections of ClO

Continuation of program for ground level monitoring of ultraviolet solar radiation

Continuation of 75-2

Measurement of OH in the stratosphere by laser induced fluorescence

Laboratory measurement of high resolution infrared spectra of chlorine-containing
molecules of stratospheric interest

Laboratory investigation of the heterogeneous interaction of Cl and ClO with H_2SO_4

Measurement of HCl, HF, ClO, etc., in the stratosphere by high resolution infrared
spectroscopy

Measurement of fluorocarbons and related chlorocarbons in the stratosphere by
collection and analysis

Continuation of 75-13

Construction of Fourier-transform spectrometer

Measurement of HCl and HF in the stratosphere by Fourier transform spectroscopy

Continuation of 75-32

Total world ozone level: statistical analysis

The Electron Capture Detector as a Reference Standard in the Analysis of Atmospheric Halocarbons

Investigator	Organization	Proposal Number	Contract Date	Contract Period (months)
Birks	U. of Illinois	75-1	4/17/75	15
Murcray	U. of Denver	75-13	6/16/75	12
Nicholls	York U.	75-30b	8/12/75	12[a]
Howard	NOAA-Boulder	75-47	11/13/75	12[a]
Rasmussen	Rasmussen Associates	75-84	11/29/75	3[a]
Rasmussen	Washington State U.	75-71	12/3/75	5[a]
Lovelock	Private	75-67	12/4/75	12[a]
Davis	U. of Maryland	75-73	2/5/76	8[a]
Berger	Temple U.	75-62	2/5/76	12
Rasmussen	Washington State U.	75-79	2/13/76	9[a]
Davis	U. of Maryland	75-87	2/24/76	12
Murcray	U. of Denver	75-92	3/4/76	12
Martin	Aerospace Corporation	75-81	3/22/76	12
Girard	ONERA-France	75-88	3/29/76	12
Ridley	York U.	76-102	3/31/76	12[b]
Murcray	U. of Denver	86-101	4/1/76	12[b]
Buijs	Bomem, Inc.	75-90	4/1/76	12[b]
Buiks	Bomem, Inc.	75-98	4/9/76	5[a]
Sze	ERT, Inc.	76-115	6/18/76	11
Parzen, Pagano	Frontier Science and Technology Research Foundation, Inc.	76-106	7/15/76	12
Lovelock	Private	76-120	8/16/76	12

(cont'd)

TABLE 10.5(b) (cont'd.)

Program
Continuation of 75-11
Continuation of 75-1
Studies of Heterogeneous Reactions
Studies of Compounds of Sulfur, Oxygen, and Chlorine
Meteorological and Multi-Dimensional Modeling Considerations Relating to Atmospheric Effects of Halocarbons
Absolute Calibration of Fluorocarbon Measurements
Photochemical and chemical kinetic measurements of stratospheric importance with respect to the fluorocarbon issue
Total Chlorine Measurements in the Troposphere and Stratosphere
Laser Magnetic Resonance Study of HO_2 Chemistry
Photochemistry of small chlorinated molecules
Stratospheric measurement of ClO and OH
Climatic effects of fluorocarbons
Reactions of the HO_2 radical studied by laser magnetic resonance
Millimeter wave observations of chlorofluoromethane byproducts in the stratosphere
Lower stratospheric measurement of non-methane hydrocarbons
Interlaboratory comparisons of fluorocarbon measurements

[a]Contract extended   [b]Estimate

Investigator	Organization	Proposal Number	Contract Date	Contract Period (months)
Nicholls	York U.	75-11-II	8/18/76	12
Birks	U. of Illinois	76-117A	8/26/76	12
Birks	U. of Illinois	76-117B	9/29/76	12
Kaufman	Emory U.	76-126	10/7/76	12
Cunnold, Alyea, Prinn	CAP Associates	76-122	10/8/76	12
Stedman	U. of Michigan	76-132	10/21/76	2[a]
Timmons	Catholic U.	76-129	10/29/76	12
Eggleton	AERE Harwell	76-116	11/2/76	4
Howard	NOAA-Boulder	76-100	11/6/76	6
Wiesenfeld	Cornell	76-128	11/8/76	12
Murcray	U. of Denver	76-135	12/9/76	4
Cunnold, Alyea, Prinn	CAP Associates	76-122S	12/10/76	12
Thrush	U. of Cambridge	75-58 II	Pending	12
Solomon	SUNY Stony Brook	76-120	Pending	12
Rasmussen	Private	76-140	Pending	4
Rasmussen	Private	76-142	Pending	3

TABLE 10.6   Publications Resulting from Work Supported by
Fluorocarbon Manufacturers

Alexander Grant & Company

1. Environmental Analysis of Fluorocarbons FC-11, FC112, and FC-22, February 5, 1976.

Allied Chemical Corporation

1. Statistical Modeling of Total Ozone Measurements with an Example Using Data from Arosa, Switzerland. W.J. Hill and P.N. Sheldon, *Geophys. Res. Lett.*, 21 (12), 541-4 (1975).

2. Analyzing Worlwide Total Ozone for Trends. W.J. Hill, P.N. Sheldon and J.J. Tiede, *Geophys. Res. Lett.*, accepted.

J.W. Birks, University of Illinois

1. Four-Center Reactions Involving Dichlorine Monoxide. J.W.B., B. Shoemaker, T.J. Lock and D.M. Hinton, draft ms.

2. Studies of Reactions of Importance in the Stratosphere. I. Reaction of Nitric Oxide with Ozone. J.W.B., B. Shoemaker, T.J. Lock and D.M. Hinton, draft ms.

M. J. Campbell, Washington State University

1. Halocarbon Decomposition by Natural Ionization. M.I.C., *Geophys. Res. Lett.*, 3 (11), 661-4 (1976).

D.M. Cunnold, F.N. Alyea and R.G. Prinn, Massachusetts Institute of Technology

1. The Impact of Stratospheric Variability on Measurement Programs for Minor Constituents. R.G.P., F.N.A. and D.M.C., *Bull. Am. Meteorol. Soc.*, 57 (6), 686-94 (1976).

F.I. du Pont de Nemours & Company, Inc.

1. Atmospheric Stability of Fluoroalkanes — Implications for Ozone Depletion. R.L. McCarthy and J.P. Jesson, Symposium on Fluorine Chemistry, Kyoto, Japan, August 26, 1976.

2. Measurement of the Reaction Rate of CF_3Cl with Atmosphere-Like Ions. B.G.Hirsch, *Atmos. Environ.*, 10 (9), 703-5 (1976).

3. Laboratory Microwave Spectrum of $ClONO_2$ and Evidence for the Existence of ClONO. R.D. Suenram, D.R. Johnson, L.C. Glasgow and P.Z. Meakin, *Geophys. Res. Lett.*, 3 (10), 611-14 (1976), 3 (12), 758 (1976).

4. The Fluorocarbon-Ozone Theory. I. Production and Release, World Production and Release of CCl_3F and CCl_2F_2 (Fluorocarbons 11 and 12) through 1975. R.L.

TABLE 10.6   (cont'd.)

McCarthy, F.A. Bower and J.F. Jesson, draft ms.

5.   The Fluorocarbon-Ozone Theory. II. Tropospheric Lifetimes, An Experimental
     Estimate of the Tropospheric Lifetime of CCl_3F. J.P. Jesson, P. Meakin and
     L.C. Glasgow, draft ms.

J.E. Lovelock, University of Reading

1.   Atmospheric Halocarbons and Stratospheric Ozone. J.E.L., *Nature*, 252, 292-4
     (1974) (11/22/74).

2.   Long-range Transport of Photochemical Ozone in Northwestern Europe. R.A. Cox,
     A.E.J. Eggleton, R.C. Derwent, J.E.L. and D.H.Pack, *Nature*, 255, 118-21 (1975
     (5/8/75).

3.   Natural Halocarbons in the Air and Sea. J.E.L., *Nature*, 256, 193-4 (1975)
     (7/17/75).

4.   Photochemical Oxidation of Halocarbons in the Troposphere. R.A. Cox, R.F.
     Derwent, A.E.J. Eggleton and J.E.L., *Atmos. Environ.*, 10 (4), 305-8 (1976).

5.   Halocarbon Behavior from a Long Time Series. D.M. Pack, J.E.L., C. Cotton and
     C. Curthoys, draft ms.

6.   The Electron Capture Detector Theory and Practice. J.E.L., *J. Chromatogr.*,
     99, 3-12 (1974).

D.G. Murcray, University of Denver

1.   Simultaneous Stratospheric Measurements of Fluorocarbons and Odd Nitrogen
     Compounds. W.J. Williams, J.J. Kosters, A. Goldman and D.G.M., draft ms.

2.   Statistical-Band-Model Analysis and Integrated Intensity for the 10.8 μm Band
     of CF_2Cl_2. A. Goldman, F.S. Bonomo and D.G.M., *Geophys. Res. Lett.*, 3 (6),
     309-12 (1976).

3.   Measurements of Stratospheric Fluorocarbon Distributions Using Infrared
     Techniques. W.J. Williams, J.J. Kosters, A. Goldman and D.G.M., *Geophys. Res.
     Lett.*, 3 (7), 379-82 (1976).

4.   Measurement of Stratospheric Mixing Ratio Altitude Profile of HCl Using Infra-
     red Absorption Techniques. W.J. Williams, J.J. Kosters, A. Goldman and D.G.M.,
     Geophys. Res. Lett., 3 (7), 383-5 (1976).

5.   Statistical Band Model Analysis and Integrated Intensity for the 11.8 μm Band
     of $CFCl_3$. A. Goldman, F.S. Bonomo and D.G.M., *Appl. Opt.*, 15 (10), 2305-7
     (1976).

R.W. Nicholls, York University

1.   The Absorption Cross-Sections and f-Values for the v" = 0 Progression of Bands
     and Associated Continuum for the ClO ($A^2\Pi_i \leftarrow X^2\Pi_i$) System. M. Mandelman and

TABLE 10.6 (cont'd.)

R.W.N., *J. Quant. Spectros. Radiat. Transfer*, accepted.

J.N. Pitts, Jr. and O.C. Taylor, University of California at Riverside

1. Fluorocarbons in the Los Angeles Basin. N.E. Hester, E.R. Stephens and O.C.T., *J. Air Pollut. Control Assoc.*, 24 (6), 391-5 (1974).

2. Relative Rate Constants for the Reaction of $O(^1D)$ Atoms with Fluorocarbons and N_2O. J.N.P., H.L. Sandoval and R. Atkinson, *Chem. Phys. Lett.*, 29 (1), 31-4 (1974) (11/1/74).

3. Reactions of Electronically Excited $O(^1D)$ Atoms with Fluorocarbons. H.L. Sandoval, R. Atkinson and J.N.P., *J. Photochem.*, 3 (4), 325-7 (1974).

4. Tropospheric and Stratospheric Chemical Sinks for Commercial Fluorocarbons. J.N.P. and R. Atkinson, *Trans. Amer. Geophys. Union*, 55 (12), 1153 (1974).

5. Mechanisms of Photochemical Air Pollution. J.N.P. and B.J. Finlayson, *Angew. Chem., Int. Ed. Engl.*, 14 (1), 1-15 (1975).

6. Fluorocarbon Air Pollutants, II. N.E. Hester, E.R. Stephens amd O.C.T., *Atmos. Environ.*, 9 (6-7), 603-6 (1975).

7. Fluorocarbon Air Pollutants, Measurements in Lower Stratosphere. N.E. Hester, E.R. Stephens and O.C.T., *Environ. Sci. Technol.*, 9 (9), 875-6 (1975).

8. The Photostability of Fluorocarbons. S.Japar, J.N.P. and A.M. Winer, draft ms.

9. Background and Vertical Atmospheric Measurements of Fluorocarbon-11 and Fluorocarbon-12 over Southern California. L. Zafonte, N.E. Hester, E.R. Stephens and O.C.T., *Atmos. Environ.*, 9, 1007 (1975).

10. Rate Constants for the Reaction of OH Radicals with CHF_2Cl, CF_2Cl, $CFCl_3$, and H_2 Over the Temperature Range 297-434K. R. Atkinson, D.A. Hansen and J.N.P., *J. Chem. Phys.*, 63 (5), 1703-6 (1975) (9/1/75).

11. Tropospheric and Stratospheric Sinks for Halocarbons: Photooxidation, $O(^1D)$ Atom and OH Radical Reactions. R. Atkinson, G.M. Brewer, J.N.P. and H.L. Sandoval, *J. Geophys. Res.*, accepted.

12. Fluorocarbon Air Pollutants. III. Fluorocarbon Measurements from the Lower Stratosphere. N.E. Hester, E.R. Stephens and O.C.T., draft ms.

R.A. Rasmussen, Washington State University

1. Halocarbon Measurements in the Alaskan Troposphere and Lower Stratosphere. E. Robinson, R.A.R., J. Krasnec, D.Pierotti and M. Jakubovic, draft ms.

2. Detailed Halocarbon Measurements Across the Alaskan Troposphere. E. Robinson, R.A.R., J. Krasnec, D. Pierotti and M. Jakubovic, *Geophys. Res. Lett.*, 3 (6), 323-6 (1976).

TABLE 10.6 (cont'd.)

C. Sandorfy, University of Montreal

1. Vacuum Ultraviolet a: Photoelectron Spectra of Fluorochloro Derivatives of Methane. J. Doucet, P. Sauvageau and C.S., *J. Chem. Phys.*, 58 (9), 1708-16 (1973) (5/1/73).

2. Vacuum Ultraviolet Absorption Spectra of Fluoromethanes. P. Sauvageau, R. Gilbert, P.P. Berlow and C.S., *J. Chem. Phys.*, 59 (2), 762-5 (1973) (7/15/73).

3. Vacuum Ultraviolet Absorption Spectra of Chlorofluoromethanes From 120 to 65 nm. R. Gilbert, P. Sauvageau and C.S., *J. Chem. Phys.*, 60 (12), 4820-4 (1974) (6/15/74).

4. Vacuum Ultraviolet and Photoelectrom Spectra of Fluoroethanes. P. Savageau, J. Doucet, R. Gilbert and C.S., *J. Chem. Phys.*, 61 (1), 391-5 (1974) (7/1/74).

5. On the Hydrogen Bond Breaking Ability of Fluorocarbons Containing Higher Halogens. T. DiPaolo and C.S., *Can. J. Chem.*, 52 (21), 3612-22 (1974).

6. Fluorocarbon Anaesthetics Break Hydrogen Bonds. T. DiPaolo and C.S., *Nature*, 252, 471 (1974) (12/6/74).

7. Photoelectron and Far-Ultraviolet Absorption Spectra of Chlorofluoro-Derivatives of Ethane. J. Doucet, P. Sauvageau and C.S., *J. Chem. Phys.*, 62 (2), 355-9 (1975) (1/15/75).

8. Photoelectron and Far-Ultraviolet Spectra of CF_3Br, CF_2BrCl, and CF_2Br_2. J. Doucet, R. Gilbert, P. Sauvageau and C.S., *J. Chem. Phys.*, 62 (2), 366-9 (1975) (1/15/75).

9. Photoelectron and Vacuum Ultraviolet Spectra of a Series of Fluoroethers. A.H. Hardin and C.S., *J. Fluorine Chem.*, 5 (5), 435-42 (1975).

10. Ultraviolet Absorption of Fluorocarbons, a Review. C.S., *Atmos. Environ.*, 10 (5), 343-51 (1976).

D.H. Stedman, University of Michigan

1. Measurement Techniques for the Ozone Layer. D.H.S., *Res./Dev.*, January, 1976, pp. 22-4, 26.

N.D. Sze, Environmental Research & Technology, Inc.

1. Measurements of Fluorocarbons 11 and 12 and Model Validation: An Assessment. N.D.S. and M.F. Wu, draft ms.

PAPER 11

Activities in the Field of Stratospheric Ozone, including Possible Geophysical and Biological Consequences of Ozone Depletion

Presented by ICSU/SCOPE

INTRODUCTION

ICSU has been involved institutionally in stratospheric ozone studies since 1930 when a Sub-Commission on Ozone was established within IUGG. During the last two decades, the activities have expanded greatly, and now encompass:

- experimental and theoretical investigations of stratospheric composition and processes;
- studies of the effects of ozone depletion on climate;
- studies of the effects of ozone depletion on the biosphere;
- reviews of relevant socio-economic studies.

EXPERIMENTAL AND THEORETICAL INVESTIGATIONS OF STRATOSPHERIC COMPOSITION AND PROCESSES

The main ICSU activities in this field are within:

IAMAP

The International Commission on Atmospheric Ozone, the International Commission on the Upper Atmosphere, and the International Commission on Atmospheric Chemistry and Global Pollution all sponsor international symposia which relate to the ozone depletion problem. A major reference work is the Proceedings of the 1974 International Conference on the Structure, Composition and General Circulation of the Upper and Lower atmospheres and Possible Anthropogenic Influences, available from IAMAP for US $10 (2 volumes).

Joint IAMAP/IAGA Symposia

A symposium of Stratospheric Ozone was held in September 1976, sponsored by NASA with IAMAP/IAGA cosponsorship. Information on that meeting can be obtained from Dr D.H. Hunten, NASA, Washington, D.C.

Another joint symposium is to be held in Seattle, Washington in August, 1977: "Minor neutral constituents in the middle atmosphere: Chemistry and Transport

COSPAR Working Group 6

Spectrophotometric measurements are being taken on Nimbus 6 at the 9.6 micrometer
wavelength. These measurements will be repeated on Nimbus G in 1978. The COSPAR
W/G 6 is encouraging the processing of these data to derive global ozone distribution, making several years of satellite ozone observations available to the scientif
community for research. This will be particularly useful for models of the three-
dimensional patterns of stratospheric ozone.

STUDIES OF THE EFFECTS OF OZONE DEPLETION ON CLIMATE

The effects of Ozone depletion on climate are being studied within the ICSU-WMO GARF
Programme. The Joint Organizing Committee of GARP should continue to provide the
leadership for the worldwide programme.

STUDIES OF THE EFFECTS OF OZONE DEPLETION ON THE BIOSPHERE

The ICSU body with a direct interest in the biological effects of ozone depletion
is the International Photobiology Association, which is a Commission of IUBS. A
symposium of the biological effects of ozone depletion was held in Rome in September
1976. The British Photobiology Association, an affiliate of IPA, is sponsoring a
symposium in June 1977, and other meetings will be organized as appropriate.

REVIEWS OF RELEVANT SOCIO-ECONOMIC STUDIES

ICSU-SCOPE has organized an interdisciplinary institute, MARC, Chelsea, the Monitoring and Assessment Research Centre at Chelsea College, London, England. This Centre
has the capability to undertake reviews of socio-economic studies related to environmental management. One such review is the report (in press) by Dr L. Machta on
the ozone depletion problem.

Other relevant reviews could be undertaken upon request.

NATIONAL ACTIVITIES

ICSU provides a focal point for international coordination and exchange of results.
The main research studies are carried out within countries, through universities,
academies of science, research councils and government departments.

The programmes are too numerous to be summarized here.

CONCLUSIONS

ICSU represents a very broad spectrum of scientific interests and capabilities
through its adhering Unions, National Academies and Research Councils. The ICSU
activities in the field of ozone depletion, including possible geophysical and
biological consequences, are illustrative of the ways in which ICSU provides a
mechanism for international cooperation amongst scientists.

ICSU is both initiative and responsive. ICSU continues to press forward the boundaries of knowledge on ozone depletion, while at the same time providing expert
scientific opinion and advice upon request to governmental and intergovernmental
bodies. The ICSU-WMO relationships are exemplary.

APPENDIX 1

LIST OF ICSU CONTACTS

Mr F.W.G. Baker, ICSU Secretariat, 51 Boulevard de Montmorency, 75016 Paris, France.

Dr V. Smirnyagin, SCOPE Executive Secretary, 51 Boulevard de Montmorency, 75016 Paris, France.

Dr P. Melchior, Secretary General of IUGG, Observatoire Royal, Avenue Circulaire 3, 1180 Brussels, Belgium.

Dr S. Ruttenberg, Secretary of IAMAP and Secretary of COSPAR W/G 6, NCAR, Boulder, Colorado, U.S.A.

Dr N. Fukushima, Secretary of IAGA, Geophysics Research Laboratory, University of Tokyo, Tokyo, Japan.

Dr P.H. Bonnel, IUBS Executive Secretary, 51 Boulevard de Montmorency, 75016 Paris, France.

Dr R.B. Setlow, Immediate Past President, IUBS Commission on Photobiology, Brookhaven National Laboratory, Long Island, N.Y. U.S.A.

APPENDIX 2

ACRONYMS

COSPAR	Committee on Space Research
GARP	Global Atmospheric Research Programme
IAGA	International Association of Geomagnetism and Aeronomy
IAMAP	International Association of Meteorology and Atmospheric Physics
ICSU	International Council of Scientific Unions
IPA	International Photobiology Association
IUBS	International Union of Biological Sciences
IUGG	International Union of Geodesy and Geophysics
SCOPE	Scientific Committee on Problems of the Environment

APPENDIX 3

ICSU ACTION PROPOSALS

(1) Coordination

If UNEP decides to establish a coordinating committee, ICSU would be willing to provide members or observers (an atmospheric physicist and a biologist).

(2) The Meteorological Problem

(a) At its General Assembly in Canberra, 1979, IUGG would organize:

- an overview symposium on the atmospheric problem; and

- a specialist symposium on the air-sea exchanges of relevant trace gases as N_2O and halocarbons;

(b) The International Ozone Commission would organize in 1980, an inter-
     national Symposium on Ozone, with special emphasis on the problem of
     "greenhouse" effects and of ozone monitoring.

(c) At its meeting in Austria in 1978, COSPAR would organize a workshop on
     rocket/satellite monitoring of ozone, including ground-truth require-
     ments.

(3) The Biological Effects Problem

(a) With appropriate financial support, ICSU would be prepared to develop
     an International Research Centre for Photobiology. This could follow
     the institutional pattern of the ICSU/SCOPE Monitoring and Research
     Centre, Chelsea College, London, England. Such a centre would have two
     functions, namely:

     - to act as a coordinating and information processing center for UV
     photobiology pertaining to the ozone problem; and

     - to initiate and sponsor appropriate research programmes at university
     or other research centres around the world.

(b) The International Association for Photobiology would organize a three
     day specialist symposium on "Effects of UVB radiation in biological
     systems" in 1979. The proceedings of the meeting should be published.
     A one day follow-up workshop on UV and skin cancer would be organized
     at the International Congress on Photobiology, France, 1980.

PAPER 12

Stratospheric Ozone Depletion - An Environmental Impact Assessment
Presented by ICSU/SCOPE

INTRODUCTION

The Problem

The stratospheric ozone layer is important for two reasons: it shields the surface
of the earth from the ultra-violet (UV) rays of the sun; and it is an integral part
of the world atmosphere system.

A widely discussed topic at the present time is the degree of fragility of the
ozone layer. Mankind is becoming capable of environmental disruption on a global
scale; trace substances are being introduced into the atmosphere in ever increasing
amounts and some of these materials are reaching the stratosphere where they may
upset the natural balance of the ozone layer. The resulting perturbations are not
yet detectable but a slow depletion in ozone concentrations is predicted in future
years if the chemical releases to the atmosphere continue in increasingly large
amounts.

Because of the great natural variability in stratospheric ozone from place to place
and from year to year, it will be difficult to determine the extent of ozone depletion in the years ahead. In fact, one of the special aspects of the ozone problem
is that if mankind waits until there are experimental data confirming a downward
trend in ozone concentrations, it will already be too late to prevent an extended
period of ozone depletion.

In this paper, an environmental impact assessment is given of the problem
of stratospheric ozone depletion.

An Historical Perspective**

The stratospheric ozone layer has been studied for many decades. As early as 1880,
Hartley suggested that atmospheric ozone was absorbing UV sunlight. However, the
first accurate ground-based measurements of total ozone were not made until 1920
when Fabry and Buisson determined the amount of ozone over Marseilles during a

---

*Prepared for ICSU/SCOPE by the Scope Monitoring and Assessment Research Centre,
 Chelsea College, University of London (15 January 1977). Edited by R.E. Munn.

**For a more detailed historical review and full citation of references in this
  section, see Dobson (1966).

period of 14 days and suggested that the amount was equivalent to that of a layer
of pure ozone about 0.3 cm thick at sea level.

In 1929, Götz realized that from measurements of total ozone at two wavelengths
and at several sun elevations on a clear day, it was possible to infer the average
height of the ozone layer and to provide a general indication of the vertical distribution of ozone. This technique is known as the *umkehr* method and it provided
a stimulus for many investigations in the 1930s and 1940s.

The first of the modern ozone spectrophotometers was built in 1929. About 110 of
these so-called "Dobson" instruments were manufactured in the next few years and
most of them are still in service. A puzzling feature of the early measurement
series was that the highest ozone values occurred in spring and in the far northern
latitudes.

The international scientific community became involved institutionally in 1930 when
a Sub-commission on Ozone was established within ICSU-IUGG.[1] After the war, a
separate International Commission on Atmospheric Ozone was organized and its main
task in 1948 was the "the organization of an ozone survey for Western Europe, while
at the same time assisting in the establishment of ozone stations in other parts of
the world, as opportunity presented itself".

In the 1950s, chemical methods for measuring ozone were developed, making it
possible to obtain *in situ* values from aircraft, balloons and rockets. An intensive monitoring programme was undertaken during the IGY[2] in 1956-57 and the WMO[3]
published the results. Subsequently, the WMO established the World Ozone Network;
and Canada, on behalf of WMO, began regular publication of the data. Today, remote
sensing of stratospheric ozone from satellites opens even more avenues for monitoring and research.

The theoretical problems thrown up by each new set of measurements have always been
interesting. In addition, there has been a continuing possibility that a better
understanding of the ozone layer might lead to an improved capability in weather
forecasting.

The classical photochemical explanation of the ozone layer given by Sidney Chapman
in 1930 has had to be modified several times since 1960 with the recognition of
additional chemical and meteorological mechanisms. In particular, the importance
of the wind fields in transporting ozone from one latitude to another, and from one
height to another, is now understood: this has cleared up one of the early riddles
– the explosive warming (and increase in ozone) that takes place in polar regions
in late winter and spring. The developments of the last few years, beginning in
1970 with the concern about supersonic aircraft emissions, and increasing in 1974
with publication of the paper by Rowland and Molina (1974) on halocarbons, are well
known, and need not be described here.

In summary, there is a tradition of experimental and theoretical investigations of
the unperturbed ozone layer. Physicists and chemists with specialist training in
this field are therefore available for study of the man-perturbed stratosphere.

Given the scenarios for ozone depletion in the stratosphere and for UV increases at
the surface of the earth, there is a need for predictions of the biological effects.
This task has been taken up by photobiologists and epidemiologists. Much of the
work can be undertaken by individual scientists; where international co-ordination

---

[1]   International Council of Scientific Unions-International Union of Geodesy and
      Geophysics
[2]   International Geophysical Year
[3]   World Meteorological Organization

is required, it can be achieved through UNESCO/MAB[4], FAO[5] and WHO[6] on the governmental side and ICSU-IUBS[7] on the non-governmental side.

THE STRATOSPHERIC OZONE LAYER: THE NATURAL STATE

The structure of the atmosphere in the Northern Hemisphere is shown schematically in Fig. 12.1*. The layer containing the dots is called the stratosphere; it is virtually cloudless and its temperature increases with height.

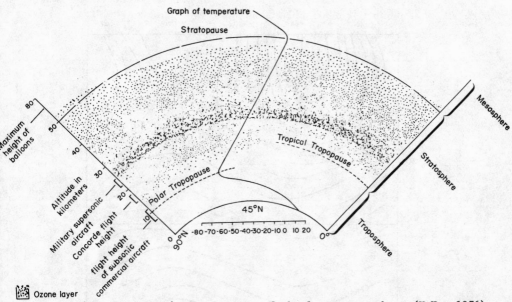

Fig. 12.1. Physical structure of the lower atmosphere (U.K., 1976)

The UV rays from the sun cause stratospheric oxygen O_2 to dissociate and to form ozone O_3. Ozone is also created near the ground on sunny days, particularly over large cities such as Los Angeles. In this case, there are photochemical reactions between oxides of nitrogen and hydrocarbons. However, the surface layers contribute only insignificant amounts to the total ozone in the atmosphere.

Ozone is destroyed by contact with the surface of the earth, clouds and suspended particles, through photochemical reactions with trace gases (particularly the oxides of nitrogen NO_x and the halocarbons) and, in the stratosphere, by absorption of UV and other wavelengths of solar radiation.

The concentrations of ozone are 10 to 100 times greater in the stratosphere than in the troposphere; and approximately 90 per cent of the total ozone lies in that upper region. The observations reveal variations of the order of 25 per cent associated with migratory weather systems and lasting only a few days. Climatologically,

---

4   United Nations Scientific Cultural and Educational Organization/Man and
    the Biosphere
5   Food and Agriculture Organization
6   World Health Organization
7   International Union of Biological Sciences

*The atmosphere in the Southern Hemisphere is almost symmetric with that in the
 Northern Hemisphere.

total ozone generally increases poleward and is greater in spring than autumn.
These patterns are shown in Fig. 12.2 (London and Bojkov, 1976), which displays
average values by latitude and season. The data also suggest biennial and 11-
yearly cycles. The latter is less certain and a proposed association with the
sunspot cycle is not universally accepted. The periodicity of the biennial and
11-year cycles and the amplitudes of these as well as the annual cycles are
variable. In fact, the natural variability is so large that man-induced changes
in ozone, if any, will have to be at least 5 per cent and persist for many years
to be distinguishable from natural trends.

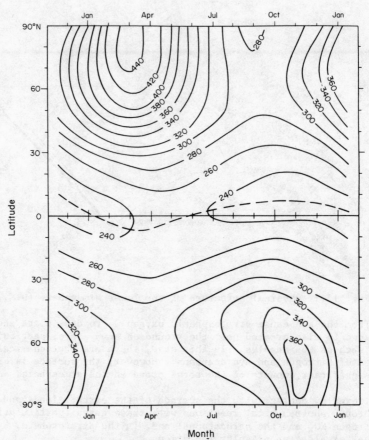

Fig. 12.2. Average total ozone isopleths by latitude and
season, in units of m atm cm. (London and Bojkov, 1976)

During the 1960s, the total ozone measured at Northern Hemisphere monitoring
stations increased at most stations by almost 5 per cent but the trend reversed in
1970 and for a few years the data suggested a decrease of a few per cent. After
about 1975 the decrease became less distinct. These trends reflect analyses based
on monitoring stations primarily over continents and in more developed countries.

Global patterns have been obtained for limited periods from satellite observations,
and this technique will assist in future trend analysis. The satellite data may
also help identify layers in the stratosphere contributing to trends in the vertical distribution of ozone.

In summary, much is known about the natural formation, destruction, transport and global distribution of ozone. Unfortunately the knowledge is inadequate to distinguish between natural and man-induced components unless the latter become large.

MAN'S IMPACT ON THE STRATOSPHERIC OZONE LAYER

Proposed Mechanisms for Disruption of the Stratosphere

Several natural and man-induced mechanisms for perturbing the stratosphere have been postulated.

Natural mechanisms. Volcanoes; extra-terrestrial events (solar flares, explosions of supernovae, etc.) perturbing the concentration fields of the oxides of nitrogen (NO_x).

Man-induced mechanisms. Aircraft emmissions; emissions of halocarbons* at the earth's surface reaching the stratosphere; emissions of N_2O from fertilizers and/or nitrogen-fixing vegetation; nuclear bombs.

Stratospheric disruptions could be caused by photochemical reactions involving a large number of trace substances, but special attention has been given to the oxides of nitrogen and the halocarbons. The stratospheric concentrations of these two classes of substances could be increased in the following ways:

Sources of Oxides of Nitrogen

Fertilizer denitrification. Table 12.1 provides an estimate of the amount of nitrogen fixation in 1974 in megatons of nitrogen. Natural processes fix nitrogen in the soil and air but man's contribution to the total is increasing (24 per cent of the total in 1974). In 1850, no industrially produced fertilizers were used; in 1950, 3.8 MT or 2 per cent of the total came from industrial fixation. If the growth in nitrogen fertilizers continues, the amount of nitrogen fixation by fertilizers alone may increase to between 100 and 200 MT/yr by the year 2000 AD.

However, it is denitrification which produces nitrous oxide, N_2O, a gas capable of reaching the stratosphere. During denitrification in the soil, only a small and uncertain fraction (about 7 per cent) is believed to be converted to N_2O; the remainder becomes molecular nitrogen, N_2. The processes of denitrification in the soil and the sea are not well understood. In particular, there is a lag between application of nitrogen fertilizer and denitrification which has been estimated by Liu et al., (1976) as possibly lasting hundreds to thousands of years.

It may be noted that large amounts of nitric oxide, NO, and nitrogen dioxide NO_2, are produced by industry, home heating, etc. A relatively small amount of N_2O is also created by combustion. But chemical transformations and rain scavenging prevent significant amounts of NO or NO_2 from reaching the stratosphere. On the other hand, tropospheric losses of N_2O are smaller although there is current uncertainty about the rates.

Aircraft. The high temperature combustion chambers of aircraft create NO and NO from atmospheric oxygen and nitrogen. Since, in the cruise mode, these oxides of nitrogen are returned to the atmosphere within or just below the stratosphere, they can, in theory, destroy stratospheric ozone.

---

*Chlorofluoromethanes (CFM)

TABLE 12.1. Estimates of Total Nitrogen Fixation for the Year 1974. (CAST, 1976)

Location or Mechanism of Fixation	Nitrogen Fixed	
	MT/yr	% of Total
Agriculture		
Legumes	35	14.8
Non-legumes	9	3.8
Grasslands	45	19.0
Forests	50	21.1
Unused land	10	4.2
Ocean	1	0.4
Atmosphere (lightning, etc.)	10	4.2
Combustion	20	8.4
Industrial ammonium fixation		
use as fertilizers	39	16.5
industrial uses	11	4.6
inefficiencies (losses)	7	3.0
Total	237	100.0

Table 12.2 provides emission rates of four types of aircraft which can fly in the upper troposphere or lower stratosphere. The proposed United States SST* which would cruise at an average altitude of 19.5 km would have an emission index (gNO_2 per kg fuel consumed) of about the same value as the Concorde/TU-144.

It has been argued that the redesign of jet aircraft engines could reduce the emission of NO_x by a large amount but many years of engineering and considerable investment must precede their operational use.

Sources of Halocarbons

Halocarbons include the following gases which are released in large amounts and which largely survive tropospheric removal processes: chlorofluorocarbons F-11 (CCl_3F) and F-12 (CCl_2F_2); and carbon tetrachloride (CCl_4). The source strengths of CCl_4 are uncertain although some analysts ascribe the atmospheric content and growth rates entirely to industrial production.

Figure 12.3 presented the estimated releases of F-11, F-12 and F-22** between 1945 and 1975. The spectacular growth ceased in 1974 with the economic recession

*Supersonic transport

**F-22 is believed to be destroyed in the troposphere and is therefore a possible alternative for F-11 and F-12.

in several industrialized countries and public reaction to the ozone depletion problem. The estimates of industrial emissions of CCl_4 by Galbally (1976) suggest annual values of between about 50 to 100 x 10^9 grams/year since about 1935.

TABLE 12.2.   Aircraft NO_2 Emmission Rates. (COMESA, 1975)

Aircraft Type	Cruise Fuel Flow per Aircraft (kg/hr)	Cruise NO_x Emission Index (g/kg) as NO_2	Cruise NO_x Emission Rate per Aircraft (kg/hr) as NO_2
Narrow-Bodied, Four-Engine Subsonic	5,400	8	43
Wide-Bodied Three-Engine Subsonic	7,500	18	135
Wide-Bodied Four-Engine Subsonic	10,000	18	180
Concorde, TU-144 First-Generation SST	21,600	20	430

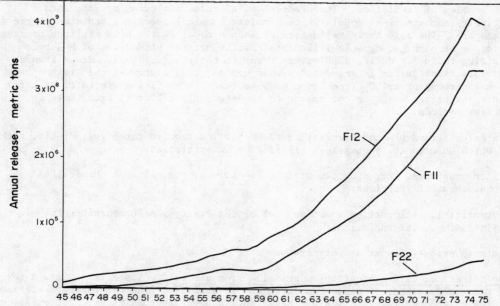

Fig. 12.3. The annual world production of chlorofluorocarbons. (MCA, 1976)

Other Sources of Oxides of Nitrogen and Halocarbons

The intense heat from the detonation of nuclear devices produces NO. Several investigators have estimated that each megaton of nuclear explosion produces $0.9 \pm 0.6 \times 10^{32}$ molecules of NO (all NO_x expressed as NO) or $45 \pm 30 \times 10^9$ grams of NO. Thus, during the atmospheric testing of nuclear devices when an estimated 510 megatons of explosive were detonated, about 2.3×10^{13} grams of NO were, in theory, introduced. Some sources predict a 4 to 5 per cent decrease in total ozone due to the 1961–62 nuclear tests when 1.5×10^{13} grams of NO were thought to have been added to the air. Most analysts have failed to detect such a change in the observed values of total ozone following the 1961–62 tests. However, the predicted decrease (4 to 5 per cent) may have been so small as to be lost in the natural ozone variability.

Volcanic eruptions can inject hydrogen chloride, HCl, into the stratosphere. The chlorine oxides, which are the reaction products of HCl in the stratosphere, can destroy ozone. As was the case with the nuclear bomb tests, no evidence for a decrease in total ozone attributable to a major volcanic eruption stands out, e.g., after the 1963 eruption of Mt Agung, Bali. Large amounts of HCl are also produced by industrial activities but this HCl is rapidly removed from the troposphere and little of it is believed to enter the stratosphere.

Finally, there have been suggestions that solar flares sufficiently intense to disrupt the ozone layer could occur about once every thousand years; and that the explosion of a nearby supernova which could increase the NO_x concentration by an order of magnitude, is not an impossibility (Reid et al., 1976; Whitten et al., 1976).

Predictions of Ozone Depletion

The problem of prediction. The predictions of ozone depletion are obtained entirely from numerical models which simulate chemical reactions and atmospheric transport. The only known validation of such a model is the high altitude decrease in ozone at high geomagnetic latitudes from the natural production of NO_x by ionizing radiation during a polar cap event in early August 1972. Ozone trends and variations during recent decades when man may have disturbed the stratosphere by halocarbons or by NO_x from large nuclear tests, high flying aircraft, or large-scale fertilizer usage do not appear to be detectably different from those in earlier periods.

The calculations of ozone generation and depletion require three ingredients, each of which unfortunately possesses significant uncertainties:

- rate constants, including temperature dependence, for relevant chemical and photochemical reactions;

- quantitative information about several of the stratospheric constituents (including solar radiation);

- atmospheric transport and dispersion.

The effects of oxides of nitrogen emission from aircraft. Table 12.3 lists a few of many calculations of ozone reduction due to aircraft. The percentage depletions in Table 12.3 and later tables are based on steady-state conditions. The emissions are assumed to be constant and continuous into the indefinite future at the indicated rate.

In addition to the calculations of Table 12.3, there are some additional estimates of ozone depletion due to types of aircraft other than the Concorde/TU–144 (CIAP,

1975):

- 400 wide-bodied subsonic aircraft flying at 10.5 km altitude produce an ozone reduction of 0.082 per cent;

- 100 "proposed" large U.S. type supersonic transports flying at an altitude of 19.5 km yield an ozone reduction of 3.27 per cent.

TABLE 12.3.   Estimates of Percentage Ozone Depletion Due to the Injection of 10^{11} grams of NO_x per Year at 16-18 Kilometers (100 Concordes/TU-144 Flying 7 Hours per Day)*

Calculated by	Global Average Ozone Depletion** Percentage
U.S. Climatic Impact Assessment Program (CIAP, 1975)	0.34
U.S. National Academy of Sciences (NAS, 1975)	0.57
U.K. COMESA (1975)	0.11

The effects of oxides of nitrogen from fertilizer denitrification. N_2O from the denitrification of fixed nitrogen spread as fertilizers and from other man-made sources can reach the stratosphere. Two of several estimates of the ozone reduction are given in Table 12.4. Note that the delay between nitrogen application to the soil and denitrification may be very long.

TABLE 12.4.   Estimates of Percentage Ozone Depletion Due to the Use of 200×10^{12} Grams of Nitrogen per Year as Fertilizer***

Calculated by	Global Average Ozone Depletion	Remarks
CAST (1976)	3%; 23%	Two analyses yield the two values of ozone; both numbers assume $N_2O/N_2 = 0.07$ during denitrification.
Crutzen (1976)	less than 1.8%	Upper limit since factors favourable to ozone depletion were optimized; $N_2O/N_2 = 0.07$ during denitrification.
Liu et al. (1976)		Argue that the response time for ozone depletion following fertilizer application is hundreds to thousands of years.

*There are differences in the assumed aircraft NO_x emission rates among the three reports. The injection rate of 10^{11}g NO_x/year corresponds to different numbers of flying hours. The table heading uses the COMESA estimates.

**The largest ozone depletion is likely in the 30-60°N band and might be about 1.8-2.0 times as great as these global average values.

***This annual level of application may occur in 2000 AD if the growth rate in usage is maintained as 6 per cent/year.

The effects of chlorofluorocarbons. Scenarios of the effects of emissions of F-11 and F-12 are normally made by projecting constant mixtures and quantities of the two gases into the indefinite future. Several of the resulting estimates of ozone depletion are listed in Table 12.5.

The time required for these reductions to reach half of their ultimate steady-state depletions is about 40 to 50 years (NAS, 1976).

TABLE 12.5. Estimates of Percentage Ozone Depletion Due to the Release of 2×10^{11} Grams of F-11 per Year and 3×10^{11} Grams of F-12 per Year. (Derwent et al., 1976)*

Calculated by	Global Average Ozone Depletion Percentage
Crutzen (1976)	6.5-7.0
Rowland & Molina (1975) Wofsy, McElroy & Sze (1975)	13-18
Turco & Whitten (1975)	10-18
IMOS (1975)	7
Derwent, Eggleton & Curtis (1976)	8
NAS (1976)**	7 (with a 2 to 20% uncertainty range)

POSSIBLE EFFECTS OF OZONE DEPLETION ON CLIMATE

Ozone plays a major role in determining the radiative balance of the atmosphere, particularly the stratosphere. A reduction and/or redistribution of ozone with altitude could alter the distribution of temperature and other weather elements, although there are two difficulties in evaluating the effects. First, climate simulation by high-speed computers or by any other technique is in a very primitive state and little confidence can be assigned to the predictions. Second, climate has varied on all time scales in the past and there is every reason to expect these oscillations to continue. Man-made changes might well be lost within this natural variability.

Ozone absorbs solar radiation so that a reduction in stratospheric ozone may be expected to reduce the temperature in the stratosphere and to allow more solar radiation to reach the ground. Using existing numerical models of climate, the effects of stratospheric ozone depletion can be simulated. The results show that even a simulated 25 to 50 per cent ozone depletion might produce only comparatively small tropospheric weather changes. The surface temperature does increase slightly in some simulations but recent calculations (Ramanathan et al., 1976; COMESA, 1975) suggest cooling by small amounts.

*The releases correspond to those actually occurring in 1973.

**The NAS (1976) estimate was not included in Derwent et al.'s paper, being published later.

In addition to the possible climate impact of a decrease in stratospheric ozone, there are other potential modifications by the ozone-depleting substances. Greatest attention has been directed to the "greenhouse" warming of the lower atmosphere by fluorocarbons, carbon tetrachloride and other trace gases (Ramanathan, et al., 1976; Wang et al., 1976). The warming effects of these gases add to the warming by CO_2. Atmospheric carbon dioxide concentrations are increasing due to the rising consumption of fossil fuels and this constitutes the more important greenhouse problem.

In addition to the trace gases, man is introducing aerosols into the stratosphere and these may also disturb the radiative balance (CIAP, 1975; COMESA, 1975). High-flying aircraft emit water vapour and sulphur dioxide which become aerosols in the stratosphere. An increase in aerosol content and the formation of ice clouds (possible in only certain parts of the stratosphere) would reduce the amount of sunlight reaching the ground and would tend to cool the lower atmosphere, although most estimates of these changes are small.

Mucray et al., (1975) has detected the absorption of solar energy due to F-11 and F-12 (almost exclusively of anthropogenic origin) but this is the only definitive evidence of man-made chemicals disturbing the radiative balance; natural levels of other gases and aerosols are insufficiently well established to determine the changes due to human activities.

Finally, mention should be made of the possibility that a decrease in stratospheric ozone would increase the UV sunlight near the ground, and thus might increase the photochemical smog over large cities. However, it can be shown that the impact is trivial, amounting to an increase in urban ozone of less than 1 pphm.

BIOLOGICAL CONSEQUENCES OF STRATOSPHERIC OZONE DEPLETION

Introduction

Ozone absorbs sunlight significantly in the near ultra-violet wavelengths (290 to 330 nanometers, called the UV-B part of the solar spectrum) (see Fig. 12.4). It is the energy in these wavelengths that produces sunburning, skin cancers and cellular and sub-cellular damage. Fortunately, the much more biologically damaging UV-C region (240-290 nm) is so strongly absorbed that a negligible amount would reach the earth's surface even with a very substantial reduction in the ozone shield (Caldwell and Natchwey, 1975). The increase in UV-B (when averaged over the wavelengths that cause biological effects) has been calculated as being about twice (2.0) the percentage ozone reduction in mid-latitudes under clear-sky conditions (or in some cases under highly idealized cloudy conditions). Observations suggest a lower value, perhaps 1.5 (Machta, 1976). For example, a 6 to 8 per cent reduction in stratospheric ozone levels would result in an increase in UV-B reaching the ground of approximately 9 to 16 per cent.

Present knowledge is insufficient for a quantitative assessment of the biological impact to be expected from any predicted level of increased UV-B, although in qualitative terms we know that a substantial increase in UV stress is harmful to many organisms, resulting among other things in reduced photosynthesis, depressed cell division, increased chromosome abnormalities and mutation rates and, ultimately death. The damaging effect is not likely to be a simple function of the total UV-B because there would be a relatively greater increase in the shorter wavelength UV-B (Fig. 12.4) which, for most responses, is more harmful to biological systems. Based on the action spectrum* for unrepaired damage to DNA (the material which

---

*An action spectrum is a curve showing the degree of biological effect as a function of wavelength.

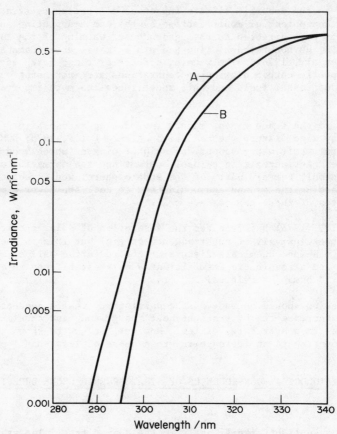

Fig. 12.4. Total direct and diffuse solar UV-irradiance at
60° solar altitude for different total atmospheric ozone
concentrations: A; 0.16 atm cm B; 0.28 atm cm.
(Caldwell and Natchwey, 1975)

carries the genetic information of living cells), the relative effect per quantum
increases from approximately 0.03 at 320 nm to 160 at 290 nm, when normalized to
a value of 1000 at 265 nm (NAS 1975). The increase in biologically damaging UV-B
can only be assessed from a knowledge of the wavelength response curve for any
particular damaging effect. As a general rule, however, the functions relating
deleterious effects to wavelength are poorly known, if at all, so that we cannot
weight the total UV increase for its potential damaging action. Although there is
good reason to believe that most, but not all, of the harmful effects of UV-B on
living organisms arise from its action on nucleic acids and proteins, the action
spectrum for damage to an organism is not a simple function of the absorbance of
the damaged molecule; this is because there are complicated interactions with
screening pigments, photorepair mechanisms, and photosensitization effects. However, a comparison of the relative effectiveness of UV wavelengths for damaging
DNA (Fig. 12.5) with the expected changes in UV radiation reaching the ground
when the ozone shield is reduced, reveals why a small spectral shift in these
regions could be of considerable biological significance.

The Photobiology of Damage by UV Light

The bactericidal and erythemal (skin burn) actions of solar UV have been known

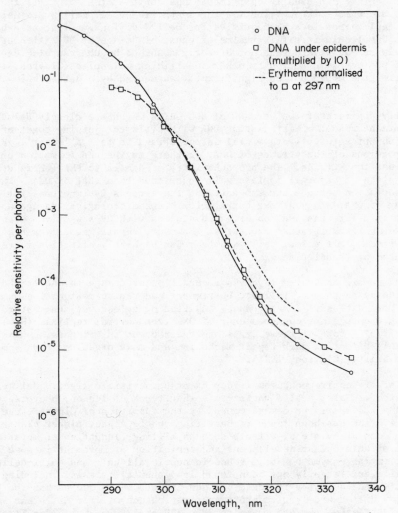

Fig. 12.5. Relative action spectra for damage to DNA and
DNA under epidermis.

since the 19th century and it is now recognized that UV causes a wide range of
deleterious reactions. Unfortunately, most of the basic photobiological research
has been carried out with wavelengths which are not present in the solar spectrum
reaching the ground. This has largely arisen because of the ready and cheap
availability of low-pressure mercury lamps with most of the emission at 254 nm
which is close to the absorbance peak for DNA and RNA* at 260 nm. Research has
been concentrated on effects of UV absorbed by nucleic acids and proteins and has
led to a good understanding of the photochemistry and photobiology of DNA damage.
Effects on RNA and proteins are less well understood (Clayton, 1971). For single-
celled organisms, the action spectrum for several responses such as killing, delay
of cell division and increased mutation rates suggests that nucleic acids are the
main targets for damage, with damage to DNA being a major cause of the harmful
effect. This is to be expected because if a molecule of protein or RNA is

*RNA is involved in translating the DNA information into the synthesis of
 proteins.

destroyed, another can usually be formed. DNA, on the other hand, is the primary
repository of genetic information for the synthesis of all cell proteins; in addition, DNA replication is a prerequisite for cell divisions. Light absorbed by the
nucleic acid molecule itself is the major cause of damage by UV. Although visible
light and longer UV wavelengths not directly absorbed by DNA can also cause damage
because of photodynamic action by other sensitizing molecules (Clayton, 1971),
these wavelengths would not be significantly augmented by a decrease in the ozone
shield.

DNA normally consists of two strands of sub-units which are closely associated.
The predominant result of irradiating DNA with UV is the joining together of two
adjacent sub-units (usually thymines) in one strand to form a 'dimer'. Breaks can
also occur in one of the strands of DNA, or there may be the formation of cross
links between two strands. The photochemistry of damage to RNA (where uracil
occurs in place of thymine) is less well understood than that of DNA. The nature
of the association between the sub-units in the strands is important for the proper
functioning of DNA, and a strong correlation between biological damage and the
production of dimers has been observed in studies with UV-sensitive mutant bacterial cells which lack 'repair' processes. However, repair mechanisms are possessed
to some degree by all normal cells and they can be highly effective in reducing the
damage caused by irradiation with UV.

One type of repair mechanism involves recognition of damage to one strand of two-
stranded DNA by an enzyme, followed by removal and replacement of the damaged
section. More extensive damage can be repaired by a less well understood process
which can piece together one good copy of DNA from partial replicas (Howard-
Flanders, 1975). Both of these types of repair occur in darkness. Human cells
have a highly effective dark repair mechanism and most organisms have some dark
repair capability (Clayton, 1971).

Another type of repair mechanism is dependent upon visible light. Unlike dark
repair, which can also heal a variety of other types of lesion, photoreactivation
appears *only* to repair the damage caused by the formation of dimers (Rupert, 1975).
Photoreactivation has been found in bacteria, fungi, algae, higher plants and
animals and can alleviate UV effects such as killing, induction of mutations,
inhibition of chloroplast development and inhibition of movement in some algae.
Photoreactivating enzyme has been found in nearly all uni- and multi-cellular
forms of life and recently has been found in placental mammals, including humans.
Photoreactivating activity is found between about 300 and 500 nm but the action
spectrum may differ widely with photoreactivating enzyme from different sources.
For *Escherichia coli*, the peaks are at 325, 350 and 380 nm while for *Streptomyces
griseus*, the peak is at 435 nm. Such a difference could modify the action spectrum
for damaging effects to DNA. A similar light-dependent mechanism apparently works
in some plant cells to repair RNA (Gordon, 1975). Lower intensities of light are
needed for repairing than for damage so that recovery from full sunlight damage
could occur in a shaded site.

In single cells, which have unusually effective DNA repair mechanisms, reactions
of proteins acquire relatively greater significance; light near the protein absorbance peak of 285 nm may be just as effective for killing as light at 260 nm
(Clayton, 1971). Ultra-violet irradiation of proteins causes a poorly characterized assortment of reactions in their component amino-acids. The most sensitive
targets are the aromatic amino-acids (especially tryptophane) because they have
the greatest ability to absorb UV. Many enzymes have been shown to be inactivated
by UV light. In laboratory experiments, these effects on proteins in single cells
are usually relatively minor, being overshadowed by DNA damage, especially with
254 nm light.

Higher organisms show the same principal kinds of photochemical damage to nucleic acids and proteins as micro-organisms with the same kinds of repair processes. Although other different types of reaction (such as the photodegradation of the plant hormones) (CIAP, 1975) could contribute to the loss of cell function under natural conditions, it is certain that photochemical changes in nucleic acids and proteins are important. Therefore, any increase in UV-B merits attention. The widespread existence of DNA repair processes prevents simple predictions of the effect of a given radiation exposure, however, even when quantitative photochemical yields are known. Any type of normal cell has its own combination of repair processes for DNA and the final effect depends on how much damage is unrepaired when the cell needs the impaired function. The amount of net unrepaired damage, unlike the initial damage, may not be proportional to the UV exposure. The amount of repair can vary with the condition of the cell; and repair systems may function more efficiently (i.e., repair a larger fraction of the damage) at lower radiation levels (Blum, 1959; Harm, 1968).

Approaches to Predicting the Biological Consequences of an Increase in UV at the Earth's Surface

The extrapolation of results obtained in the laboratory at 254 nm is too tenuous to be of much value in predicting the quantitative changes that might be expected to result from an increase in UV-B outdoors. The results, however, contribute to basic understanding of how cells are damaged by UV light.

The effects of UV radiation on intact higher organisms include both acute damage from a single large dose (e.g., sunburn in man) and chronic effects resulting from repeated or long-term exposures (e.g., skin cancers in man and animals and diminished growth in plants). Action spectra are known for some acute radiation effects. Between 290 and 320 nm, these sufficiently resemble the nucleic acid absorption spectrum to make this a reasonable guide for predicting damage (Caldwell, 1971). Practial difficulties prevent the determination of similar action spectra for effects due to repeated or long-term exposures but it is reasonable to argue that, as damage to nucleic acids and proteins is a major cause of biological damage, similar action spectra can be inferred (Setlow, 1974). They may, of course, be distorted by screening or sensitizing effects of pigments in the outer tissues of the organism.

In order to be able to make useful predictions about possible damage we need to know not only the relative effectiveness of different wavelengths but also the exact dose-response relationships. Do low intensities for a long time give the same result as higher intensities for a short time? In many simpler systems there are synergistic and/or antagonistic effects between UV-B and longer wavelengths (Clayton, 1971). We need to know whether similar interactions are present in the responses of higher organisms and, if so, what effect they would have under natural conditions.

The straightforward approach to predicting the effect of ozone depletion is to irradiate biological systems with solar simulators that mimic sun exposure as a function of time and various levels of ozone. Lamp filter combinations simulating changes in terrestrial solar radiation resulting from depletion of ozone have been described by Sisson and Caldwell (1975) and have been used in experimental pro-grammes with plants and small animals at a simulated increase in UV equivalent to a 35 to 50 per cent reduction in stratospheric ozone. Continued and extended use of such solar simulators over a range of latitudes with different local climates should provide data for refining our predictions, even though the natural UV varie with solar angle, fog, dust, cloud cover, etc., whereas the simulators impose a constant UV load. A modulated lamp system to provide a constant simulated frac-tional ozone reduction during the day would require a substantial degree of tech-nical development (Caldwell and Natchwey, 1975).

It is not possible to use solar simulators to look at effects on man, nor in large
ecosystems. Here predictions must be based on a knowledge of the basic photobiology, which is still far from complete in the UV-B region, and on correlations
between damage and spatial or temporal changes in incident UV. The latter approac
based on the increase in solar UV with decrease in latitude, has been used to
predict possible increases in skin cancer consequent on ozone depletion. However,
it is important to recognize that most of the epidemiological data for skin cancer
incidence and mortality rates are for locations which differ in many ways; for
example, they may vary in average UV received during the year, maximum intensity
of UV received, total amount of visible light, life styles, clothing, exposure
patterns and also often different ethnic and occupational backgrounds. Without
data from many more places and an understanding of the impact of the many variables
including the weighting factors for different wavelengths in the UV-B region, the
predictions will remain speculative.

The sizeable natural variations in stratospheric ozone could provide a limited bas
of experience of biological effects to be expected from ozone change. Irregular
changes of 20 to 40 per cent lasting for a few days occur in middle latitudes whil
seasonal variations of 20 to 40 per cent also take place in these latitudes. Thes
changes modify the variations in monthly average UV-B radiation which result from
the seasonal changes in solar altitude. All species exposed to ordinary sunlight
must be adapted to these variations.

More nearly resembling the depletions that might be expected from continued CFM
release at somewhat more than the 1973 rates are the longer-term variations in
ozone of about 10 per cent which are experienced over about an 11-year cycle.
When weighted for biological effectiveness (based on nucleic acid damage), these
must result in changes of about 20 per cent in damaging UV. No striking effects
on plants or animals have ever been shown to be correlated with these longer-term
cycles of ± 5 per cent in ozone level but variations in other climatic factors
could obscure any effects directly due to changes in incident UV-B (NAS, 1975).
Any additional reduction of ozone caused by man would be superimposed on this
natural fluctuation and could result in the exposure of ecosystems to levels of
UV-B flux not yet experienced and, therefore, whose effect is unknown.

Biological Effects of UV-B Radiation

Man. The known effects of UV-B on man include sunburn and snow-blindness, both of
which can be avoided by suitable protection, and the production of vitamin D_3,
which is beneficial. Ageing and wrinkling of the skin also result from continuous
exposure to the UV-B component of sunlight (NAS, 1976; CIAP, 1975). In addition,
there is a substantial body of evidence that links UV-B with skin cancer and indicates that an increase in solar UV-B would increase the incidence of these cancers
in man. Skin cancers associated with solar UV are diseases of less pigmented
peoples and are, therefore, of major concern in North America, Europe and Australi
Changes in individual DNA molecules are thought often to be a key step in cancer
production and sun-induced cancers are, therefore, probably caused by the action
of sunlight on DNA (Setlow, 1974); as already mentioned, it is the UV-B portion of
sunlight which is potentially damaging. The available evidence indicates that the
spectral sensitivity for skin cancer induction in man is similar to the action
spectrum for erythema production or to the spectrum for damaging DNA (the spectra
are similar but not identical, Fig. 12.5). Work with laboratory animals is consis
tent with this conclusion (Blum, 1959).

The most common skin cancers (non-melanoma) are only locally invasive and respond
to treatment. They occur most frequently on habitually exposed areas of the body
and show a strong correlation of incidence with latitude. These observations,
together with differences linked with occupational exposure, point clearly to sun-

light as prime cause of cancer induction and to increased incidence as a consequence of increase in UV-B. Generally disseminated skin cancer (melanoma) is
much rarer but has a very high mortality rate and, therefore, represents a significant health hazard. The evidence linking ultra-violet radiation with melanoma
is not so strong as for non-melanoma. The latitude dependence is well established
but melanoma is not concentrated on the habitually exposed areas of the body.
However, it is rarely seen in the regions which are least exposed (e.g., those
usually covered by a bathing suit) and most cases are found in those areas which
are lightly covered or occasionally uncovered. Differences in patterns of location between sexes correlate with differences in exposure to sun, with more on the
chest in males and on the legs in females (NAS, 1976). It has also been found that
the latitudinal variation in Norway is more pronounced for malignant melanoma of
the male neck-trunk and female lower limb (Magnus, 1976). Changes in mortality
and incidence with time are large and can reasonably be associated with changes in
exposure patterns (NAS, 1976). There is, therefore, a strong likelihood that
solar UV contributes to the induction and/or to the development of melanoma and
that an increase in UV-B reaching the ground would result in an increase in melanoma incidence and mortality. Where UV contributes to the induction of cancers,
the dose-response relationships are complex and unknown and quantitative predictions can only be speculative. A percentage increase in non-melanoma about equal
to the percentage increase in damaging UV (i.e., weighted for the presumed action
spectrum) has been predicted (NAS, 1976), but there is a large uncertainty factor
(Machta, 1976). For melanoma the situation is more complex and predictions become
even more speculative. If melanoma induction is related to solar UV, a depletion
of ozone by about 7 per cent might be expected to result in an increase in melanoma
deaths of the order of about 15 per cent (NAS, 1976); an absolute maximum may be
about 2,000 per year (Machta, 1976). A decrease in the uncertainty associated
with these predictions requires not only a better basic understanding of the origin
and growth of skin cancers but also better epidemiological data.

Mattingly (1976) has presented a theoretical analysis of the latitudinal changes
in the total annual sunburn dosage that would result from a postulated 10 per cent
decrease in stratospheric ozone. Assuming that the sky is free of aerosols and
clouds, the percentage increase in sunburn dosage ranges from 16 per cent near the
equator to slightly more than 30 per cent in polar regions. However, the absolute
value of the increase is largest in the tropics, decreasing rapidly poleward. As
an example, moving from 50°N to $42\frac{1}{2}^{\circ}$N would increase the annual sunburn dosage by
50 per cent, while moving from 50°N to $32\frac{1}{2}^{\circ}$N would bring an increase of 100 per
cent. The effects of aerosols and clouds on these estimates have not yet been
investigated.

Plants. Most of the early experimental work with plants was carried out at 254 nm;
only relatively recently have possible effects of UV-B received attention. Degradation of chlorophyll, depressed rates of photosynthesis, and cell death, which
are among the observed responses to 254 nm UV light, are assumed to be caused by
damage to nucleic acids and proteins. Longer wavelengths of UV may also destroy
some plant hormones and induce the release of a related substance, ethylene (CIAP,
1975). Experiments with lamp-filter combinations simulating the increase in UV-B
which would be associated with a 35 to 50 per cent decrease in ozone have been
carried out in glasshouses, controlled environment rooms and, to a limited extent,
in the field. These studies (CIAP, 1975) have revealed a great range of sensitivity to damage. Many plants are adapted to grow in full sunlight and mechanisms
to cope with potentially damaging UV-B have developed. These include elaboration
of effective screening pigment systems in the outer layers which protect the photosynthetic apparatus, temporal separation of UV-sensitive reactions such as cell
division so that they occur mainly during darkness, repair systems, and possibly
some degree of avoidance by positional movements of leaves or organelles*. It was

---

*Organelles are structural components of cells, e.g. chloroplasts.

found that some plants tolerated an increase in UV-B such as may accompany a 50 per cent depletion of ozone while others showed signs of stress with regimes that are considered normal fluxes of UV-B. In sensitive species, the UV supplement inhibited growth and development, depressed photosynthesis and biomass, and enhanced somatic mutation rate in stamen hairs of *Tradescantia*. Pollen germination was also found to be inhibited. Depressions in growth rate and biomass have been documented in the field for some species in Utah and in glasshouses with the roof removed in Florida (CIAP, 1975). However, there is a great heterogeneity of response depending on species and environmental conditions; and a greater degree of damage is observed more commonly and to an enhanced degree under growth room conditions than in the field. This is possibly because of the greater amount of light available for photorepair under the latter conditions. The importance of photorepair for plants growing in the open has been demonstrated for some alpine plants where one day of exposure to normal levels of solar UV without visible light resulted in photochemical damage (Caldwell, 1968). It is possible that some higher plants may be able to adapt to a gradual increase in UV-B, e.g., by increased pigmentation, alteration of surface wax layers, etc.

In most experiments solar UV has been supplemented to correspond roughly to a 35 to 50 per cent reduction in ozone at 60° solar altitude; the effects of exposing sensitive plants to lesser degrees of ozone depletion have not yet been explored. The dose-response relationships have not been well investigated but certain types of damage, such as the depression of photosynthesis, appeared to be cumulative (Caldwell, 1976). Other types of damage, for example, the inhibition of leaf expansion, seemed to be more affected by peak dose. In general, exposing plants to UV-B causes damage similar to that found in the shorter wavelength UV-C region. Photosynthesis in particular appears to be sensitive to UV irradiation which results in ultra-structural changes in the chloroplasts and an overall loss of photosynthetic capacity (CIAP, 1975).

Terrestrial animals. The effects of increased UV-B radiation on animals other than man have generally received little attention as it is usually assumed that UV-B is essentially absorbed by their outer coverings. Hair, feathers, shells, scales and pigments protect the critical organs of many animals from UV-B. A few experiments (CIAP, 1975) showed that in some species, sufficiently large doses of UV-B given at the appropriate stage of development could lead to abnormalities which might impair the animals' competitive ability.

In latitudes where long periods of high natural solar UV are prevalent, three effects attributable to UV have been noted in farm animals (U.K., 1976; CIAP, 1975). These are cancer eye, a form of ocular carcinoma of cattle; pink-eye, a bacterial infection of cattle which is aggravated by UV; and photosensitization, a complex reaction which results in hypersensitivity following ingestion of certain sensitizing agents (often found in plants) together with UV exposure. Carcinomas are also found in goats, sheep and horses, and in most cases they are found in parts of the body lacking protective melanin, such as eyelids and genitals.

Most animals will not expose themselves to full sunlight unless they have a way of preventing excessive light from reaching sensitive areas (e.g., by the presence of UV absorbing pigments) but farm animals may not have protective cover available to them.

Although UV-B is within the visible spectrum of many insects and might affect their behaviour, it is not yet possible to assess the relative importance of the UV-B component compared with other factors. Preliminary experiments with simulated increases in UV-B have so far revealed no adverse effects on pollination behaviour in which UV-dependent perception (mostly UV-A) of flower markings (honey guides) is often important. Little information is available on growth and development

responses to UV; to date only the larvae of a few insect species have been found
to be sensitive to prolonged irradiation with large UV doses (CIAP, 1975).

Ecosystems. A broader view of the possible biological impact of increased UV-B
includes the non-agricultural aquatic and terrestrial ecosystems which are important to man for timer, recreation, pest and predator balance, and as a food
resource. No effects of the 11-year cycle of stratospheric ozone variation have
been detected. However, experiments with simulated solar sources (CIAP, 1975)
have revealed considerable differences in sensitivity to UV between species; additional UV might therefore influence the ability of some species to compete in a
particular natural ecosystem. Changes in plant metabolites could affect the
animals that feed on them.

Aquatic ecosystems might also be affected because solar UV has been shown to
penetrate to considerable depths in clear waters. The attenuation of UV-B largely
depends on the amount of organic matter present: 10 per cent of the UV penetrated
to a depth of 23 m in clear waters of the west equatorial Pacific but only to 20 cm
in Douglas Lake, Michigan (see Zaneveld, 1975). Calculations indicated that about
2 per cent of the euphotic zone (the zone with sufficient light for net photosynthesis) might be affected by UV-B with a stratospheric ozone reduction of 25
per cent resulting in 3 to 10 per cent of the phytoplankton of the world's oceans
receiving more UV-B than is now received at the surface. Preliminary experiments
indicate that phytoplankton may be quite sensitive to UV (CIAP, 1975) and consequences to the aquatic food chain might, therefore, be far-reaching.

The prediction of effects in large and complicated ecosystems require knowledge of
the penetration of UV into the canopies, the sensitivity and behaviour patterns of
the component species and the species interactions. At present we lack the basic
data and the models that would enable us to make reasonably confident predictions
for complex natural ecosystems.

The Reduction of Harmful Effects

Some of the effects discussed above could certainly be minimized. Ability to
recognize persons at risk together with better health educational programmes about
exposure to sunlight could prevent much of the predicted increase in skin cancer
incidence. Because solar UV is likely to increase only gradually over decades,
breeding programmes for crops would inevitably include the increased UV stress as
a component of the environment in which the genotypes were being screened for yield.
Conscious effort could also be made to test new cultivars for UV sensitivity or to
breed for increased UV resistance. However, the time scale almost completely pre-
cludes breeding programmes for long-lived crops such as trees, and such plants
would be expected to be at greater risk from cumulative effects of UV-B exposure.

In natural ecosystems nothing can be done to minimize the impact of increased
solar UV-B. Some organisms may have spare capacity to endure an increased UV load
in their repair mechanisms, in their ability to replace damaged cells, and/or in
the effectiveness of their screening. Adaptive (UV-induced) increases or redistribution of screening pigments may occur in some animals as it does in man. However,
laboratory tests have shown that some organisms are deleteriously affected by even
small increases in UV-B and presumably are already near their tolerance levels
(CIAP, 1975). Indeed, even current levels of solar UV have been linked with mutations, delayed cell division, depression of photosynthesis in phytoplankton and
lethality in some bacteria and aquatic invertebrates (Caldwell and Natchwey, 1975).
Under natural conditions organisms near their UV tolerance can prevent excessive UV
doses by behavioural reactions which avoid sunlight. Correlations have been
observed in some cases between skin pigmentation and behavioural avoidance of
exposure to sunlight: many snakes and reptiles which are diurnal in habit have

more UV-absorbing skin pigments than those which are nocturnal (CIAP, 1975).

As has been pointed out in previous reviews (Caldwell and Natchwey, 1975), most
animals do not see UV-B radiation; thus the behavioural cue for UV-avoidance is
the intensity of visible light. Because an ozone-dependent increase in UV-B would
not be accompanied by parallel increases in the visible spectrum, the relationship
between the cue for avoidance (visible light) and the unfavourable factor avoided
(UV-B) would be changed. Thus the effectiveness of behavioural avoidance might be
lowered. However, this point may have been overemphasized for relatively small
changes in UV. Because the transmittance of UV through leaves is virtually zero
and their reflectance is less than 10 per cent (Caldwell, 1971; Gates et al., 1965;
CIAP, 1975), penetration into plant communities is largely dependent on the geometry of the vegetation stand (CIAP, 1975). The most important parameters are the
leaf area per unit ground area (leaf area index), the distribution of leaves within
the canopy (leaf area per unit volume, or leaf area density) and the leaf angles
(CIAP, 1975; Anderson, 1970; De Wit, 19765; Warren-Wilson, 1965). Less UV penetrates into dense canopies, and relatively more penetrates into vegetation stands
where the leaves are erect than into those where the leaves are largely horizontal.
The effectiveness of canopies in providing shade from UV-B for animals and other
plants is, therefore, a function of canopy geometry. Shade will be almost complete
where there is a dense, broad-leaved canopy in which the leaves are largely horizontal as in many tropical forests.

The UV-B sensitive species of green plants are likely to be the most vulnerable
organisms in terrestrial ecosystems because they cannot avoid exposure (except to
a limited extent by leaf and organelle movements) and because photosynthesis, their
only source of food, is a process known to be highly sensitive to UV damage. In
the lower strata of forests and woodlands, plants would be strongly protected by
the canopy above, and, therefore, it is the canopy species themselves which are
likely to be most affected. Furthermore, the canopy trees are usually long-lived
species which would be most subject to the cumulative effects of exposure to UV-B.
In grasslands, the vegetation stands consist largely of species with upright leaves
and, therefore, a greater penetration of UV-B into the canopy would be expected.
Thus most of the leaves of the component species would be exposed to an additional
UV-B load, although the more erect leaves of grasses would tend to reduce the area
exposed to direct light when the sun is at high elevations. Changes in the component species might occur because of varying levels of UV-B tolerance and/or differences in ability to develop resistance. Species with short lifespans and a more
rapid turnover of their gene pool could become more abundant. Individual plants
of some grasses are known to have long lifespans and, therefore, would be more
subject to the cumulative effects of UV exposure.

In aquatic environments a shade habitat may be simply an increase in depth, and
the light attenuating properties of water are quite different from those of plants:
UV-B light has been shown to penetrate to a considerable depth into clear waters
with a 50 per cent attenuation of UV-B not being acheived until about 4 m (Caldwell
and Natchwey, 1975; CIAP, 1975). Where plants or organic matter are present, the
UV-B is attenuated rapidly and does not penetrate to any great depth. For animals,
therefore, the effectiveness of UV-B attenuation is highly dependent on the organic
matter content of the water, although many find shelter in rock crevices. The
visual spectrum of many marine animals does not extend into the UV-B region and
thus visual cues for increased avoidance are lacking. Some aquatic plants, such
as phytoplankton, could avoid an increase in UV-B exposure and consequent damage
to their photosynthetic apparatus by moving to greater depths, but the resultant
attenuation of photosynthetically active visible light would lower their photosynthetic ability.

Conclusions

Research to date indicates that likely consequences of increased UV-B are: (1) an increase in the incidence of skin cancers with a probable increase in mortality; (2) an increase in carcinoma and perhaps in other UV-related conditions in cattle, leading to some economic loss; (3) a reduction in crop and timber yields, which could have social and economic consequences; and (4) a decrease in plankton production with consequences for the aquatic food chain. There might also be changes in the competitive ability among organisms which could alter entire ecosystems, although at present these are impossible to predict. A benefit of increased UV-B is the increased production of Vitamin D, although Daniels (1975) concludes that "any hope that an increase in UV might be beneficial should be tempered with caution".

At present, biologists and epidemiologists are being asked to make quantitative predictions from insufficient data. The only predictions that have so far been made are those for the increase in incidence of certain types of skin cancer; the results are summarized in Table 12.6 (Machta, 1976). The nature of the inherent problems has been stressed in this report. We know that the main targets for UV damage are the important biological macromolecules, proteins and nucleic acids, although in plants destruction of other molecules (e.g., the plant hormones) may add to the damaging effect and be a cause of depressed growth rates. However, extrapolation from laboratory data obtained with single-celled organisms or cell cultures to multi-cellular organisms under natural conditions is immensely complicated by screening effects of the outer layers, by behavioural characteristics, by antagonistic and synergistic effects between UV-B light and other wavelengths, and by the presence of repair mechanisms of varying degrees of effectiveness. Experiments with UV-B have now begun in many places but a realistic evaluation of the possible damage to man and biosphere that might arise as a consequence of ozone depletion to any particular level requires the input of new information, a refining of the models for predicting damage to individuals and to ecosystems, and a programme of long-term research.

SOCIO-ECONOMIC CONSIDERATIONS

Introduction

Decisions relating to environmental protection require an understanding of the penalties paid for the decisions. If no action is taken, then an environmental risk is possible; if a restriction is imposed in order to reduce the risk, an economic or social penalty is often paid.

The following sub-sections contain brief accounts of the socio-economic considerations relating to aircraft emissions and to halocarbons. These are as yet no studies on benefit-disbenefit relationships associated with nitrogen fertilizers.

Aircraft Emissions

There are several steps which could be taken by the aircraft industry to reduce emissions. First, advanced technology could reasonably be expected to produce about one-sixth of present-day emissions of oxides of nitrogen from supersonic aircraft (even lower values are potentially possible). Development costs for a new engine combustor are difficult to estimate; the development of a new engine to accommodate the modified combustor might cost U.S. $250,000,000 (CIAP, 1975).

The reduction of sulphur in fuel will proportionately reduce the amount of sulphate aerosol in the stratosphere created by high-flying aircraft. Reduction of sulphur

TABLE 12.6.   Estimated Incidence of and Mortality from Skin Cancers from Steady-State Release of Selected Chemicals* (Machta, 1976)

Chemical	Input (g/yr)	Altitude of Input (Km)	Source	Dose Commitment		Harm Commitment Additional Cases per Year**		
				% Ozone Depletion	% Increase UV-B	Incidence		Mortality
						Non-Melanoma Skin Cancers	Melanoma Skin Cancers	Melanoma Skin Cancers
NO_x	156×10^{11}	10.5	400 wide-bodied subsonic aircraft	0.082	0.12	2,000	40	10
NO_x	1×10^{11}	16-18	100 Concorde TU-144 aircraft	0.11-0.57	0.17-0.86	2,500 -	50 - 250	10 - 50
NO_x	164×10^{11}	19.5	100 large supersonic aircraft	3.27	4.91	75,000	1,500	300
N_2O	$2 \times 10^{14}(N_2)$	0	Fertilizers in 2000 AD and constant thereafter	<1.8 - 23	<2.7 - 35	<40,000 -	<900 -	<150 -
F-11 F-12	$\{2 \times 10^{11}, 3 \times 10^{11}\}$	0	Aerosols, refrigerants, etc. at 1973 rate	6.5 - 18	9.8 - 27	150,000 400,000	30,000	600 - 1,500

* Estimates assume a constant world light-skinned population of 10 persons.

**Based on current non-melanoma skin cancer incidence of 150/100,000 light-skinned persons per year; melanoma incidence of 4/100,000 melanoma mortality of 1.5/100,000.

in fuel is technologically feasible. Costs to achieve this reduction vary from
U.S. $0.01 to about $0.30 per barrel of fuel depending upon the initial and final
sulphur contents.

Lowering the flight altitude of aircraft would reduce the ozone depletion. Since
this altitude change would adversely affect aircraft performance, this alternative
is unlikely to receive serious attention.

As long as hydrocarbons are used as aircraft fuels, it is unlikely that the water
vapour component of aircraft emissions can be reduced.

Thus, there exists an expensive technological fix to reduce the oxides of nitrogen
but this would take many years to implement. The reduction of sulphates is easier,
less costly, and more quickly undertaken. There are no procedures for reducing
water vapour emissions, other than by using a non-fossil fuel engine.

The Halocarbons

Some general remarks. The raw materials for the production of fluorocarbons are
carbon tetrachloride, hydrofluoric acid and chloroform. In the U.S.A. (1974), 80
per cent, 40 per cent and 56 per cent, respectively, of those materials produced
ended up as fluorocarbons.

Over 40 companies (considering the same company name in two countries as distinct
since the degree of international management is unknown) make chlorofluorocarbons
in about 23 countries. Virtually every country in the world uses F-11 and F-12 in
some product; the U.S.A. alone shipped over 20 million kilograms to over 70 countries in 1974.

Table 12.7 provides the U.S.A. chlorofluorocarbon-dependent employment and U.S.
dollar value of products. The use of F-11 and F-12 in aerosol spray cans and in
air-conditioning and refrigeration are well known. By foaming is meant the blowing of F-11 in the manufacture of rigid and flexible polyurethane and F-12 in the
manufacture of polyurethane and polystyrene. The employment and dollar value
figures for the refrigeration and air-conditioning industry are the largest. In
the U.S.A. (1973) about 5 per cent of refrigerant gas was F-11 and 65 per cent
F-12. The remaining 30 per cent was primarily F-22 which is not presently under
suspicion as a cause of ozone depletion.

TABLE 12.7. U.S.A. Dollar Value and Employment in 1974 for Uses of
Fluorocarbons*.   (Machta, 1976)

Use	U.S. $ Millions	Employment 1000s
Aerosol propellants	1,873	27.5
Refrigeration and Air-conditioning	22,769	495.5
Foaming	840	57.0
	PLUS	
Chemical Manufacturers		
Basic Chemicals**	339	1.3
Fluorocarbons**	551	2.7

* Source – U.S. Department of Commerce "Economic Significance of Fluorocarbons",
  December 1975.

**Include production of some fluorocarbons besides F-11 and F-12.

Aerosols. Over 75 per cent of the aerosol cans in the U.S.A. (and possibly for the whole world) which used F-11 and F-12 as propellants in 1973 were for hair care, antiperspirants and deodorants.

Substitute fluorocarbons to replace F-11 and F-12 in aerosols suffer from several disadvantages and problems, as noted in Table 12.8. Hydrocarbons, now used extensively as paint spray propellants, are cheaper than F-11 and F-12 but are highly inflammable (Table 12.8). Compressed gases, particularly carbon dioxide, suffer mainly from an inferior delivery spray (Table 12.8). Finally, chloro-carbons are also cheaper than F-11 and F-12 but have other disadvantages (see Table 12.8).

TABLE 12.8. Alternative Aerosol Propellants and Their Characteristics*.
            (Machta, 1976)

Propellant Types	Price U.S.$ per kg	Relevant Characteristics
Fluorocarbons		Highly favourable as a propellant; non-inflammable, non-toxic, favourable boiling points and vapour pressures.
F-11	0.77	
F-12	0.90	
F-114	1.07	Safety must be assured (they also contain chlorine which may destroy ozone in stratosphere). Higher cost. Some are inflammable, irritate the skin, or have undesirable boiling points or vapour pressures, or are poor solvents.
F-115	3.30	
Others (such as perfluoro-carbons)	4.4-13.2	
Hydrocarbons		Highly inflammable but other properties favourable as a propellant; cheap.
Propane	0.30	
Isobutane	0.26	
Butane	0.26	
Compressed Gases		Failure to deliver fine spray with 1975 technology. Nitrogen has pour stability. Nitrous oxide presents a personal hazard because of the strong oxidizing power. They are cheap. Carbon dioxide and nitrogen offer no environmental problems and even nitrous oxide may have negligible environmental hazard.
Carbon dioxide	0.11	
Nitrous oxide	0.41	
Nitrogen	0.06	
Chlorocarbon		May also threaten stratospheric ozone. Problems of inflammability, toxicity and skin irritation require mixtures with other propellants. Cost is less than for F-11 and F-12.
Methylene chloride	0.37	

*Adapted from A.D. Little, Inc. Report.

Mechanical or other (roll-on) delivery systems are currently being sold but public preference for aerosol dispensers in many countries is evident. Typical pump costs for mechanical delivery systems are about $0.10-$0.15, but a cap to minimize leakage adds slightly to the cost. For comparison, valves for F-11 and F-12 aerosol cans cost $0.03-$0.04 and filling about $0.88-$0.99 per kg. including filling losses. The fluorocarbon propellant costs are estimated to be about 30 per cent of the direct production cost. The cost of mechanical substitutes are less expensive than aerosol cans using F-11 and F-12.

Technological developments may overcome the drawbacks of certain delivery systems, especially compressed carbon dioxide and mechanical ones. This development may be accelerated by threats, in some countries, of restrictions in F-11 and F-12 usage. The delivery system must, of course, be treated as a whole and in some cases the product may have to be redesigned for alternative delivery.

Refrigerants. A 1972 Handbook lists 78 refrigerants. But F-11, F-12 and F-22 have largely displaced all others despite their higher prices. Their advantages are non-toxicity, chemical inertness, and good performance. For example, ammonia can be explosive if it leaks while sulphur dioxide, although non-inflammable, is highly toxic.

F-12 appears to have very few, if any, substitutes as the working gas in common household refrigeration for thermodynamic reasons. F-22 can replace many refrigeration and air-conditioning products but only after major redesign of equipment.

Other types of cooling devices such as thermoelectric absorption, or vapour pressure are either in an experimental stage or much more expensive than using fluorocarbons in either their capital and/or operating costs.

The present pessimistic outlook for substitute working fluids for cooling devices may be balanced by the likelihood that relatively simple procedures and modifications can reduce the emissions from devices now using F-11 and F-12. It appears that more than 60 per cent of all refrigerant emissions could be eliminated with improved service procedures and with relatively minor design changes. Another 25 per cent could be recovered at the point of product disposal.

Foam blowing. The blowing agents form the cellular structure in the manufacture of foams made from polyurethane, polystyrene, and polyolefins (very small amounts used for this latter purpose).

Table 12.9 lists some suggested substitutes for F-11 and F-12 in plastic foam manufacture. Each of these substitutes has some minor or major disadvantage over the currently used halocarbons. For example, substitution of water/CO_2 blowing agent systems for halocarbons in rigid polyurethane foams would result in the loss of an economic advantage over other insulating systems due to higher heat conduction.

In some cases, the large percentage losses of halocarbons to the air can be reduced by halocarbon recovery or destruction during manufacture or fabrication.

Some medical applications. Aerosol products have special benefits to medicine. In hospitals, fluorocarbons are used as propellants to sterilize and sanitize operating rooms, etc. For example, ethylene oxide is valuable in the sterilization of surgical instruments, bandages, etc. However, it is a highly inflammable and explosive gas. When mixed with fluorocarbons, its inflammability and explosiveness is controlled but sterilization is not inhibited. In pharmaceutical applications, inhalation units are used to relieve asthma, and fluorocarbon sprays are employed to apply bandagees.

Time required to implement alternatives. Table 12.10 provides estimates for the time required to introduce substitutes for the present spray cans using F-11 and F-12. Up to five more years should be added if a new chemical propellant must be identified and developed.

TABLE 12.9. Halocarbon Usage and Substitutes in Plastic Foam Manufacture*.
(Machta, 1976)

Halocarbon Typically Used	Type of Plastic Foam	Quantity of Halocarbon Used U.S.A. 1973 (millions of kg.)	Typical Product Application	Alternative Blowing Agent
F-11, -12 -13	Low density rigid polyurethane	20.4	Insulation	Water/CO_2
F-11	High density rigid polyurethane		Furniture parts	Water/CO_2
F-11 Methylene Chloride	Flexible polyurethane	16.1	Cushioning, bonded foam fabrics	Water/CO_2
F-12 Methyl Chloride	Extruded polystyrene	3.8	Food trays	Pentane
F-11, -12	Expanded polystyrene	0.2	Packaging, insulation	Pentane
F-114, -12	Polyolefin	0.5	Packaging, insulation	Azodicarbamide aliphatic hydrocarbons, nitrogen

*Adapted from A.D. Little, Inc. Report.

TABLE 12.10. Time Required for Introducing Products to Personal Care Market**.
(Machta, 1976)

Expand production of existing non-aerosol products	1 - 3 years
Introduce compressed gas aerosol products	1 - 3 years
Introduce liquefied gas aerosol products	2 - 4 years after new propellant identified

**From Arthur D. Little, Inc., estimates based on aerosol product industry sources.

Table 12.11 suggests very uncertain estimates for the time needed to improve
existing cooling equipment or to substitute new refrigerants.

Table 12.12 shows the estimated times to develop and install emission control
equipment and to substitute a new blowing agent. Up to 6 more years should be
added if new substitute chemicals must be identified and developed.

TABLE 12.11. Time to Improve Existing Air-Cooling Equipment or to Introduce
Products with New Refrigerants or Cooling Methods*. (Machta, 1976)

Product Categories	Reduce or Eliminate Losses in Products Using F-11 & F-12	Convert to New Refrigerant** or New Cooling Method
Appliances (e.g. home refrigerators)	2 - 3 years	3 - 6 years
Mobile air-conditioners	3 - 4 years	3 - 5 years
Room (home) air-conditioners	2 - 3 years	3 - 5 years
Commercial refrigeration	2 - 3 years	3 - 6 years
Commercial air-conditioners (chillers)	2 - 3 years	1 - 5 years

* Adapted from A.D. Little, Inc. Report.

**After a new refrigerant is identified and assuming it is available in sufficient
quantities. This time may be another five years.

TABLE 12.12. Time to Install Control Equipment and Substitute Other Blowing
Agents***. (Machta, 1976)

Develop and install vapour recovery equipment	2 - 3 years
Substitute blowing agent in rigid foam applications	3 years
Substitute blowing agent in flexible foam applications	6 months

***From A.D. Little, Inc., estimates based on industry sources.

Consequences of various strategies for restricting F-11 and F-12. Because the
current uses of F-11 and F-12 exist in a price- and performance-competitive
situation in most countries, one can generally contend that for a given product
performance, the current uses of F-11 and F-12 are the least costly way of satis-
fying the need as perceived by the consumer. Restrictions would shift the consumption to the next best alternatives which would either be more expensive or
perform less satisfactorily. However, as already noted, the alternative need not
be more expensive.

Net employment changes resulting from restrictions may be either upward or downward although an adverse job impact on some employees is almost certain.

Several studies of the economic impact of possible restrictions in the U.S.A. have
been published but they must be considered preliminary. The results have been
summarized by Machta (1976, pp. 25-33). These studies need to be expanded and to
be internationalized.

RESEARCH REQUIREMENTS

The recommendations for future work to resolve unanswered questions about the ozone depletion problem fall into four categories: (1) predictions of the amount of ozone depletion given a certain chemical input into the environment; (2) the effects of ozone depletion on people and other forms of life; (3) climate effects of ozone depletion and associated chemicals; and (4) economic and social effects of restrictions on emissions of potential ozone-depleting chemicals.

Since predictions of ozone depletion are based on theory, and all of the biological consequences and the need for restrictions depend on their validity, this aspect deserves major emphasis. A monitoring programme to keep abreast of global values of ozone itself and of the threatening chemicals such as F-11, F-12, CCl_4, N_2O and NO_x in the stratosphere can avoid unexpected surprises. Additionally, support should be given to a programme to help improve and validate the ozone depletion theory by measuring those related substances including naturally occurring radicals which take part in stratospheric photochemistry, (e.g., OH, H_2O_2, HO_2), and certain products of the man-made chemicals (e.g., CLO_x, N_2O_5). Laboratory experiments are required to confirm or determine for the first time, certain rate reactions which are used in the theory. Better estimates of atmospheric transport, especially in the stratosphere, are also needed. Finally, an expanded modelling effort to add realism to simulations of ozone depletion predictions by man-made chemicals should incorporate up-dated atmospheric and laboratory information. The recommendations in this paragraph are already being implemented within the WMO Global Ozone Research and Monitoring Project.

The climatic effects of certain ozone depleting chemicals (or those derived from the same technology) or from reduced or redistributed ozone are unlikely to be predicted with confidence for many years to come. A programme of climatic predictions must continue for a long time.

Perhaps the most serious knowledge gap lies in the biological consequences of ozone depletion. The programme should encompass further epidemiological surveys in places where the solar UV radiation is known as well as studies of the effects of enhanced ultra-violet radiation on agriculture, animals and organisms in the seas. The recommendations made at a recent workshop are given in Tables 12.13 - 12.16, below.

The economic and social consequence of restrictions on ozone-depleting chemicals have been studied for fluorocarbons and aircraft. Little has been done in the fertilizer field and for chemicals other than F-11, F-12, and NO_x. Even for fluorocarbons and aircraft emissions, changing technology modifies the penalties for restrictions so that updating of economic analyses should be undertaken.

In general, each area of proposed study uses personnel from different disciplines. Further efforts in one area are independent of those in other areas. Thus, research in all four areas can be undertaken simultaneously, but interdisciplinary and international co-ordination of activities is desirable.

TABLE 12.13. The Major Effects of Ultra-Violet Radiation and Ozone Depletion on *Human Health*. (Setlow, 1976)

What is known: UV-B (280-320 nm) is carcinogenic

	(a) *dose-response ?*	(b) *reciprocity ?*	(c) *action-spectrum ?*	(d) *synergisms ?*
Precancerous lesions basal and squamous cell carcinoma malignant melanoma	Epidemiologic data not good enough. For a 10 per cent O_3 fall there is a rise of 40 per cent (90 per cent confidence limits 5 to 150 per cent)	Uncertain even in mice.	Unknown. Erythema and DNA in cells used. Mouse data may be consistent with them.	Mouse data uncertain. Effects of visible uncertain. Some in bacteria. UV affects ability of mice to accept grafts (a systematic immunological effect).

What is needed:

1. Data on erythema: To identify high risk segments of the population; a human model for photobiological responses.

    Answers to a, b, c, d, above are needed for
      (i) different skin types
     (ii) tanning
    (iii) different erythemal end points
     (iv) individuals with skin cancer

2. Epidemiolgoical data: To sort out the many variables; can one use data on precancerous lesions as epidemiological indices of skin cancer?

    Need:
      (i) data for ∿ 100 localities using a random sample at each locality with questions about life style, etc.
     (ii) repeat the survey in 5 to 10 years to determine how base line changes
    (iii) dosimetry (average and variability) at these locations
     (iv) dosimeters on people to obtain average and variability

3. Data for extrapolation to people:
      (i) animal model for melanoma
     (ii) probes of accumulated damage in people (connective tissue?) (immunological detection of photoproducts in skin?)
    (iii) answers to a, b, c, d, for bacteria, animal cells and animals so that we have sufficient understanding of endpoints such as mutations, chromosome aberrations, neoplastic transformation to make the extrapolation bacteria→animal cells→animals ---→people

TABLE 12.14. The Major Effects of Ultra-Violet Radiation and Ozone Depletion on *Agricultural and Native Plants* (Setlow, 1976)

What is known: High levels of UV-B are deleterious; there is a large variation among species and between experiments

(a) *dose-response ?*	(b) *reciprocity ?*	(c) *action-spectrum ?*	(d) *synergisms ?*
Not known; if linear a 10 per cent O_3 fall would give a 1-10 per cent fall in producti- vity (such effects are detectable in growth chambers but not in the field)	Not known except it holds for photosyn- thesis in the one sensitive plant tested	Only guessed at	Not clear, visible may be antagonistic to UV-B

What is needed:

1. Better simulators for enhanced UV in the field and laboratory

2. To extrapolate to field predictions

   (i)   effects of visible light on UV effects
   (ii) growth chamber data and rules for prediction
   (iii) keys to the identification of sensitive plants (transmission through cuticle?)
                                    (effects on auxin and other growth regulators?)
   (iv) cellular or tissue indicators of effects on plants

3. Predictors of damage

   (i)   chloroplast damage quantitatively related to dose
   (ii) answers to a, b, c, d, for cells in culture
         mutations in plants
   (iii) interactions between other environmental variables
         (temperature, H_2O, salts, ...) and UV

4. Degradation of pesticides by UV-B. Are the degradation products safe or harmful for humans, animals and plants?

TABLE 12.15. The Major Effects of Ultra-Violet Radiation and Ozone Depletion on *Animals other than Man.* (Setlow, 1976)

What is known:	Little

cancer eye in cattle
insect behaviour
other animals (not mice)   } no large
fish   anticipated
                                                     effects

What is needed:

1. Fish eggs: answers to the four key questions a, b, c, d, in Table 12.13.

2. Birds' eyes (any bad effects are difficult to observe in the wild because they would lead to death). Look at lenses of old birds (especially sea birds)

TABLE 12.16. The Major Effects of Ultra-Violet Radiation and Ozone Depletion on *Aquatic and Terrestrial Ecosystems.* (Setlow, 1976)

What is known: Little or nothing

some field studies and laboratory studies on algae and protozoa
no ocean field studies
community structures can change rapidly

What is needed: 1. On plankton

     (i)   answers to the four key questions a, b, c, d, in Table 12.13. for representative species
    (ii)   and under field conditions
   (iii)   the penetration of UV-B into plankton-rich waters
    (iv)   build up the ecosystem from its components

2. On terrestrial systems

     (i)   any useful data
    (ii)   are some trees sensitive?

REFERENCES

Anderson, M.C. (1970) Radiation climate, crop architecture and photosynthesis.
    Proc. IBP/PP Tech. Meeting, Trebon, PUDOC, Wageningen, 71-78.

Blum, H.F. (1959) *Carcinogenesis by ultra-violet light.* Princeton University
    Press, Princeton, New Jersey.

Caldwell, M.M. (1968) Solar UV radiation as an ecological factor for alpine
    plants. *Ecol. Monogr. 38,* 243-268.

Caldwell, M.M. (1971) Solar UV radiation and the growth and development of higher
    plants. *Photophys. 6,* 131-177.

Caldwell, M.M. (1976) The effects of solar UV radiation on higher plants;
    implications of ozone reduction. Abstract, VII Int. Congress on Photo-
    biology, Rome.

Caldwell, M.M. and D.S. Natchwey (1975) Introduction and overview. Ch. 1, CIAP
    Monograph 5, U.S. Dept. of Transportation, Washington, D.C.

CAST (1976) Effect of increased nitrogen fixation on stratospheric ozone. Rep.
    No. 53, U.S. Council for Agr. Sc. and Tech., Washington, D.C., 33 pp.

CIAP (1975) CIAP Monograph 5, U.S. Dept. of Transportation, Washington, D.C.

Clayton, R.W. (1971) *Light and living matter. Vol. 2: The biological part.*
    McGraw Hill, Inc., New York.

COMESA (1975) The Report of the Committee on Meteorological Effects of the
    Stratosphere. U.K. Meteorological Office, Bracknell, U.K.

Crutzen, P.J. (1976) Upper limits on atmospheric ozone reductions following
    increased application of fixed nitrogen to the soil. *Geophys. Res. Letters
    3,* 169-172.

Daniels, F., Jr (1975) Miscellaneous detrimental and beneficial effects of UV on
    man. CIAP Monograph 5, U.S. Dept. of Transportation, Washington, D.C.

Derwent, R.G., A.E.J. Eggleton and A.R. Curtis (1976) A computer model of the
    photochemistry of halogen-containing trace gases in the troposphere and
    stratosphere. AERE-R-8325, Harwell, U.K.

De Wit, C.T. (1965) Photosynthesis of leaf canopies. Agr. Res. Rep. 665, PUDOC,
    Wageningen.

Dobson, G.M.B. (1966) Forty years research on atmospheric ozone at Oxford - a
    history. Clarendon Lab., Univ. of Oxford, England, 50 pp.

Galbally, I.E. (1976) Man-made carbon tetrachloride in the atmosphere. *Science
    193*: 573-576.

Gates, D.M., H.J. Keegan, J.C. Schleter and V.R. Weidner (1965) Spectral
    properties of plants. *Appl. Opt. 4,* 11-20.

Gordon, M.P. (1975) Photorepair of RNA. In *Molecular Mechanisms for Repair of
    DNA* (P.C. Hanawalt and R.B. Setlow, eds.), pp. 115-121. Plenum Press,
    New York.

Harm, W. (1968) Effects of dose fractionation on ultra-violet survival of *E. coli*. *Photochem. Photobiol. 7,* 73-86.

Harm, W. (1969) Bioloigcal determination of the germicidal activity of sunlight. *Rad. Res. 40,* 63.

Howard-Flanders, P. (1975) Repair by genetic recombination in bacteria: overview. In *Molecular Mechanisms for Repair of DNA* (P.C. Hanawalt and R.B. Setlow, eds.), pp. 265-274. Plenum Press, New York.

IMOS (1975) Fluorocarbons and the environment. Inadvertent modification of the stratosphere. U.S. Government, Washington, D.C.

Liu, S.C. et al. (1976) Limitations of fertilizer induced ozone reduction by the long lifetime of the reservoir of fixed nitrogen. *Geophys. Res. Letters 3,* 157-160.

London, J. and R. Bojkov (1976) Atlas of total ozone 1957/1966. NCAR Tech. Note, Boulder, Colorado.

Machta, L. (1976) The ozone depletion problem. MARC Chelsea Report, 33 pp.

Magnus, K. (1976) Epidemiology of malignant melanoma of the skin in Norway with special reference to the effect of solar radiation. Abstract, VII Int. Cong. Photobiology, Rome.

Mattingly, S.R. (1976) Spatial and temporal variation of solar UV sunburn dosage. *Atm. Env. 10,* 935-939.

MCA (1976) Worldwide annual fluorocarbon production figures. Manufacturing Chemists Association.

Murcray, D.G., F.S. Bonomo, J.N. Brooks, A. Goldman, F.H. Murcray and W.J. Williams (1975). Detection of fluorocarbons in the stratosphere. Geophysical Reference Copy 2, pp. 109-112.

NAS (1975) Environmental impact of stratospheric flight: Biological and climatic effects of aircraft emissions in the stratosphere. National Academy of Sciences, Washington, D.C.

NAS (1976) Man-made impacts on the stratosphere. National Academy of Sciences, Washington, D.C.

NAS (1976) Halocarbons: environmental effects of chlorofluoromethane release. National Academy of Sciences, Washington, D.C.

Ramanathan, V. et al. (1976) Radiative transfer within the earth's troposphere and stratosphere: a simplified radiative-convective model. *J. Atm. Sc. 33,* 1092-1112.

Reid, G.C., I.S.A. Isaksen, T.E. Holzer and P.J. Crutzen (1976) *Nature 259,* 177-179

Rowland, F.S. and H.J. Molina (1974) *Nature 249,* 810-812.

PAPER 13

Atmospheric Exchange Processes and the Ozone Problem*

Presented by the Institute for Environmental Studies, Toronto

TERMS OF REFERENCE

This study was commissioned by United Nations Environment Programme as
"A state-of-the-art review of the dynamics of production and distribution of stratospheric ozone, which should include

    (i)   the man-made factors that may affect the resulting balance and distribution of ozone;

    (ii) the effects - both climatological and biological - that may be expected from the depletion of stratospheric ozone."

Since, however, the World Meteorological Organisation has also prepared a detailed analysis of these problems, we have tried to avoid duplication of effort. The principal author of the present study (F.K. Hare) attended the WMO working group meetings as a UNEP observer, and tried to judge the most useful form the study could take.

What we have done is to prepare a review that summarizes existing knowledge, and tries to address the general environmental issues of interest to UNEP. We have written for the non-meteorological scientist, and have included a basic introduction to atmospheric structure and behaviour, as a background to the discussion of stratospheric pollution.

We have been greatly assisted in this work by the President of the Commission for the Atmospheric Sciences and the Chairman of the Working Group on Air Pollution and Atmospheric Chemistry of WMO, Warren L. Godson and R.E. Munn, both of whom happen to be members of the Institute's working group. The text has been written mainly by F.K. Hare, but H.I. Schiff (of York University) contributed much of the section on stratospheric chemistry.

THE ISSUES

Ozone is an intensely toxic gas that attacks man's respiratory system to such an extent that exposure to concentrations above one part in two million by volume is

---

*Prepared by the Working Group on Climate and Human Response, Institute for Environmental Studies, University of Toronto.

unacceptable. It also attacks a wide variety of industrial materials, notably rubber and nylon. It is present in significant concentrations (of order 5 to 10 ppmv*) in the stratosphere, chiefly between 10 and 35 km above sea-level. If such air were not purged of its ozone by compression systems, lethal concentrations woul occur in the cabins of aircraft flying in the stratosphere. Yet this toxic gas is essential to the survival of man and other living organisms (the biota). Ozone is created at high levels by ultraviolet radiation** from the sun. Much of the ultraviolet is absorbed, and hence fails to reach the ground. Since ultraviolet radiation attacks most living tissues, this ozone screen is a necessary part of the natural environment.

The total amount of ozone in the atmospheric column (abbreviated as "total ozone") is quite small. If extracted and compressed to standard sea-level pressure and temperature it would form a layer about 0.3 cm deep, subject to significant variations in time and space. Thin though it is, this layer, together with oxygen, filters out of the sun's rays ultraviolet radiation below a wavelength of 295 nm*** and weakens the ultraviolet (wavelength 295-400 nm) that does reach the earth's surface. Since the time when life first colonized the land areas of the earth, thi screen must have been present though occasional, catastrophic thinnings may have threatened the biota by allowing ultraviolet to reach the surface at greater intensities.

The ozone layer exists in its present form because of *a sensitive balance between natural processes of creation and destruction.* Human activities now pose threats to this balance. We are powerless to alter the incoming ultraviolet that *creates* the ozone, but our technology is releasing significant quantities of substances tha *attack* it. We may be accelerating the destruction of ozone without affecting the creation, leading to falling concentrations. The technologies responsible include high altitude flight, stratospheric nuclear testing, refrigeration, air conditioning, use of spray-cans, soil fertilization, and possibly others still unidentified.

Since these issues became apparent, beginning in 1970, several countries have mounted programmes of research aimed at determining their importance. Results so far suggest that the dangers are real, and that drastic international action will be required. At the third session of the Governing Council of UNEP in 1975 it was decided that priority should be given to the study of these potential threats. The present review, and the WMO working group report, are direct outgrowths of this decision.

SOURCES

We have relied on the combined experience of our members, and on guidance from WMO, OECD and other intergovernmental agencies. Because of the speed with which the review has been written, we have been unable to take adequate account of work in various countries. We have also relied heavily on the documents listed in Appendix I. These include reviews by the U.S. National Academy of Sciences, the Climatic Impact Assessment Program (CIAP) of the U.S. Department of Transportation (which involved workers in several countries) and the report of the U.S. Federal Task Force on Inadvertent Modification of the Stratosphere (IMOS). Equally valuable were the proceedings of the Anglo-French Symposium at Oxford, in September 1974. This was sponsored by the U.K. Committee on Meteorological Effects of Stratospheric Aircraft (COMESA) and the Comité d'Etudes sur les Conséquences des Vols Stratosphériques (COVOS). We have made much use of individual reviews by F.S.

---

*   parts per million by volume

**   ultraviolet radiation is solar electromagnetic radiation resembling visible light, but shorter in wavelength (in practice below 400 nm) ***

***   nanometre, i.e., one billionth (10^{-9}) of a metre.

Rowland, M. McElroy, P. Crutzen, R.E. Munn, B. Bolin, L. Machta, R.J. Murgatroyd and their associates. R.E. Dickinson helped us with privately communicated results.

We have taken account of the WMO statement on modification of the ozone layer due to man's activities (WMO/No. 315) issued on January 6, 1976. We are in full agreement with the statement. Our treatment also takes account of the final report of the WMO-CAS Working Group on Stratospheric and Mesopheric Problems, based on its Geneva meeting of September 8-11, 1975.

Inevitably we have been too much influenced by work in North America, and much too little by research in other continents. This reflects the extreme pressure under which we worked and not a lack of interest or appreciation.

We make no claim for this report other than that it is the discussion draft of an extended review, trying to snythesize a vast body of diverse research by numerous research workers. Our own contribution to this research has been small, but large enough to give us some feeling for the results obtained. We have cited only the major references, since the report is addressed to non-specialists.

BASIC ATMOSPHERIC STRUCTURE AND BEHAVIOUR

In order to understand the ozone problem, it is necessary to understand the basic structure and behaviour of the atmosphere. Hence we begin with an examination of its thermal and dynamical properties.

Thermal Structure

The lower atmosphere (surface to about 85 km) is permanently made up of three layers defined by distinctive thermal properties (Fig. 13.1).

TABLE 13.1 Thermal Structure

Height above Sea-Level	Approx. Pressure (mb)	Layer or Surface
		Thermosphere (extensive ionization)
∿85 km	∿.01 ----------------	Mesopause - (cold)
		Mesosphere (strong decrease of temperature with height)
∿50 km	∿.1 --------------	Stratospause - (warm)
		Stratospause (usually a slow increase of temperature with height)
9-17 km	∿100-300------------	Tropopause - (cold)
		Troposphere (strong decrease of temperature with height)
0-1 km	∿.1000	Planetary boundary layer (between earth and atmosphere)

We are dealing here with the troposphere and stratosphere, which together contain 99.9 per cent of the atmosphere's mass. The ozone layer is in the stratosphere. Above about 30 km many chemical reactions are brought about by ultraviolet radiation. These photochemical reactions* are slow or absent in the lower stratosphere and troposphere, except in the planetary boundary layer, where pollutants often interact photochemically. Much of the atmospheric ozone is stored *below* the photochemically active part of the stratosphere, chiefly in high latitudes.

The troposphere, containing about 85 per cent of the mass of the atmosphere, is turbulent, windy, cloudy and stormy, and hence well mixed – i.e., tends towards uniform composition. It contains most of the water vapour, and nearly all the clouds. It is the layer in which most flying takes place. We live in its basal kilometre, the planetary boundary layer, into which our industrial pollutants are released. Average temperature decreases strongly with height to a cold surface called the tropopause. This surface is of great importance, since air does not readily cross it. The air above the tropopause is sharply different in history and properties from the air of the troposphere.

The tropopause is in practice complex (Fig. 13.2). Over the polar caps of both hemispheres, and equatorward to about 55 degrees latitude, it tends to be a single, distinct surface at some level near 10 km, subject to small seasonal variations and day-to-day disturbances. Its temperature is in the -40C to -70C range. From 35 degrees latitude in the summer hemisphere to about 20 degrees in the winter hemisphere the tropopause is again a single, sharp surface, almost constant in temperature near -80C. Between this *tropical tropopause* and the two *polar tropopauses* are complex zones corresponding to the belts of westerly jet-streams, in which the tropopause is often multiple. Only rarely do polar and tropical tropopauses join. This *tropopause gap* is important because some of the mixing between tropospheric and stratospheric air takes place through it.

The stratosphere is much less turbulent than the troposphere. Its thermal structure – at most times a slow increase of temperature with height – tends to damp down vertical exchanges of air (i.e., convection). Even in the polar night area, where coldest temperatures (below -80C) lie well above the tropopause, in the 25-30 km layer, vertical exchange is discouraged. There are strong westerly or easterly wind systems, some of which are shared with the troposphere below, some with the mesosphere above. The stratosphere is very dry and almost cloudless. Much high altitude jet flight is in the basal stratosphere.

The upper limit of the stratosphere, the *stratopause,* is a level of high temperatures, much like those at ground level. The warmth of the upper stratosphere and lower mesosphere is due mainly to the absorption of solar radiation by ozone. About one per cent of the solar radiation is used to heat this layer. The ozone also absorbs infrared radiation from below.

Chemical Composition

The gases of the troposphere and stratosphere are largely molecular, though at all levels there is a small dissociation into electrically charged ions. Above 85 km ionization becomes more general, leading to the term "ionosphere" widely used for the upper atmosphere, in which temperature rises rapidly with height.

The lower atmosphere is very uniform in composition, although there are significant differences between troposphere and stratosphere. It is a mixture of gases,

---

*A photochemical reaction is a chemical change induced by the absoprtion of electromagnetic radiation, chiefly ultraviolet in the present case. Photosynthesis, also a photochemical process, requires visible light.

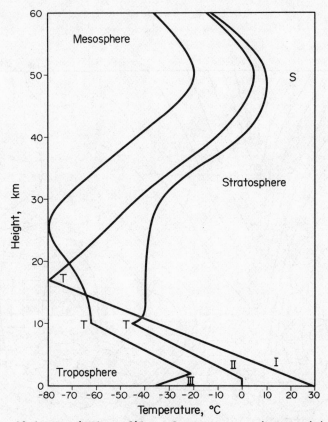

Fig. 13.1. Vertical profiles of temperature (schematic) for regime as
follows: I, equatoral region, year-round; II, polar regions, summer;
III, polar night region (cold regime). T = tropopause, S = stratopause.

together with an exceedingly large number of suspended particles, some liquid, some
solid, some electrically charged ions. The permanent or fixed gases are virutally
constant in proportion throughout the troposphere and stratosphere. They include,
in descending order of abundance,

Nitrogen   , N_2 , ... 78.1 per cent by volume*

Oxygen   , O_2 , ... 20.9 "   "   "   "

Argon   , Ar , ... 0.93 "   "   "   "

Neon (Ne), helium (He), krypton (Kr), xenon (Xe)~0.00014 "

Oxygen and nitrogen interact cyclically with the earth's surface, mainly with the
biota. Helium is released from the crust and ultimately escapes to space. The
remaining gases are inert, permanent parts of the atmosphere.

---

*Of dry air. These are fractional concentrations or mixing ratios by volume,
 called mole fractions by chemists. Meteorologists normally restrict the term
 mixing ratio to fractions by mass. Ozone concentrations are more often given in
 these mass mixing ratio terms.

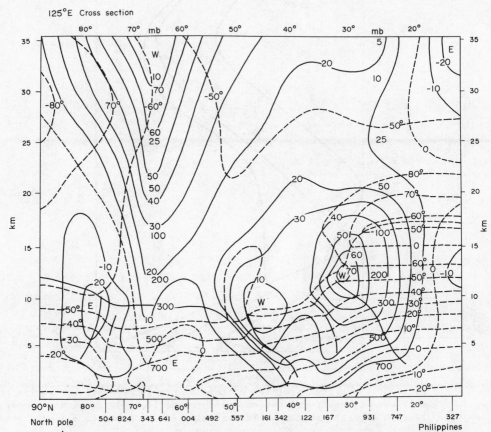

Fig. 13.2. Cross-section from north pole to Philippines along 125E, Dec.
4, 1959, showing typical tropopauses (dashed heavy line), jet streams
(W, from west, E from east) and fronts. In this case the polar
tropopause is shown extending to the sub-tropical jet streams in 30N,
where it almost joins the tropical tropopause, which is very high and
cold. An intense, separate polar night westerly jet is shown in 65N
at 33 km, but winds were even stronger aloft. "Downfolding" of the
tropopause may be occurring below the two lower level westerly jets.
Dashed lines are isotherms (C), thin solid lines isotachs of zonal wind-
speed, positive for westerlies, m s^{-1}. Numbers on abscissa are WMO
station identifiers. (McGill University, AMRG)

The stratosphere and troposphere also include certain gases that vary in time and
space. Although thin by comparison with the heavy blanket of the fixed gases, these
variable constituents play major environmental roles. They include

         Water vapour   , H$_2$O ,........................... <3 per cent by
                                                                 volume*

―――

*Of the moist atmosphere. Again these are volumetric mixing ratios, or mole frac-
 tions.

Carbon dioxide , CO_2 , ... 325 ppmv (increasing at
                                               about 0.8 ppmv per annum)

Methane   , CH_4 , ... 2 ppmv

Ozone   , O_3 , ... 0 to 10 ppmv

Hydrogen   , H_2 , ... 0.5 ppmv

Nitrous oxide   , N_2O , ... 0.3 Ppmv

Carbon monoxide, CO   , ... 0.1 ppmv

In addition the atmosphere includes many other trace gases, some natural, some due
to pollution. These include sulphur dioxide, SO_2 (up to 1 ppmv), other oxides of
nitrogen, and a variety of synthetic species such as the chlorofluoromethanes
(e.g., $CFCL_3$, CF_2Cl_2). Small concentrations (of order parts per billion) of various
organic compounds also occur naturally, notably carbon tetrachloride, CCl_4, methyl
chloride, CH_3Cl, and hydrocarbons (terpenes). At stratospheric levels our knowledge of these concentrations is very sketchy, because of lack of observations.
Even water is not known with precision.

Most of these variable gases interchange cyclically with the earth's surface. In
fact the entire set of ecosystems depends on the continuous cycling of nitrogen and
its oxides, oxygen, water, carbon dioxide, carbon monoxide and methane between sea,
soil, biota and atmosphere.

In addition to the gases, the atmosphere supports in finely-divided, colloidal form
(aerosols) prodigious numbers of particles (of order 10^8 to 10^{12} per cubic metre).
These are often solutions of sulphates, nitrates and chlorides, but also include
clay particles, organic particles, spores or pollen grains, and meteoric dust,
chiefly metallic compounds. Most of the particles are in the lower troposphere,
where they interact with the earth's surface, but there is a distinct layer of
particles (largely sulphates) in the lower stratosphere on either side of the 20 km
level. Haze layers are often visible in the lower stratosphere from aircraft flying
at polar tropopause level, and have recently been detected above the tropical
tropopause. The tropospheric particles play key roles in the precipitation process,
and in the earth's radiative balance. Significant particle emissions occur from
aircraft in flight, and this may influence their concentration at high levels.

In the upper stratosphere many molecular species are broken up (photolysed) by
ultraviolet radiation. These photochemical changes lead to an active chemistry.
Most ozone (O_3) originates at these levels, following the photolysis of oxygen (O_2).
This chemically active layer involves certain gases that originate at ground level,
and that then diffuse upwards into the upper stratosphere. The list of such upward
diffusing gases includes water vapour, methane, nitrous oxide, methyl chloride and
certain pollutants - notably the chlorofluoromethanes and carbon tetrachloride
(CCl_4).

Environmentally it is important that some of the gases involved in vital ecosystem
functions - the growth and decay of living tissues - are also critical to the
chemistry of the ozone layer. *In other words the earth's living cover and soil
interact chemically with the trace gases of the stratosphere.* There are essential
links between the earth's surface and the middle levels of the atmosphere.

ATMOSPHERIC EXCHANGE PROCESSES

It follows from the conclusion of the last section that there must be an efficient
process whereby gases can be exchanged between the earth's surface and the
chemically active layer about 35 km above. The extensive horizontal uniformity of

atmospheric composition proves, moreover, that there are efficient ways of distributing materials around the earth. In fact the latter - horizontal transport and
redistribution - is more rapid and more effective than the vertical, at least on
the larger scales. In this section, we describe these exchange processes, vertical
and horizontal.

Convection (Vertical Exchanges)

At root, the motion of the atmosphere is a gigantic system of convection, set up by
the unequal receipt of solar energy. We recognize a wide variety of types and
scales of convection. Some are *forced*, being created mechanically by the interaction between moving wind and obstacles on the earth's surface, or between layers
of air moving at different velocities. Others are *free*, meaning that they arise
from the natural buoyancy of air that is less dense than its surroundings. When
warm air rises and cold air sinks, kinetic energy is created. This last principle
embodies the process whereby the overall motion of the atmosphere is maintained
against the endless braking action of friction.

Small-scale convection includes the gusts (or turbulence) associated with winds
near the ground, the individual bumps affecting aircraft in rough air, and the
individual small cumulus clouds of a summer's day. The convection consists of
rising and falling parcels of air within the general wind. If there is a gradient
upwards or downwards of some substance or property, such convective currents will
bring about an upward or downward transport of the entity concerned. Net transport
upwards of water vapour, carbon dioxide, nitrous oxide, methane and pollutants
through the planetary boundary layer is largely achieved in this way.

In the rest of the troposphere and stratosphere, upward transport also takes place
on much larger scales. There are three broad mechanisms, mesocale (thunderstorm)
convection, synoptic scale cyclone-wave activity, and mean meridional circulations.

Mesoscale convection consists of clusters (10-200 km across) of large cumulus and
cumulonimbus clouds that contain strong updraughts extending from the planetary
boundary layer to the upper troposphere at 9-12 km. Some of the highest cumulonimbus towers may reach above 20 km, and at their crest may penetrate several
kilometres into the stratosphere. Much of the water that they transport upwards is
at once precipitated as snow or hail. As the clouds disperse, however, they leave
behind at high altitudes a little water vapour, plus many other particles and gases
brought up rapidly from below. Such *penetrative convection* is one of the ways in
which mixing occurs between troposphere and stratosphere across the tropopause.
Mesoscale convection is the dominant upward transporting process in the troposphere
over most humid, low latitude areas. Penetration of the tropopause is believed
especially frequent in the severe thunderstorms of the U.S. south-west, the Indian
summer monsoon, and probably the excessively wet areas of S.E. Asia and parts of
South America - but reliable statistics (derived from satellite measurements of
cloud top height) have not yet been compiled on a world basis.

Vertical motion associated with synoptic-scale cyclone waves* is on a much larger
geographical scale, though it is much slower - of the order of a few centimetres
per second. In this case the vertical motion is slanted at very low angles -
typically one kilometre of ascent while the air is travelling several hundred or
more kilometres horizontally, usually towards the north-east,** with corresponding

---

*the travelling cyclones and anticyclones of the surface weather map, plus the wave
 disturbances of the circumpolar westerlies (see below) that overlie them. The term
 "synoptic-scale" comes from the expression "synoptic-chart," and means systems
 typically 25 to 70 degrees of longitude in east-west extent.
**
  ascent in south-eastward and descent in north-eastward in the Southern Hemisphere.

rates of descent (subsidence) in south-eastward* moving currents. Such storm
systems - for cyclones are the main bringers of rains and strong winds in middle
and high latitudes - can achieve considerable north-south (meridional) as well as
upward transport.

The cyclone and anticyclone waves of the westerly belts thus achieve a major amount
of stirring of the troposphere along all three spatial directions (for they
typically move eastwards). They extend vertically well into the lower stratosphere,
since they are disturbances of the entire westerly current; the latter normally
extends to about 20-23 km (Fig. 13.3). The strong vertical motion is associated also with divergence or convergence** of the wind systems. Since these occur in the
parts of the lower stratosphere containing strong vertical gradients of ozone
concentration, where there is also strong horizontal convection, they cause considerable day-to-day variation in total ozone content of the air columns above regions
traversed by the cyclones. Similar effects are caused by disturbances of higher
level wind systems.

The remaining vertical transfer mechanisms are the so-called mean meridional
circulations, which are vast, slow, planetary overturnings in which general uplift
in some latitude is balanced by descent in another. The Hadley cells,*** for
example, consist of uplift over near-equatorial latitudes balanced by descent in
subtropical latitudes of both hemispheres. These particular cells are mainly
tropospheric. Obviously there has to be equatorward horizontal motion near the
ground (the trade-winds) and poleward motion at high levels - in this case mainly
in the 10-15 km layer.

The mean meridional circulations affecting the stratosphere, which might bring
about transport of substances upwards, downwards and along north-south lines, remain
imperfectly known. They cannot be observed directly. Instead we rely on two
sources of evidence - time averaging of observed horizontal winds at all levels,
which is prone to error, and the predictions of numerical models of the atmosphere's
dynamics and heat balance. We also take account of the displacement of trace
substances. Fig. 13.4 sketches a recent estimate of the overturnings that actually
matter in this report.

Scavenging and Fall-Out

Not all vertical transport depends, however, on convection - in which the air
carries some entity bodily. Some particles are large enough to descend at signif-
cant speeds usually fall out in this fashion in a few hours or, at the most, days.
Meteoric dust or deliberately planted tracers injected at high altitudes in the
upper atmosphere similarly fall under gravity until air resistance plus turbulence
slows down their passage.

The removal of gases and small particles, however, depends on scavenging, chiefly
by rain, snow or hail - i.e., precipitation. Certain particles that fall out from
the stratosphere may carry gaseous molecules with them, either in combination or
adsorbed. But scavenging is primarily a tropospheric sink for particles and gases.
Falling rain in particular will carry down in solution many sulphates, nitrates,
and chlorides - including the corresponding acids. The tropospheric residence

---

*   ascent in south-eastward and descent in north-eastward in the Southern
        Hemisphere.
**   divergence is a net fractional outflow from a volume. Convergence is an in-
        flow. Meteorologists apply this idea two-dimensionally, meaning a net hori-
        zontal inflow to or outflow from an <u>area</u>, i.e. $\frac{\partial u}{\partial x} + \frac{\partial v}{\partial y}$, where u and v are
        wind components towards x and y.
***   so-called because their existence was predicted *in 1735* by George Hadley, F.R.S.

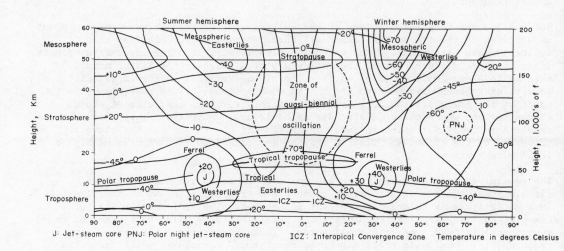

Fig. 13.3. Schematic generalized cross-section from pole to equator,
for summer (left) and winter (right) hemispheres. Isotherms in C,
isotachs of zonal wind in m s^{-1}. (WMO)

time* of water vapour itself is only of the order of ten days. Turbulence tends to
carry water upwards, thereby equalizing the mixing ratio, but rising air is generally cooled below its frost- or dewpoint, with cloud formation resulting. These
clouds in turn precipitate water or ice back to earth, carrying with them soluble
materials and particles encountered in descent. Soluble species hence tend to be
washed out before they reach the stratosphere. The chemically active oxides of
nitrogen, for example, are rained back to earth, and it is only the inert, feebly
soluble nitrous oxide, N_2O, that is able to reach the stratosphere. Chlorides, too,
which abound in the particle load of the lower troposphere, are quickly scavenged.

Advection (Horizontal Exchanges)

The atmosphere is in ceaseless motion, the winds being organized into great systems
familiar to everyone - the trade winds of the tropical oceans, the mid-latitude
westerlies of both hemispheres, and the seasonal monsoons of Africa and Asia.
These winds are as much convective as the vertical currents just described, but
meteorologists conventionally use the word *advection* for the horizontal transports
of air, particles and pollutants.

Figure 13.5 shows in chart form time-averaged pressure distribution over the earth
at two levels - the 300 mb pressure surface (about 10 km), and the 30 mb surface
(about 24 km), the surfaces being at polar tropopause level and in mid-stratosphere
respectively. The charts show contours of the height above sea-level of the pressure surface. The time-averaged wind blows nearly parallel to these isobars or
contours at a speed proportional to the pressure gradient (slope of the pressure
surface), in such a sense that lower pressure (or height) is to the left of an

---

* i.e. the average length of the time spent by a molecule or particle in some
   specified reservoir, in this case the troposphere.

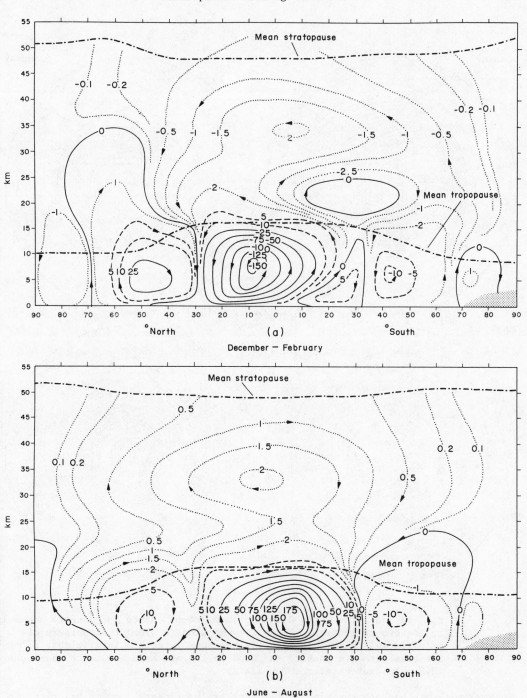

Fig. 13.4. Mean meridional circulation over earth, showing flow-lines in units of 10^{12} g s^{-1} mass transport. Diagram (a), for December-February, is more typical of the year than Diagram (b), June-August, where the Hadley cells are modified by the summer monsoons of the northern hemisphere. (J.F. Louis, CIAP Monograph 1)

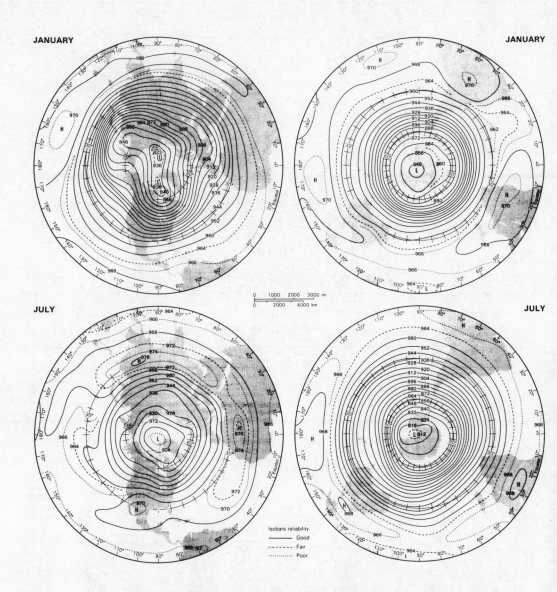

Fig. 13.5(a). The above diagrams show 300-mb charts for January and July in the two hemispheres. Mean flow of wind is along the contours of geopotential height, so that low pressure is to the left (right) of an observer with his back to the wind in the northern (southern) hemisphere. (H.L. Crutcher et al.— 300-mb data)

observer with his back to the wind in the northern hemisphere, and to his right in the southern. We have not included a 30 mb chart for the southern hemisphere.

These charts show that the dominant wind systems are zonal (i.e., west-east or the reverse), and that the zonal currents encircle the earth. The two hemispheres

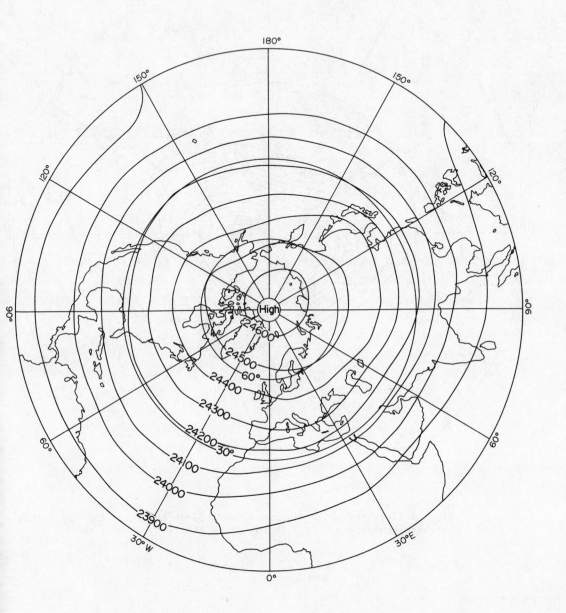

Fig. 13.5(b). The above diagram shows a 30-mb chart for July in the northern hemisphere. (R.A.Ebdon, 30-mb charts)

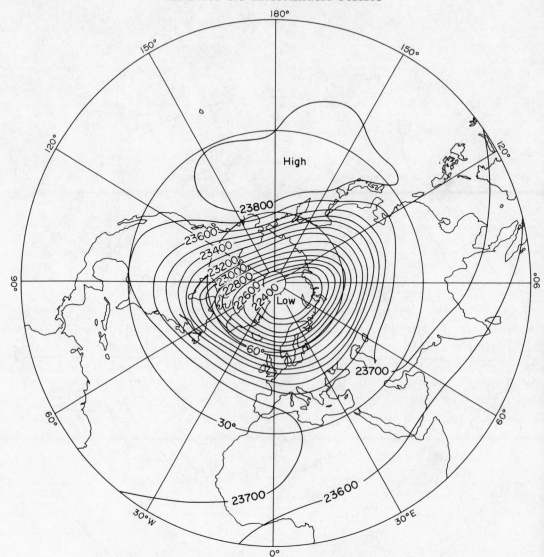

Fig. 13.5(b). The above diagram shows a 30-mb chart for January in the northern hemisphere. (R.A.Ebdon, 30-mb charts)

look as though they function separately, with only small and localized exchange
between them. But this is partly illusory. Certain quasi-permanent eddy systems,
such as strong flow from south to north over eastern equatorial Africa in the
northern summer, bring about exchange of air between the hemispheres. It has been
calculated that it takes about 12 months to complete such an exchange. Since most
particles fall out in a few days, the particle loads of the two hemispheres are
essentially separate.

The rule connecting wind to pressure stated above, the so-called *gradient wind law*,
is not perfectly obeyed. There is some flow of air across the isobars or contours,
chiefly (but not exclusively) towards lower pressure. This happens in the planetary
boundary layer because of frictional retardation near the ground. It also happens
in fast moving disturbances like cyclones, and in the trade-wind easterlies of both
hemispheres. Many of these cross-contour flows balance one another, but others are
not balanced by return flows. In the trade-wind belts, for example, there are net
flows towards the atmospheric equator (the Intertropical Convergence Zone) over
which the innumerable rising towers of convection are not quite balanced by equivalent descending currents. Together these net cross-latitude flows, with their
accompanying uplift and subsidence zones, constitute the mean meridional circulations
of the last section.

To bring this home we refer again to Fig. 13.3, showing vertical cross-sections of
the troposphere and stratosphere. In this meridional plane (i.e., a vertical plane
extending along a meridian from pole to pole) one can show, by means of lines of
constant speed (isotachs), the average structure of the easterly or westerly winds
blowing through the meridional plane. In addition one can sketch flow lines showing the direction and relative strength of the mean meridional circulations, not,
however, attempted in this discussion draft.

The tropospheric circumpolar westerlies of the two hemispheres clearly have strong
cores of maximum speed at 10-13 km above sea-level. There are typically two or
three jet-streams at these levels, each lying above a strong front separating cold
air on the poleward side from warm on the equatorward. In the stratosphere the
westerlies decrease with height - to zero in summer, and to about 20 m sec-1 in
winter - at the 20-23 km level. They are narrower and less strong in the summer
hemisphere. Except for a belt about 20 degrees wide on either side of the equator,
therefore, the lower stratosphere (where most of the ozone resides) is dominated -
forced in the dynamical sense - by zonal currents whose real sources of kinetic
energy are in the troposphere below.

On the 300-millibar chart (Fig. 13.5) - whose shape is typical of most levels from
3 to 16 km - it can be seen that these westerlies encircle the earth. If a parcel
of air remains near the core of the currents it may circumnavigate the globe in one
to two weeks. Constant-level balloons released into the southern westerlies have
indeed done so, though at slower speeds, since they did not remain in the fastest
moving air. In the northern westerlies, however, there is much entrainment of air
from either side in some longitudes, and much ejection of air in others, so that
objects or substances being transported by the westerlies have a good chance of
being thrown out on either side, and hence mixed through a wide latitude belt.
The southern westerlies behave similarly, but are wider and more intense than the
northern. In both hemispheres natural or artificial additions to the atmosphere
in mid-latitudes will be quickly transported round the globe, and widely distributed
geographically.

North-south - meridional - mixing of this sort depends on the disturbances to which
the westerlies are highly prone. As Fig. 13.5 shows, there are large-scale roughly
horizontal waves in the currents, especially in the northern hemisphere. These
planetary waves are slow-moving, and have certain preferred longitudes. Hence they

are visible even on time-averaged charts. They are of considerable importance in transporting heat. Not visible in the charts, however, are the cyclone and anti-cyclone waves already discussed. These are transient systems, moving rapidly east-ward, usually with a poleward component. We have already seen that the vertical motion created by these large systems - 25-70 degrees of longitude wavelength - brings about much vertical transport. They are also responsible for north-south transfers of water vapour, particles and pollutants. Moreover, in their late stages cyclones often move slowly across the north Polar basin, aiding in the horizontal mixing of the entire hemisphere. Model calculations suggest that uniform meridional mixing of the northern hemisphere is approached in two to three months. Still more critically these waves bring about some of the actual exchanges of mass, including trace gases, between troposphere and stratosphere.

The low-level intertropical regime is quite different. On either side of the equator the zonal flow is a light easterly called the trade winds - north-east at low levels in the northern hemisphere, south-east in the southern. The low-level trades converge on an axis called the Intertropical Convergence Zone that is most often on the northern side of the geographical equator. The trades are associated with general but rather shallow cumulus-scale convection. They are quite shallow - well below 10 km in depth. Very large disturbances are absent (except for the pro-found rearrangements over the Indian Ocean, South-East Asia, West Africa and the western Pacific due to monsoonal currents), but there are numerous mesoscale or larger areas of shower and thunderstorm activity that are effective upward trans-porting agents, and in summer and autumn small but intense cyclonic storms - hurricanes, typhoons - occur in oceanic areas.

The dominant feature of the stratospheric flow outside the tropics is a colossal, annually reversing system of circumpolar zonal winds - westerlies in the winter hemisphere, easterlies in the summer, with maximum speeds in both cases at the stratopause or in the mesosphere. As in the troposphere, substances injected at these levels will be rapidly carried round the earth. Poleward or equatorward dispersion, however, is much less rapid than in the troposphere, since there are fewer disturbances at these high levels.

The winter stratospheric westerlies, blowing from September through March, are strong, variable from year to year, and rather different in the two hemispheres. In the northern hemisphere they are systematically deformed into standing waves by a persistent area of high pressure over Alaska. In the southern they never flow west-east. They appear as strong westerlies as low as 20 km in latitudes 55-75 of both hemispheres, where they are known as *polar-night westerlies*. They are subject to vigorous disturbances that produce strong uplift and subsidence through great depths. These give rise to strong temperature changes. The associated subsidence creates dramatic "sudden warmings," sometimes of 20° to 30°C in a day, which quickly destroy the westerly system. Large increases in total ozone occur with each warming, and there is no doubt that these systems account for the accumulation of much ozone in the lower polar stratosphere late in winter. Figure 13.6 illustrates some of their effects.

The summer stratospheric easterlies also descend to about 20 km, but are light and almost disturbance-free below 30 km. Slow moving waves, with some vertical motion, occur at higher levels in middle and low latitudes.

The remaining area of the stratosphere is between the 30th parallels. In this broad intertropical zone - half the area of the earth - there are two non-annual cycles at work, superimposed on the average zonal winds shown on Fig. 13.3. These are a six-month cycle, very variable from year to year, with maximum effect at about 40 km, and a roughly two-year cycle - actually with a variable period averaging near 26 months - with maximum effect at 20-35 km. Figure 13.7 shows that the result of the latter is to produce a remarkable quasi-periodic reversal of wind

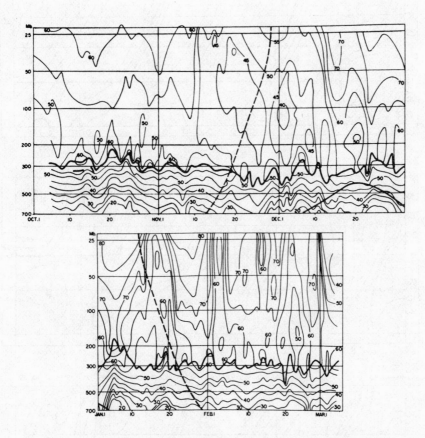

Fig. 13.6. Time-pressure diagram showing variation of temperature (C)
and tropopause height (heavy black line) in 1958-59 at Resolute,
Canada, in latitude 75N. Note rapid variation of temperature in the
troposphere and basal stratosphere, caused by passage of synoptic disturbances, which also caused the rapid variation of tropopause level. The
slower, very deep temperature waves in the middle and upper stratosphere
are associated with disturbances of the polar night westerlies. Both
sets of disturbances produce short-term variations in total ozone.
(F.K. Hare)

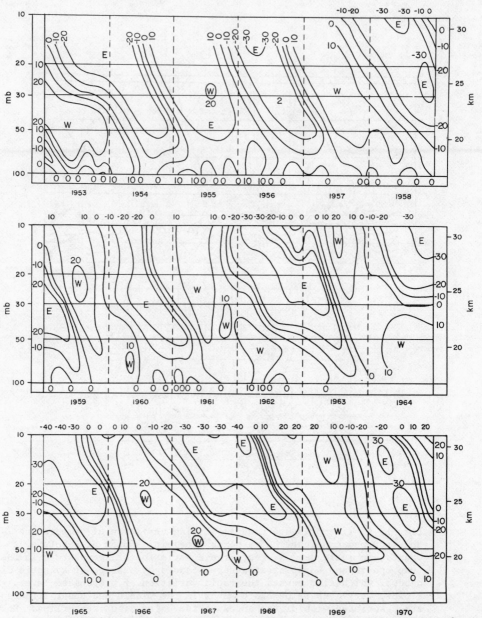

Fig. 13.7. Time-height diagram of mean monthly zonal winds (m s^{-1}) over
Canton Island (1953-1967) or Gan (1967-70), both near equator, showing
the quasi-periodic alternations of west and east winds due to the
26-month oscillation. Note the slow descent of each wave. (Kriester,
CIAP Monograph 1)

from west to east and back. This is the layer into which much of the nuclear
debris was injected in early U.S. testing over the Pacific.

Troposphere-Stratosphere Mixing Processes

We said above that the lower stratosphere is resistant to vertical motion, and
hence to the exchange of substances and properties with the troposphere below. In
other words the tropopause appears to act as a lid to the troposphere. Obviously
it is an imperfect lid. Many substances must cross it to balance the various
budgets. What mechanisms are there for such transfer?

At least six possibilities can be recognized. These include, (i) transfers through
the tropical tropopause at 17 km by the upward and downward arms of the Hadley
mean meridional circulation cells; (ii) penetrations of the tropopauses by tall
thunderstorms; (iii) roughly horizontal transfers through the tropopause gap in
mid-latitudes; (iv) down-folding of the tropopause in strong frontal zones,
especially when cyclonic disturbances are active; (v) transfers by other mean
meridional circulations not yet positively identified; (vi) fall-out, possibly with
some scavenging. To these we must add upward-moving gravity waves, which are
thought to play a rather special role in the lower stratosphere.

Figure 13.4 shows the mean meridional circulations, according to a very recent
analysis. The powerful Hadley cells of the tropics are visible in all seasons
except June-August (when they fuse into a single, powerful cell centred just south
of the equator, mainly because of the great Asian and African monsoonal systems).
The flow lines of the cell extend into the lower stratosphere, though at very low
mean rates. They imply an upward drift of air near the equator, and a gentle
subsidence near the 30th parallels of latitude. Above them there is the suggestion
of a slow but enormous overturning involving the entire planet, with slow uplift
in the heated summer stratosphere, and descent in the cooled winter polar caps.
There is also a net transfer of mass from the summer to the winter hemisphere
estimated to be in the range 5 to 8 x 10^{13} tonnes.

A more complex arrangement may be typical of the lower stratosphere in winter, below 30 km, and one that differs between hemispheres. In the northern hemisphere
at these levels general ascent has been postulated for both equatorial and polar
latitudes, with subsidence in mid-latitudes (where the lower stratosphere is
warmest). In the southern hemisphere a three-cell arrangement has been suggested,
with slow uplift over the equator and in high-non-polar latitudes, with subsidence
in mid-latitudes and over Antarctica. There is doubt about these overturnings,
however; Brewer (private communication) believes that they are inconsistent with
the observed lack of water vapour transport into the stratosphere.

Clearly if these mean meridional circulations are accurately portrayed - and this
is not yet certain in the stratosphere - they must play a significant role in
transporting trace substances, particles and water vapour between troposphere and
stratosphere, and also between hemispheres. But difficulties remain. What, for
example, is the actual process whereby air passes upwards or downwards through the
tropical tropopause, which is highly permanent, and which cuts across the Hadley
cells? This is a cold surface, at about -80C, and only significant forcing could
make air rise through it into the warmer stratosphere above. Is this in practice
an intermittent process due to some species of unidentified disturbance capable of
providing the forcing? Is the net *upward* transfer due to the penetrative convection
mentioned earlier, whereby numerous thunderheads penetrate the tropopause? These
are in part unanswered questions.

The gap between the polar and tropical tropopause offers an obvious route for
roughly horizontal mixing. A current moving poleward from the tropical zone at,

say 16 km (about 100-mb pressure) will in fact move from the tropical troposphere into the polar stratosphere. Strong waves in the westerly jet streams, which are 2 to 5 km below this level, must often induce such transfers. Such air is necessarily dry, because it can hardly be warmer than -75C. Hence no appreciable water exchange can occur, though particles and trace gases may well be carried through the gap.

A related mechanism studied intensively by U.S. and U.K. groups is *tropopause folding* (Fig. 13.8). Immediately below the jet stream cores of the westerlies there are found strong, sloping frontal zones - airless boundaries between cool air on the poleward side and warmer on the tropical. In association with eastward moving cyclone waves of the westerly current, strong, slanting subsidence of stratospheric air occurs in the frontal zones, bringing down low humidities, radioactive tracers and ozone from the lower stratosphere. Aircraft flying through such fronts have encountered these obvious symptoms of subsidence from the stratosphere as low as 3 or 4 km above sea-level. The substances thus removed from the stratosphere are subsequently mixed into the general tropospheric reservoirs, and in due course may reach the earth's surface.

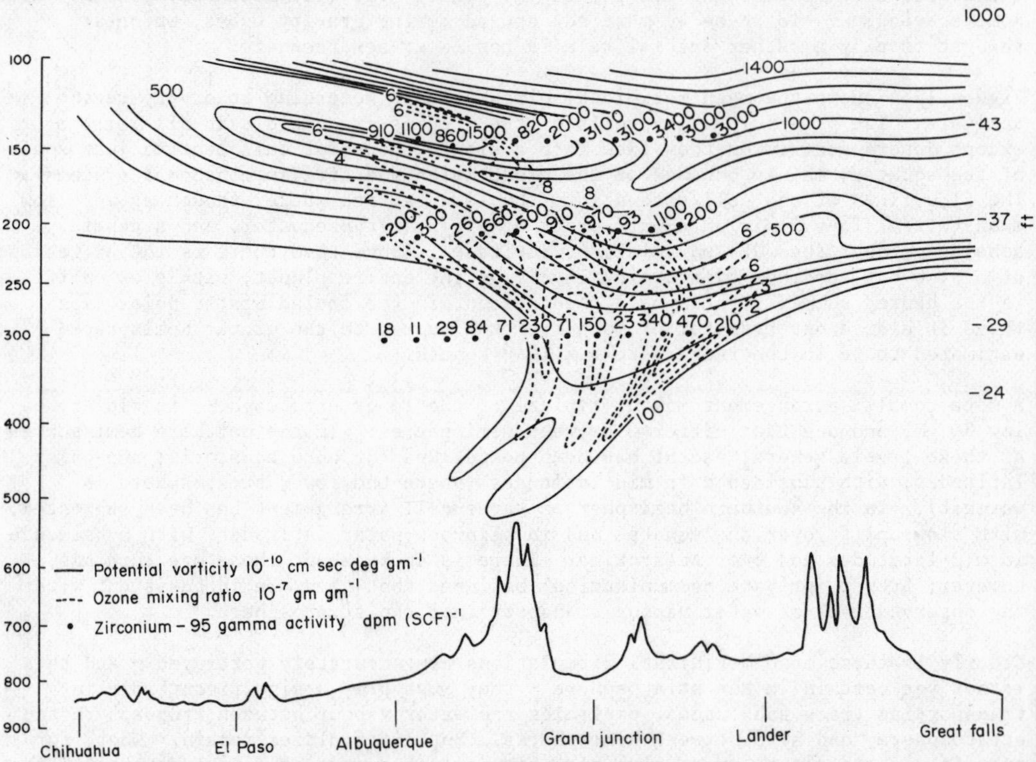

Fig. 13.8. Isopleths of potential vorticity (a thermodynamic parameter), ozone mixing ratio and zirconium-95 gamma radiation, showing penetration of troposphere by tongues of air descending along a tropopause downfold over Grand Junction. (E.F. Danielsen et al, American Geophysical Union).

We have already discussed the penetration of the basal tropical or mid-latitude stratosphere by tall thunderheads, which leave behind some water vapour, particles and trace substances as they dissipate. One recent analysis sees this as the largest source for water vapour in the stratosphere, estimating that perhaps a tenth of all Indian monsoonal thunderstorms penetrate the tropopause. There are well authenticated cases of cloud tops reaching over 20 km.

Fall-out can be dismissed briefly, since it cannot be very significant. The particles of the lower stratospheric aerosol layer at about 20 km are too small to precipitate rapidly. Some meteoric dust does presumably gravitate downwards across the tropopause. Such dust may enter into various physical and chemical processes in the stratosphere, and may hence carry some non-meteoric molecules or atoms along with it.

It is painfully clear that we have only a qualitative picture of most of these vital exchange processes. Much research remains to be done at this dynamically very active level before we can predict with reasonable certainty the rates of transfer of pollutants or natural constituents across the tropopause.

Calculation of Net Transport

Later in the report we discuss the various calculations that have been made of changes in ozone likely to result from human interference. All such calculations depend on an adequate estimate of the net transport of various gases achieved by the above exchange processes.

So far, these calculations have mainly made use of ideas based on the properties of small-scale turbulence, which dominates the diffusion process near the ground. These ideas in turn derive from analogies with molecular diffusion. In brief, it is assumed that turbulent eddies will cause gases and aerosols to diffuse from regions of high concentration to regions of low. The net transport is hence down the concentration gradient, from high to low.

The effectiveness with which turbulent eddies (of unspecified size) bring about this transport is expressed by a diffusion coefficient, properly called the *eddy diffusivity*, usually written K. The rate of net transport is proportional to the product of this coefficient with the magnitude of the concentration gradient.*

By making estimates of source strengths and reservoir capacities, one can calculate the value of the diffusion coefficient from the observed concentration gradient. Various tracers have been used for this purpose, nitrous oxide (N_2O), methane (CH_4) and certain radioactive substances (notably carbon-14) being current favorites.

Figure 13.9 shows values of the coefficient that have been typically adopted in recent calculations of the ozone cycle. In the middle and upper troposphere estimated values tend to be of order 10^5 cm^2 s-1. In the lower stratosphere - immediately above the tropopause - they fall abruptly to near 10^3 cm^2 s-1, indicating that net transport rates for a given gradient are only about one hundredth of those typical of the troposphere. At higher levels in the stratosphere values rise again to 10^4 cm^2 s-1 or more. Horizontal transfer coefficients in the upper troposphere appear to be of order 10^{10} cm^2 s-1, but these also decrease rapidly in the stratosphere.

---

*For water vapour, for example, the upward rate is presumed to be
$E = \rho K_w \, \delta q / \delta z$, where ρ is air density, q is the fractional concentration by mass of water vapour in the air (specific humidity), z is height and K_w is the diffusion coefficient for water in cm^2 s-1, the required dimensions.

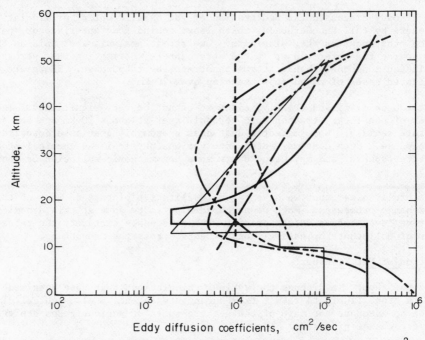

Fig. 13.9. Vertical profiles of the eddy diffusivity (K) $(cm^2 s^{-1})$ showing the wide range of values chosen in various attempts at one-dimensional models of stratospheric processes. (National Academy of Sciences)

Meteorologists feel uncomfortable about the use of these coefficients, because they do not adequately describe the large-scale exchange mechanisms just listed. Eddy diffusion coefficients are very variable in time and space. In the lowest layers of the atmosphere turbulent transfer is brought about by very numerous small rising and falling currents, and for such a regime a statistical measure like a diffusion coefficient has some physical validity. But the large-scale mechanisms often involve strong negative time correlations between velocity components, and this may actually cause transport *up* the mean concentration gradient. Sensible heat, for example, may flow *up* the average gradient of potential temperature.* The spring ozone transport in the stratosphere is often *up* the mean gradient of ozone mixing ratio. To incorporate these seemingly paradoxical results into diffusion models one has to adopt a far more sophisticated kind of calculation than is possible from present observational data.

This has been one of the major difficulties in attempting to calculate the disturbance to the ozone concentration resulting from human interference. Charac-teristically the results of such calculations are subject to large errors, some-times of the order of a factor of two or three, because we have not known how much confidence to attach to the diffusion "constants" - which, as we have just seen, are actually highly unpredictable variables.

Later in the report we deal with the various one-dimensional models from which one can calculate the effect of stratospheric trace gases on ozone concentrations. These models calculate a global average in which height and time alone vary. They

*the temperature the air would have if brought to a pressure of 1000-mb in the unsaturated state and without loss of heat.

depend on eddy diffusion coefficients calculated from very sparse data on the
vertical profiles of methane and nitrous oxide, measured in a few areas of the
world - Texas and southwestern France, for example. The observed profiles show
considerable irregularities. It is not clear how much of this is real, and how
much is "noise." The eddy diffusion coefficients calculated from these profiles
are wildly variable unless considerable smoothing is applied. The lower stratosphere regularly displays a slab structure as regards ozone and particle distribution, so it may not be legitimate to regard the irregularities in the profiles
of other tracers as noise. Nevertheless, the one-dimensional models described
below seem to do a reasonable job, *provided that they are seen as accounting for
average global conditions that do not attempt to simulate the physical processes
described above.*

Stratospheric Residence Time for Gases and Particles

When an atmospheric layer is well-mixed, as is the troposphere, it is reasonable to
talk about a *residence time** for substances stored in such a reservoir. The
residence time obviously depends on the strengths of the sources and sinks for the
substance, and the capacity of the reservoir. Thus sulphur dioxide, largely
emitted at ground-level from industrial sources, has an average tropospheric residence time of about four days, the sink to the earth's surface being efficient.
Water vapour, regularly precipitated in the form of rain or snow, has a residence
time of about ten days.

But the stratosphere's high static stability retards vertical mixing, and meridional
mixing is also slow. Substances injected at specific levels and latitudes tend to
remain there for weeks, months or even years - the time increasing with height
above the tropopause - while being distributed round the earth by strong zonal winds.
Hence the residence time is not a single figure for the entire stratosphere. Moreover there are numerous chemical sources and sinks at these levels, so that a
substance such as nitrous oxide may enter the stratosphere only to be transformed
into some other molecule. Ozone, on the other hand, is created and probably
destroyed within the layer, yet part of the annual production diffuses downwards into the troposphere. In fact, residence time in the stratosphere is usually taken
to mean the time spent in the stratosphere by substances uniformly mixed in the
tropospheric reservoir - a sort of two-reservoir exchange. Obviously this is in
many ways an artifical idea.

Our knowledge of this question comes mainly from studies of well-known tracers, both
particles and gases. The list includes ozone (which is photochemically more stable
below 25 km) and a whole list of elements, such as phosphorus-32 and beryllium-7
created by cosmic radiation, or tungsten-185, plutonium-238, carbon-14, strontium-
89, and -90, zirconium-85, rhodium-102 and cadmium-109 derived from nuclear bomb
tests or deliberate plantings. Some of this material is gaseous (e.g., the carbon-
14 occurs in carbon dioxide) but most is particulate.

From the many studies of the behaviour of these tracers we can conclude roughly as
follows:-

- Residence times for chemically stable substances in the lower stratosphere increase with height from a week or two at the north polar tropopause to several
  years at greater heights. In detail the residence time is much affected by the
  level and topography of the tropopause (see Fig. 13.2).

---

*As defined previously, this is the average time spent in a layer or reservoir by
  individual molecules or particles.

- Substances injected into the lower equatorial stratosphere (17-25 km) tend to remain there for some months, and are still present in appreciable concentrations two years after injection.

- When tropical concentrations disperse to higher latitudes they do so along downward sloping paths. When they arrive in 60 degrees latitude they are often 5-10 km below their original level. Substances injected in mid-latitudes tend to disperse upwards towards the equator and downward towards the poles along similarly sloping surfaces. Such processes require many months to complete.

- Injections at levels above the stratopause lead to the descent of particles into the high latitudes of both hemispheres, rather than the tropical belt.

- The sloping surfaces of dispersion referred to above slope poleward more steeply than surfaces of constant potential temperature; a fact that has important dynamical implications.

- It is hence impossible to define a single residence time for the stratosphere. Instead we have to reckon with a rapid increase with height of such residence times, especially above 20 km. The various exchanges processes with the troposphere discussed above are all most effective in the lower stratosphere.

- These facts have serious implications for stratospheric pollution, delt with later in detail. The chlorofluromethanes, which appear to be stable at low levels are by now well-mixed in the northern hemisphere troposphere. They are engaged in slowly penetrating the upper stratosphere, (where they are dissociated by ultraviolet) by means of the mechanisms described earlier. Because of the rapid increase of residence time with height, it will take more than a century before the process cleanses the atmosphere of the molecules already present, even if we cease manufacture now.

CLIMATOLOGY OF OZONE VARIATION

As will be shown later, stratospheric pollution threatens to alter the ozone content of the stratosphere by 10 per cent or more. It may already have lowered it by as much as 1 per cent. These changes have been and will continue to be mixed with larger natural variations, spatial, seasonal and short-term in character. One must obviously take into account these natural variations when considering the effects of pollution.

Ozone Observations

We have already seen that the total ozone content of an atmospheric column is expressed as the depth it would have in centimetres if compressed to standard surface-level temperature and pressure. Average total ozone in these units is a little over 0.3 cm. The ozone concentration at specific levels is usually given in terms of its mass mixing ratio (in micrograms - μg - per gram of air), or its partial pressure. Also widely used is the number density (molecules per cubic centimetre or cubic metre).

Direct observation of ozone has a history going back over 50 years, but most information comes from the past three decades, and is very unequally distributed about the earth, with a marked northern hemisphere bias. There is an urgent need for more stations, and for international uniformity of instruments and standards. Observations can be obtained in the following main ways:-

- Spectrophotometric observations of total ozone obtained at ground level by an instrument devised by G.M.B. Dobson of Oxford University, England. The method involves measuring the attenuation of one or two wavelength pairs of direct sunlight, but can also be used for zenith-sky light on cloudy days. Figure 13. 13 shows the network of Dobson-spectrophotometers in present or recent use.

Sixty-four such stations currently publish observations through Ozone Data for the World, published by the Atmospheric Environment Service, Canada, on behalf of WMO. The longest records are from Arosa, Switzerland (since 1925), Tromsö, Norway (since 1930) and Oxford (since 1926).

- Filter photometric observations from the U.S.S.R., also yielding estimates of total ozone, unfortunately not easily compared with the Dobson values. These are shown on Fig. 13.13.

- Various balloon-sonde or rocket-sonde observations of vertical profiles of ozone partial pressures below 35 km (with balloons) or below 70 km (with rockets). The high cost of these networks has meant that they have usually operated for only short periods.

- Satellite observations of the ozone layer from down-looking sensors detecting back-scattered ultraviolet, which yield some vertical detail, and give world-wide geographical distributions.

- The so-called "umkehr" method, in which the Dobson-spectrophotometer is used to get crude vertical profiles of ozone density. Historically this was the first method to yield such profiles, but it is not as useful as the balloon/rocket-sonde or satellite observations.

The development and operation of these observational systems has depended very largely on the enthusiasm, dedication and pertinacity of a small group of specialists, often in part at their own expense. It is urgently necessary that ozone monitoring be put on a firm international footing.

Vertical Distributions
_____

In Fig. 13.10 we show a recent estimate of the mean vertical distribution of ozone, in terms of mid-latitude number density. There is a peak just above 20 km (about 7×10^{18} molecules per cubic metre) overlying tropospheric values near 8-9 $\times 10^{17}$. There is high short-term variability below the level of maximum density, with a maximum relative standard deviation near 11-13 km, a level just above the polar tropopause, and very close to the core of the westerly jetstreams. Above the 20 km maximum, density decreases quite smoothly, but is still almost at 10^{17} molecules m^{-3} at the stratopause. If we translate these figures into mass mixing ratios (micrograms of ozone per gram of air), the relative abundance of ozone is highest in the 30-35 km layer. Fig. 13.11 shows that these levels depend on latitude. Density is highest over the equator at 25 km, but from the 60th parallels to the poles the maximum is at about 17 km, the downward trend being most marked in mid-latitudes.

In Fig. 13.12 we show sample vertical ozonesonde soundings for stations in northern Canada (where total ozone is usually highest) and in tropical America. At the tropical site the ozone layer is high, and has a smooth profile both above and below the maximum level. At the Canadian site, however, the profile *below* the maximum is marked by many shallow layers or slabs of alternating low and high values. Similar short-lived slabs occur in profiles of wind, temperature and particle concentration. Most ozonesonde ascents in the westerly belt and polar latitudes show this slab structure. It may be due to variations in the amplitude of the waves affecting the westerly current. Given that there is a strong north-south gradient of ozone density at these levels, short-term variations with height in the meridional component of wind could account for significant differences of ozone density. The fact that the slabs are short-lived, however, suggests that they may be due to smaller-scale horizontal and vertical motion induced by breaking gravity waves. These are induced in the lower stratosphere by the travel of the tropospheric westerlies over mountains, or even over strong thunderstorm areas.

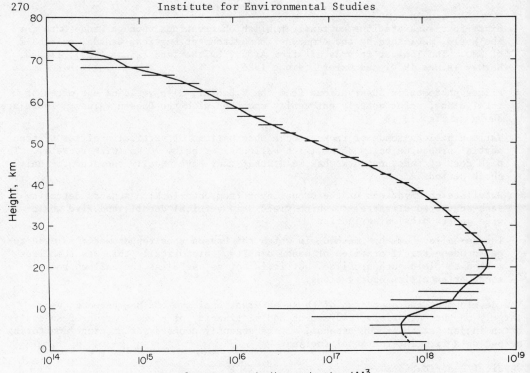

Fig. 13.10. Tentative mean annual mid-latitude ozone number density
as a function of height, with range of variation (horizontal bars).
(Kruger & Minzner, CIAP Monograph 1)

Regardless of cause, the presence of these slabs means that the vertical gradient
of ozone density in middle and high latitudes varies widely between the tropopause
and the level of maximum density (about 20 km). It may even reverse in sign
several times within this short distance. The same may at times be true for
horizontal density gradients along a north-south axis across the westerly jets.
Instantaneous vertical and horizontal gradients in these latitudes are rarely smooth,
uniform, and of consistent sign - unlike the time-averaged gradients.

Spatial Variation of Total Ozone

Figure 13.13 is a recent synthesis of long-term mean annual values of total ozone.
Given the very unequal distribution of stations on which it is based, only the
overall pattern can be trusted, and not all the details. Recent satellite observations, however, confirm the patterns shown, while displaying considerably more
transient detail.

Ozone is a maximum in sub-polar latitudes in both hemispheres, in about 70 N and
60 S. Highest values (over 0.380 cm) occur over north-east Siberia and the eastern
Canadian Arctic. In the southern hemisphere (where distribution is only sketchily
known) highest values (over 0.340 cm) are believed to occur south of the Indian
Ocean, Australia and New Zealand, but north of the Antarctic shoreline.

Minimum total ozone occurs in a belt a little north of the geographical equator.
Values are near 0.250 cm. A strong equatorward gradient extends in both hemispheres from the 50th or 55th parallels to the 20th or 25th - almost directly above
the tropopause gap. The spatial amplitude of variation about the global mean is

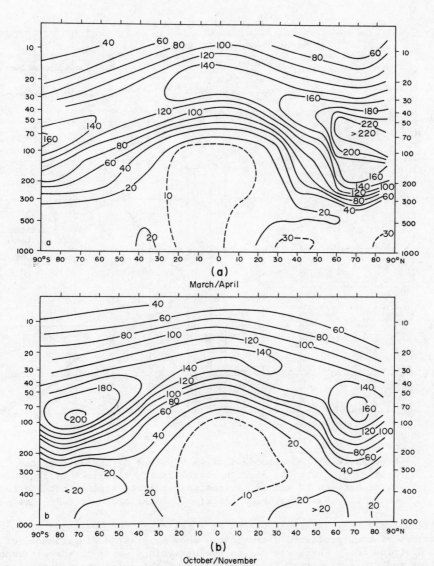

(a)

March/April

(b)

October/November

Fig. 13.11. Pole to pole cross-section of vertical ozone distribution (partial pressure, in nanobars - 10^{-3} mb), showing the characteristic altitude and topography of the ozone layer. Ordinate in total pressure (10 mb or 30 km). (H.V. Dutsch)

hence 17 to 19 per cent of that mean. It will be shown later that even short-term time variations may exceed this value.

Although this general pattern persists through the year, in March and April the northern hemisphere maximum tends to shift to the pole. The corresponding drift in the southern hemisphere maximum is less emphatic, and is delayed until November and December (i.e., the equivalent of two months later than in the northern spring).

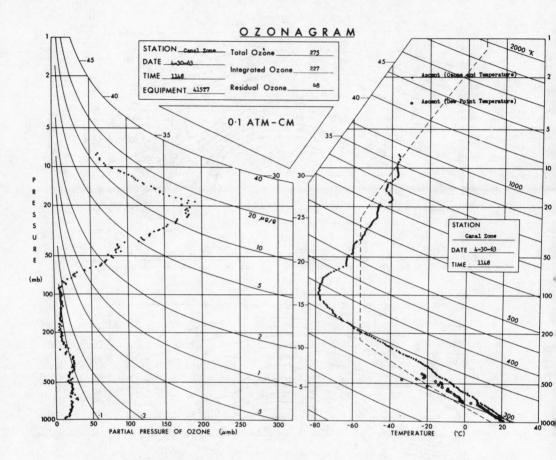

Fig. 13.12(a). Ozonesonde diagram typical of spring over the Canal Zone (9N). Note the absence of the rapid, slab-like variations of ozone between 10 and 17 km (central ordinate scale) that are found in Churchill, Manitoba, Fig. 13.12(b). Total ozone is 0.275 cm.

On individual days instantaneous values of total ozone, now revealed by satellite measurements of back-scattered ultraviolet, show many large disturbances of the average distribution. These are due to moving cyclone waves in the circumpolar westerlies and, in some cases, to disturbances of the polar night westerlies (see below).

Long-Term Variations in Total Ozone

We have very little information about long-term trends in total ozone, since even the oldest continuous records are less than fifty years old.

There is some evidence, however, of a small quasi-cyclic variation with an eleven-year period, presumably related to the similar sunspot cycle, and associated variations in solar ultraviolet or particle emissions. The range at Arosa is about 4 per cent of the long-term mean. Ozone maxima tend to follow sunspot maxima by several years. The establishment of trends is in any case difficult, because for individual stations apparent trends may actually have a dynamical origin - because of shifts in the long waves affecting the thermal structure.

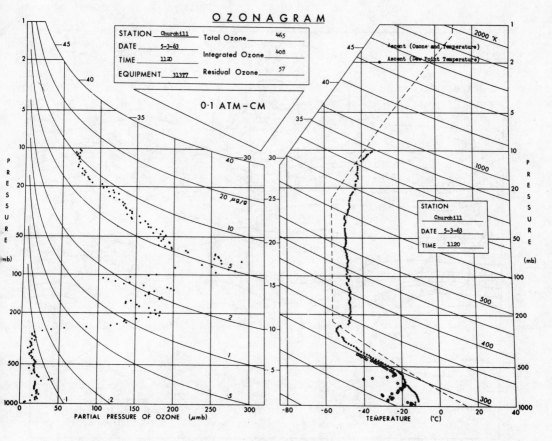

Fig. 13.12(b). Ozonesonde diagram typical of spring over Churchill, Manitoba (59N). Note the rapid, slab-like variations of ozone between 10 and 17 km (central ordinate scale). Total ozone is much higher over Churchill than over the Canal Zone (0.465 cm vs. 0.275 cm).

At many stations around the world total ozone rose slightly in the middle and late 1960s. A fall of a few per cent has occurred since 1970. These variations closely resemble earlier variations at long established stations. There is nothing to suggest that they are out of the ordinary.

The impact of stratospheric pollution to date is calculated to be much smaller than the above natural changes. If real, it is obscured in the records.

Short-Term Variations in Total Ozone

Rapid day-to-day changes in total ozone occur at stations in middle and high latitudes of both hemispheres, especially in winter and spring. These were observed fifty years ago, and it was recognized that they were strongly correlated with the passage of moving cyclones and anticylones at low levels. These are, of course, the cyclone-anticyclone waves of the circumpolar westerly currents.

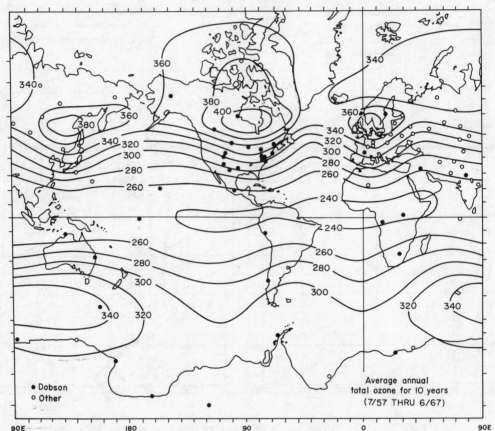

Fig. 13.13. Average annual distribution of total ozone (in thousandths of a cm), showing location of observing stations. (WMO, after Bojkov and London).

These variations are caused by advections, and ascent and descent of the air in the lower stratosphere, where much of the total ozone in these latitudes is stored. Descending currents tend to bring down higher ozone mixing ratios, and in addition are accompanied by high level convergence of ozone-rich air into the column. Such descent occurs immediately *behind* the surface cyclone. Ahead of it, over the preceding ridge of high pressure, there is ascent and high level divergence. Between them these two forced motions in the lower stratosphere produce large variations in surface measurements of total ozone - occasionally of amplitude 15 to 20 per cent of the local average. The range of such forced variations is between 0.230 cm and 0.550 cm, which exceeds the latitudinal range of mean annual values.

Other short-term variations are induced by uplift and subsidence due to disturbances of the middle and upper stratospheric polar-night westerlies, at altitudes above 15 km, with maximum effect in the 20-30 km range. At these levels the wave disturbances are mostly of long wavelength, and they move slowly, with little

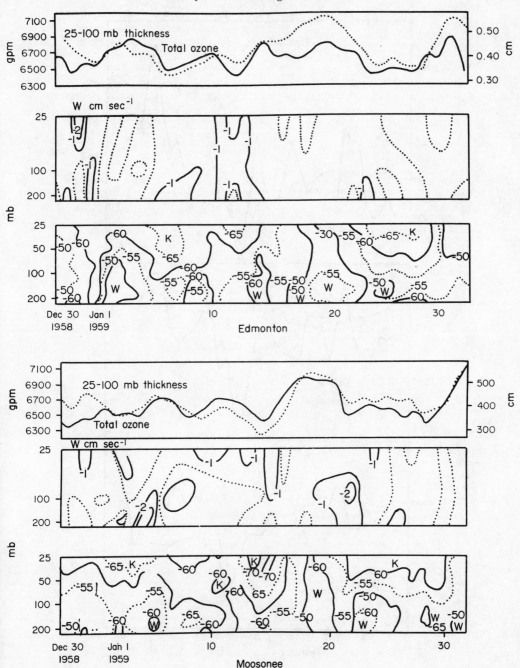

Fig. 13.14. Variations of total ozone, temperature and vertical velocity over
Edmonton and Moosonee, Canada, in January 1959. For each station top graphs
show 25-100 mb thickness (an index of mean temperature) and total ozone. Actual
temperature variations are shown in lower graphs. Obviously most of the variance
of total ozone was accounted for by the mean 25-100 mb thickness : a warm lower
stratosphere was ozone-rich, a cold was ozone-poor. (Allington, Boville and Hare)

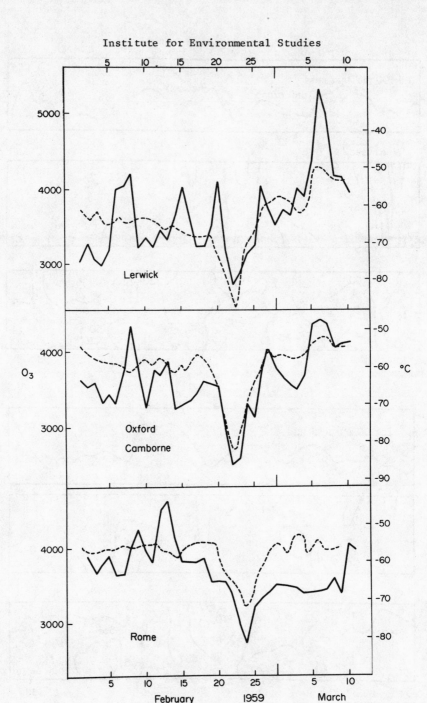

Fig. 13.15. Variation of temperature (C) (dashed curves) and total ozone at
various European stations during February and March, 1959. The rapid, short-
term ozone variations were due to the passage of cyclone-anticyclone waves.
The sharp minimum (with very cold temperatures) between February 20 and 25 was
due to the passage across Europe of a middle-stratospheric disturbance of the
polar-night westerlies. (Boville and Hare)

vertical motion. But from time to time in winter very rapid changes in the amplitude and phase of these long waves may occur almost impulsively, especially in the northern hemisphere. Strong uplight, with rapid cooling, alternates with even stronger subsidence, which produces rapid rises of temperature. In addition, deep troughs of low pressure may break away from the polar-night vortex, and travel towards mid-latitudes. These troughs are very cold, and poor in ozone, whereas the warm ridges produced by the sudden warmings are ozone-rich.

To illustrate these effects we present Fig. 13.14-15, which show the effect of both low level disturbances and events in the middle stratosphere on total ozone.

It is possibly the case that solar eruptions, which at infrequent intervals bombard the earth with showers of protons, produce significant reductions of ozone. The event of August 1972 had a measureable effect on total ozone, especially in north polar latitudes.

Summary

It is clear that there are natural variations in total ozone on short-term, seasonal and spatial scales that are at least as large, and in some scales larger, than the reared reductions in total ozone due to pollution. Since the natural variations are due primarily to rearrangements of chemically stable ozone in the lower stratosphere, well below the photochemically active layer, similar variations can be expected in other inert trace gases at these levels. The frequency of observation is not sufficient at present to bring this out.

CHEMISTRY OF ATMOSPHERIC OZONE*

As we have seen, ozone (O_3) is one of the minor constituent gases of the stratosphere. Although its concentration there is only a few parts per million, ozone plays a vital role in the existence of life on this planet. This is because it is the only substance in the atmosphere which absorbs most of the solar ultraviolet radiation in the spectral region from 240 to 320 nm. It thus prevents this radiation from reaching the surface where it would otherwise be destructive of all life forms. We deal with these effects later. Moreover, the absorption of this radiation (as well as the absorption of infrared radiation) by ozone affects the heat balance of the atmosphere and is, in fact, responsible for the upward increase of temperature that characterizes the stratosphere. Thus, any change in the amount of ozone in the stratosphere would have an effect on the biologically active radiation which reaches the surface and would also affect the atmospheric heat budget, thereby affecting global climate.

Ozone Creation and Destruction

The ozone layer is the result of a balance between processes that produce it and others that destroy it. The production processes were first identified by Chapman in 1930. They begin with the dissociation of molecular oxygen by solar ultraviolet radiation at wavelengths less than 242 nm:

$$O_2 + h\nu \rightarrow O + O \qquad\qquad (1)**$$

The oxygen atoms then combine with molecular oxygen to form ozone:

---

* Based largely on a text by H.I. Schiff.
** $h\nu$ is the product of Planck's constant with the frequency of the radiation.

$$O + O_2 + M \rightarrow O_3 + M \tag{2}$$

where M represents any gas molecule and is required to stabilize the newly formed ozone molecule by converting the chemical energy into kinetic (heat) energy.

Chapman originally proposed that the main destructive process was the attack of O_3 by another oxygen atom:

$$O + O_3 \rightarrow O_2 + O_2 \tag{3}$$

It is now known that this reaction removes only about 20 per cent of the O_3 that is formed. Another 0.5 per cent is removed by downward transport to the troposphere, and eventual destruction by contact with the ground. Since production must balance destruction on any extended time scale there must be other chemical removal processes that occur in the natural stratosphere.

Bates and Nicolet were the first to suggest (in 1950) that trace constituents in the stratosphere could provide the balance by removing O_3 in catalytic cycles in which one molecule of the catalyst could destroy many molecules of O_3. One of the first catalytic schemes that they proposed involved the free radical HO, which is formed in the atmosphere by fragmentation of water and methane, CH_4. The main catalytic scheme involving this free radical is the reaction

$$HO + O_3 \rightarrow HO_2 + O_2 \tag{4}$$

followed by

$$O + HO_2 \rightarrow HO + O_2 \tag{5}$$

It will be seen that the HO radical consumed in reaction (4) is regenerated in (5). The net effect of these two reactions can be seen by adding the equations algebraically:

$$O + O_3 \rightarrow O_2 + O_2$$

which is identical to reaction (3) in the simple Chapman scheme but, because of the catalysis by HO, occurs much more rapidly. Reactions (4) and (5) can repeat many times, consuming many O_3 molecules until, eventually, the chain is terminated by the combination of the two free radicals:

$$HO + HO_2 \rightarrow H_2O + O_2 \tag{6}$$

It is now believed that this and other catalytic cycles involving hydrogen-containing species ($HO_x = H + HO + HO_2$) contribute about 10 per cent of the ozone destruction in the normal stratosphere.

The Role of Nitrogen Oxides

Most of the remaining O_3 removal is provided by catalytic chains involving the oxides of nitrogen. The dominant processes are

$$NO + O_3 \rightarrow NO_2 + O_2 \tag{7}$$

$$NO_2 + O \rightarrow NO + O_2 \tag{8}$$

which constitute a catalytic chain since the NO is consumed in reaction (7) and reformed in reaction (8). Algebraic additions of these two reactions show that they too have the net effect of reaction (3), this time catalysed by the oxides of nitrogen. The main removal process for these nitrogen oxides is the formation of nitric acid:

$$NO_2 + HO \xrightarrow{M} HNO_3 \tag{9}$$

the nitric acid is slowly transported downward into the troposphere, from which it is removed by rain. The occurrence of reaction (9) does not immediately terminate the catalytic chain, however, since other processes occur to regenerate NO_2, the most important one being photodissociation by sunlight:

$$HNO_3 + h\nu \rightarrow HO + NO_2 \tag{10}$$

which, effectively, reverses reaction (9).

The main source of nitrogen oxides in the stratosphere is the reaction of nitrous oxide with excited oxygen atoms. Nitrous oxide is formed by bacterial action in the soil (denitrification) and slowly diffuses upwards into the stratosphere.* Current estimates are that approximately one million metric tons of N_2O are converted annually to NO on a global basis. This, then, represents the amount of nitrogen oxides which controls about 70 per cent of the O_3 budget in the normal stratosphere. In 1971 Harold Johnston pointed out that the projected fleet of supersonic aircraft to be operational by 1985 would inject about the same quantities of nitrogen oxides as is currently produced by nature. He drew world attention to the potential threat to the ozone layer from this source of pollution. The Climatic Impact Assessment Program – CIAP – of the U.S.A. devoted three years and $21 million to an intensive study of this problem, which involved more than 1,000 scientists from 10 countries. Similar, though smaller, projects – COMESA and COVOS – were launched by the United Kingdom and France, partners in the production of the Concorde SST.

The CIAP results were reviewed in depth by the National Academy of Sciences of the United States. The NAS report *Environmental Impact of Stratospheric Flight* summarizes the main results, and subjects them to a critical analysis. In general there seems to be agreement that internal combustion engines at the surface – chiefly in vehicles – or in tropospheric flight, produce chemically active oxides of nitrogen that are promptly washed back to the surface by rain. Stratospheric flight, however, must release such oxides in the layers of high ozone content. Hence they might be expected to attack the ozone and lower its concentration.

Using the CIAP experience, but incorporating later data on rates of atmospheric transport, the NAS panellists concluded that the existing and projected fleets of wide-bodied jets would produce only minor effects on stratospheric composition, chiefly because they operate below 14 km (the present generation operating mainly near 10 or 11 km). Even the present supersonic transports – Concorde and the Tupolev-144 – operating at 16-17 km will have rather small effects unless their

---

*The role of the sea in this release is still highly controversial.

numbers increase substantially. But more advanced aircraft have been planned to
fly at or above 20 km, near the level of maximum ozone density. Such planes would
release the active oxides of nitrogen and also many particles at a very crucial
altitude for the ozone balance.

The potential impact of the various possibilities can be assessed in tabular form,
in which the effects are given *per hundred aircraft*.

TABLE 13.2 Impact per Hundred Aircraft of Specified Types of Total
Ozone (NAS report, Environmental Impact of Stratospheric
Flight)

Aircraft Type	Calculated % Reduction of O_3 per 100 aircraft	Uncertainty Factor
Present subsonic transports	0.02	10
Projected subsonic transports	0.2	10
Present supersonic transports	0.7	3
Advanced supersonic transports	3	2

The advanced supersonic of this calculation was the U.S. Boeing aircraft expected
to fly at about 20 km, now abandoned. The results of similar calculations being
made by U.K. and French projects are not yet available, but are expected within a
few months.

If these calculations are correct, a fleet of 500 SSTs of the present generation
would induce a reduction of O_3 between 1.2 and 10.5 per cent, with 3.5 per cent the
most likely value. But such calculations depend critically on the assumptions made
on such matters as duration of flight (assumed to be five hours in the stratosphere
daily) and the rate at which each aircraft releases oxides of nitrogen (assumed to
be 1.8 per cent of the weight of fuel consumed). They also depend on the design of
exhaust systems, and of the engines themselves.

In the past two years the importance of this issue has receded as it has become clear
that most countries have pulled back from announced plans to develop advanced SSTs,
whose potential for stratospheric damage is much higher. It is also clear that
aircraft design can minimize the impact. Nevertheless it will always be necessary
to keep a close watch on all propulsion systems that threaten to release nitrogen
oxides in the layers above the tropopause.

Since the SST problem became widely publicized, other possible ways in which man
may be disturbing the nitrogen oxide to ozone relationship in the stratosphere have
been tentatively identified.

The increased use of nitrogeneous fertilizers on the world's arable land, for
example, must in some measure be accelerating the denitrification process, since
the rate of cycling of nitrates in cropped land under advanced farming is greater
than in natural ecosystems. Nitrogen fixation on land is brought about by soil and
root bacteria, largely attached to the roots of such leguminous plants as acacias
and vetches. Traditional agriculture substitutes other legumes, such as clover,
beans, and peas, for the same purpose. In the past twenty-five years, however,
synthetic nitrogen compounds have been applied on a rapidly increasing scale.
Preliminary calculations suggest that such application will significantly increase

the surface N_2O release — and hence ultimately ozone concentrations in the stratosphere. There is wide disagreement about the magnitude and time scale of this
effect, which needs intensive examination, particularly as regards the role of the
ocean in natural N_2O release.

Another hypothesis depends on the suggestion that the ratio of nitrous oxide (N_2O)
to nitrogen (N_2) in the output of denitrification on land depends on the acidity
of the soil. If acidity increases — and in many parts of the northern hemisphere
it is probably doing so because of acid rainfall due to air pollution — N_2O release
should form a higher proportion. McElroy has suggested that it is 8 per cent at
present, and that a doubling of the release rate would lead to a 30 per cent
reduction in total ozone. Here again uncertainty concerning the role of the oceans,
and of the reality of the supposed increase in soil acidity, has prevented agreement;
a recent analysis by Crutzen, for example, paints a less dramatic picture.

The realization that surface nitrous oxide release is the main modulator of
stratospheric ozone concentration is less than a decade old. It is hence to be
expected that for a few more years, while research progresses, uncertainties of
this sort will continue. As stated earlier, the dependence of the ozone layer, and
hence of surface ultraviolet irradiance, on soil, vegetation and ocean chemistry,
is one of the dramatic discoveries of modern times. Surface ecosystems, and hence
human use of the surface, are linked functionally with the stratosphere, and not
merely with the boundary layer.

The Role of Chlorine

More recently world attention has been focused on the possible effects of man-made
halocarbons* on the ozone layer, following a 1974 paper by M.J. Molina and
F.S. Rowland. It is well known that atomic chlorine can also catalyse the destruction of ozone by a set of reactions analogous to (7) and (8):

$$Cl + O_3 \rightarrow ClO + O_2 \qquad\qquad (11)$$
$$ClO + O \rightarrow Cl + O_2 \qquad\qquad (12)$$

Once again the net effect is identical to reaction (3), in this case catalysed by
Cl. The Cl atoms are lost by conversion into hydrogen chloride (HCl), mainly by
reaction with methane (CH_4)

$$Cl + CH_4 \rightarrow HCl + CH_3 \qquad\qquad (13)$$

followed by slow transport of HCl into the troposphere and removal by rain. The
HCl can also be reconverted to Cl, principally by the reaction

$$HCl + HO \rightarrow H_2O + Cl \qquad\qquad (14)$$

ClO may also be converted to $ClONO_2$ by the reaction

$$ClO + NO_2 \rightarrow ClONO_2 \qquad\qquad (15)$$

$ClONO_2$ which reacts only slowly with O or O_3 has recently been discovered to be
more stable than previously believed. It is reconverted to the active form of
chlorine mainly by photolysis

$$ClONO_2 + h\nu \rightarrow ClO + NO_2 \qquad\qquad (16)$$

The catalytic cycle involving chlorine, reactions (11) and (12), is about six times
more efficient in destroying O_3 than is the cycle involving nitric oxide, reactions
(7) and (8). On the other hand the fraction of chlorine atoms which is tied up as
HCl or $ClONO_2$ is greater than the fraction of nitrogen oxides that is tied up as
HNO_3; that is

---

*i.e., compounds of carbon containing chlorine, bromine, iodine or fluorine. The
    chlorofluorocarbons contain two of these elements.

$$\frac{HCl + ClONO_2}{ClONO_2 + HCl + Cl + ClO}$$

is greater than

$$\frac{HNO_3}{HNO_3 + NO + NO_2}$$

These two effects very nearly compensate one another, so that, to a close approximation, one atom of chlorine, in the form of Cl, ClO or HCl, is as effective in destroying O_3 as one atom of N, in the form of NO, NO_2 or HNO_3. A small amount of the natural ozone removal (5 per cent or less) may be due to the small amount of chlorine naturally present in the stratosphere. But the release of man-made chlorine compounds into the atmosphere, in amounts currently totalling about one million metric tons per year, is the basis of the present concern, since it represents an amount which is of comparable size, and therefore represents a comparable effect, to the naturally released nitrogen oxides, which we have seen control about 70 per cent of the natural ozone budget. The average time required for a given chlorine atom in the form of a man-made halocarbon to be returned to the surface again in the form HCl in rain may be tens of years. Therefore its effects on the ozone layer is not fully achieved until some 15 years after the halocarbon is released and the recovery time of the ozone once such release is terminated by, for example, government regulation, is of the order of 100 years.

However, to make quantitative predictions of the effects of the introduction of pollutants to the earth's ozone layer it is necessary to consider, quantitatively, a number of factors:

- the rates of release of the pollutants into the stratosphere;

- the transport by atmospheric motions of the chemically active species;

- the interaction of the pollutants with solar radiation, and chemical reactions of the products with all atmospheric constitutents;

- the removal of the reactive products from the atmosphere.

We shall summarize the present state of knowledge about each of these factors for the use of chlorine.

Sources of Chlorine in the Atmosphere

There are large quantities of chlorine-containing compounds, both natural and man-made, that are released into the troposphere. But these must enter the stratosphere and be decomposed into chlorine atoms before they can contribute to ozone destruction. Hydrogen chloride and inorganic chlorides are introduced into the troposphere by volcanic emissions, by seaspray evaporation and by human activity. Removal of these compounds by rain is so rapid that their concentrations decrease rapidly with altitude. Molecular chlorine (Cl_2) escaping from industrial processes is also rapidly removed by chemical processes in the atmosphere. There is general agreement that no significant amount of chlorine enters the stratosphere from these sources.

Halocarbons, on the other hand, are relatively inert in the troposphere and are transported into the stratosphere to a greater or lesser extent depending on their removal rate in the troposphere. Halocarbons containing hydrogen atoms or double bonds (unsaturated) in their chemical structure are partially removed in the troposphere, whereas saturated, completely halogenated halocarbons (such as chlorofluoromethanes) are very stable and will be transported into the stratosphere virtually intact. Those halocarbons that reach an altitude of about 30 km, where

the solar UV radiation is sufficiently intense, are photodissociated, releasing
chlorine atoms that then enter the catalytic cycle of ozone destruction.

The two chlorofluoromethanes,* CF_2Cl_2 (F-12) and $CFCl_3$ (F-11), are the largest
source of man-made chlorine compounds in the atmosphere. The U.S. production data
for these compounds are believed to be reliable within 5 per cent. Total non-
U.S.A. production is less well known, but is of comparable magnitude to the U.S.
figures. These compounds are mainly used as aerosol propellants and refrigerants,
which are eventually released into the atmosphere. Less than 10 per cent are used
as blowing agents for closed-cell polyurethane foam, and their eventual fate is
uncertain, though they may be stored for long periods. Recent analyses of global
production and use of F-11 and F-12 have greatly reduced the uncertainties in the
release rates of these compounds and the total amounts released through 1975 are
now believed to be known to within $+5$ per cent. Average measured levels of these
compounds in the atmosphere are consistent with the assumption that all the F-11
and F-12 that has been manufactured to date is currently in the atmosphere, although
it must be pointed out that there is a 40 per cent uncertainty in the average global
amount. So the data are also consistent with the existence of a significant
(unknown) removal process.

Carbon tetrachloride (CCl_4) has been produced in large quantities for many years
and has been formly used extensively as a solvent and dry-cleaning agent, with
eventual escape into the atmosphere. However, these uses are currently curtailed
because of the toxicity of this compound and present use is largely confined to
providing feed-stocks for other chemicals, with relatively little escape to the
atmosphere. There is some indication that a natural source for this chemical may
also exist.

Methyl chloride (CH_3Cl) is currently the most abundant halocarbon in the troposphere.
It is undoubtedly mostly of natural origin. However, it is chemically attacked in
the troposphere, so that only about 6 per cent of the amount produced at the surface
reaches the stratosphere. At present it is believed that this compound contributes
about as many chlorine atoms to the stratosphere as do F-11 and F-12 combined. The
important thing to note is that CH_3CL has reached a steady state in the atmosphere,
i.e. it is contributing as much to the stratospheric chlorine now as it is likely
to do in the future. In contrast, F-11 and F-12 have been released for a
relatively short period of time (roughly 10 years) and have therefore not yet reached
the steady state condition. In other words, if the release of F-11 and F-12 were
to cease now the amount already released would, in time, (about 15 years) contribute
a larger amount of chlorine to the stratosphere than the contribution from CH_3Cl.
If F-11 and F-12 production is continued their contribution will increasingly
exceed that from CH_3Cl. Other man-made halocarbons contribute smaller amounts to
the stratospheric chlorine burden, but their industrial production is increasing
and may become significant in the future.

Interaction of Halocarbons with Solar Radiation and Subsequent Chemistry

The halocarbons are dissociated mainly by ultraviolet light at wavelengths less
than 280 nm. Only in the middle and upper stratosphere are they unshielded from
this radiation by ozone. Laboratory measurements have shown that this photolysis
releases 2 Cl atoms from F-11 and F-12. There is laboratory evidence that the
third chlorine atom is also released from F-11. Halocarbons also react rapidly with
excited oxygen atoms in the stratosphere, which contributes to stratospheric
destruction of these compounds, amounting to about 1 per cent of that for photolysis
in the case of F-11, and 10 per cent in the case of F-12. However, these reactions
also lead to the formation of reactive forms of chlorine which destroy O_3.

---
*
Written CFMs in what follows. The terms Freon-12 and Freon-11 (F-12 and F-11)
are tradenames used by the largest U.S. manufacturer.

Since the catalytic chain for O_3 destruction involves the chain carriers Cl and ClO, the efficiency of the chain depends on the ratio of the chlorine atoms present in either of these forms (ClO_x) to those that are in molecules that do not react catalytically with ozone, i.e. on the ratio $[ClO_x]/[ClX]$ where ClX refers to all the molecular products of halocarbon decomposition containing chlorine atoms. The main unreactive form is HCl, and possibly $ClONO_2$.

The primary uncertainty in this ratio arises from uncertainties in the rate coefficients for the well established reactions that largely determine the partitioning between ClO_x and the inactive forms, HCl and $ClONO_2$.

Tropospheric Sinks

If photodissociation in the stratosphere were the only process removing halocarbons from the atmosphere, they would have a residence time in the atmosphere in the range 50-100 years. Any process that removes halocarbons at the surface or in the troposphere will lower the atmospheric residence time and consequently decrease the effect on the stratospheric ozone destruction. Since more than 90 per cent of the halocarbons are in the troposphere, a relatively inefficient tropospheric removal process can have a large effect. For example, photolysis of F-12 in the stratosphere, which leads to ozone decomposition, results in an atmospheric residence time of 100 years. If a process occurred in the troposphere (or at the earth's surface) that removed 1 per cent of the F-12 from the atmosphere per year, the net effect of F-12 on the ozone destruction would be halved. The best test for whether such a removal process occurs would be provided by a comparison between the amount of that halocarbon which has been released into the atmosphere, minus the amount which is photolysed in the stratosphere, with the measured amount found to be present in the atmosphere. Unfortunately, at this time uncertainties in the release rates (±5 per cent) and in the atmospheric measurements (±40 per cent) are too large to make such a direct comparison definitive. The best alternative available is an exhaustive examination of all possible removal processes.

Because of their low solubilities a negligible fraction of CFMs is removed from the troposphere by rain-out. Hydrolysis in water and in acid solution (characteristic of rain and most aerosols) is also very slow and will not increase the removal rate.

The oceans are potentially large reservoirs for the storage of gases. Equilibrium between surface air and the oceans is attained rapidly down to the depth of the thermocline (of order 100 m or less) but mixing times below this level are much slower ($\sim$500 years) than the atmospheric residence times of CFMs. Even if the oceans down to the thermocline were to become saturated with the respect to CFMs they would contain only 0.06 per cent of the total atmospheric burden. There is very limited evidence to suggest that CFMs may be removed from the surface waters of the oceans by some, as yet unidentified, processes. If this suggestion is substantiated it would correspond to a removal time of between 300 to 800 years and might lower the effect of CFMs on the ozone layer by as much as 20 per cent.

The possibility of entrapment in polar or glacial snow and ice has been suggested as a result of a recent observation that the gas released by melting polar snow is twenty-five times richer in CFM than is the surface air. However, even if such enrichment existed in all the permanent ice of the world it would provide only a negligible sink for CFM removal.

Most tropospheric aerosols are liquid and acidic. Solubility and hydrolysis in such aerosols are negligible and no chemical reactions are known to occur with these acidic solutions. Solid particles constitute a minor fraction of tropospheric aerosols so that adsorption of CFMs on the surface will be too small to provide a significant removal rate by dry deposition of these solid particles.

Removal of halocarbons by interaction with solar radiation in the troposphere will be extremely slow because the radiation in the wavelength region over which halocarbons show any measurable absorption is attenuated to very low values by the ozone and oxygen in the stratosphere.

Removal rates of halocarbons by reaction with neutral molecules are determined by the product of the rate constant for the reaction and the concentration of the reactant species. The magnitude of the combination of these two parameters in the troposphere is such that only reactions of HO and O_3 with the halocarbons need be considered. Such reactions are very effective in removing unsaturated halocarbons (those containing carbon-carbon double bonds). Reactions of HO with saturated halocarbon molecules containing one or more hydrogen atoms are also fast enough that only a small fraction will survive long enough to penetrate into the stratosphere. This also applies to naturally produced CH_3Cl which reacts sufficiently rapidly that only 6 per cent reaches the stratosphere. In contrast there is no evidence that HO or O_3 reacts with CFMs or with CCl_4.

Since CFMs have large cross-sections for electron capture, bombardment by electrons from cosmic rays or from natural radioactivity has been suggested as an effective removal process. However, oxygen also has an appreciable electron capture cross-section, and its atmospheric abundance is nine orders of magnitude greater than that of the CFMs, which therefore cannot compete for the electrons.

It has also been suggested that positive and negative ions present in the atmosphere might react readily with the halocarbons. Although the identity of the ions in the troposphere and stratosphere has not been subjected to direct measurement, there is sufficient laboratory information available to leave little doubt about the nature of atmospheric ions. The positive ions are undoubtedly hydrated ions of the type $H_3O^+(H_2O)_n$ and $NH_4^+(H_2O)_m$ where the degrees of hydration, m and n, have values in the range 5 to 7. These are relatively stable ions and have been shown by laboratory studies to react with halocarbons very slowly if at all. The negative ions in the troposphere are almost certain to be hydrated NO_3^-, NO_2^-, CO_3^- and O_2^- which also have been found to be too stable to react with halocarbons at a measureable rate.

Therefore, at present, there is no reason to believe that any process or combination of processes occur which remove CFMs from the troposphere at a rate approaching 1 per cent per year.

Uncertainties in Ozone Reduction

The prediction of the extent of ozone reduction resulting from a given release rate of halocarbons into the atmosphere must be made from atmospheric models. The only type of model that is sufficiently well developed at present is the one-dimensional (1-D) model. Such a model averages concentrations, motions and chemical reactions over latitude and longitude, leaving them dependent only upon height and time. The averaged transport rates depend in these models on the vertical eddy diffusion coefficient K. (see calculation of Net Transport above).

The most stringent test for any atmospheric model is its ability to account for the time and height dependence of a chemically reactive constituent other than the ones used to determine the eddy diffusion coefficient. The various 1-D models have been moderately successful in accounting for the measured height profiles of the nitrogen oxides, and quite successful in fitting the observed diurnal variations of these compounds. For example, the partitioning of NO_x between NO and NO_2 is directly dependent on the solar flux since NO is converted to NO_2 by reaction (7) and reconverted to NO by reactions (8) and (15)

$$NO_2 + h\nu \rightarrow NO + O$$

A steady state ratio is obtained between NO and NO_2 within about 2 minutes in the stratosphere. As the sun sets the rates of both reactions (8) and (15) approach zero, and NO rapidly alters to NO_2. The 1-D models have accurately predicted the observed rate of conversion. Of particular interest for the present discussion are the recent measurements of F-11 and F-12 in the mid-stratosphere, where theory predicts that photodissociation of these compounds should occur. The measured height profiles of these compounds are in excellent agreement with the model profiles.

Height profiles of HCl have also been obtained both by remote sensing and by "in situ" techniques. The results are in general agreement with the models although the number of such measurements are still rather sparse and there is some question as to whether some disagreement between model and measured profiles exists at the highest altitudes of the measurements.

The models can be used to predict various "scenarios"* of CFM release. The National Academy of Sciences has recently made a critical assessment of three 1-D models currently in use. In their assessment the sensitivities of the models to the choice of eddy diffusion coefficients and their dependence on altitude were tested by holding constant all other parameters. We can illustrate the results of this assessment by considering one scenario, in which it is assumed that CFMs are released with a 10 per cent per year growth until 1978 and then abruptly terminated. Complete cessation of CFM release does not result in an immediate drop in ozone reduction, since the CFMs already in the troposphere provide a reservoir for these compounds, and the ozone reduction will continue to become more serious for a number of years before recovery begins. We can then ask and answer three questions:

- *How many years are required (after 1978) for ozone reduction to reach its maximum value?* The models predict a value of 10 years with an uncertainty due to transport alone of +3 years.

- *At this maximum value how much does the ozone reduction exceed the value it had when CFM release was terminated?* The models predict a peak value 1.5 times the 1978 value with an uncertainty due to transport of +0.15.

- *How many years are required (after CFM release is stopped) for the ozone reduction to return to its 1978 value?* The models predict a time of 55 years with an uncertainty due to transport of +10 years.

We may now consider the uncertainties due to the uncertainties in the rate constants of the chemical reactions. The rate constants for the reactions

$HO + HO_2 \rightarrow H_2O + O_2$, $Cl + CH_4 \rightarrow HCl + CH_3$, $HO + HCl \rightarrow H_2O + Cl$, and $ClO + NO_2$
$\rightarrow ClONO_2$, are the largest sources of uncertainty at present in the factors affecting the chlorine-ozone chemistry, with smaller contributions from other reactions. Their combined effect produces a five-fold range of uncertainty.

Finally, we can consider the most probably value and the overall uncertainty in the prediction of ozone depletion. Perhaps the most useful scenario here is one of continued, *constant* release of F-11 and F-12 at the 1973 rates.

The findings of the National Academy of Sciences (U.S.A.) Study are as follows: If the effect of $ClONO_2$ on the chemistry is not included in the model, calculations show a most probable value of 14 per cent for the ozone reduction when a steady state has been reached under constant CFM in conditions. When $ClONO_2$ chemistry is included the most probable value is reduced by a factor of about 1.85 to 7.5 per cent. If the existence of an oceanic sink is confirmed the predicted value is

---

*i.e., hypothetical programmes assumed to represent future events.

further reduced by about a factor of 1/5 to 6 per cent. The overall uncertainty
in the central value based on 95 per cent confidence limits (25) is an eight-fold
range, i.e. the ozone reduction could be as low as 2 per cent or as high as 20 per
cent for this scenario. These uncertainties represent a combination of individual
uncertainties in release rates (1.05), transport (1.7) and stratospheric chemistry
(2.5). It does not include other intangible uncertainties such as the possible
existence of other, unidentified tropospheric sinks or stratospheric processes.

Other 1-D models by other groups have produced similar results. The central values
range from about 7 to 10 per cent. The effect of $ClONO_2$ by at least two groups
appears to be somewhat lower than those of the N.A.S., corresponding to a reduction
of the predicted ozone depletion of 1.4 and 1.6 respectively.

Summary of the Chlorine Effect

Suggestions have been made that action need not be taken to limit CFM release (1)
until the theory has been proven by the detection of ClO in the stratosphere; (2)
the recent discovery of the relative stability of $ClONO_2$ suggests that other
processes may be discovered that might further mitigate the ozone reduction; (3)
because no evidence exists for O_3 depletion from man-made chlorine compounds, no
action should be taken until such evidence is obtained; (4) because considerable
effort is now being expended in making laboratory and atmospheric measurements that
will decrease the uncertainties in the model predictions, action need not be taken
until the data obtained from these measurements are forthcoming.

We respond to these suggestions on the following grounds:

(1) Preliminary measurements have indeed confirmed the presence of ClO in the
stratosphere. This comes as no real surprise since there is ample laboratory
evidence to support the Cl/O_3 catalytic cycle. The measurements of F-11 and F-12
confirm that these compounds photodissociate in the stratosphere. The laboratory
evidence is overwhelming that photodissociation of these compounds produces Cl
atoms and that these atoms effectively catalyse O_3 destruction.

(2) The discovery that $ClONO_2$ is more stable than originally believed was indeed
somewhat of a surprise to atmospheric chemists. However, inclusion of its
chemistry in the models has not changed the general conclusions that continued
injection of man-made halocarbons into the atmosphere will lead to an appreciable
reduction in O_3. The maximum decrease in the effect is believed to be a factor of
1.85 and recent attempts to observe this compound in the stratosphere suggest the
effect may be smaller than this. The possibility that other unknown or ill-defined
chemical processes may play a significant role cannot, of course, be definitively
ruled out but it is, becoming increasingly unlikely. There are a finite number of
such possibilities and the intensive efforts of the N.A.S. and the scientific
community have failed to reveal any during the past two years.

(3) The difficulty inherent in waiting for an observed reduction in ozone arises
from the natural short- and long-term fluctuations in ozone concentrations discussed
in Climatology of Ozone Variation, pp. 22-39 above. An improved network of ozone
monitoring stations making measurements for the next several decades *might* be able
to discern a 3 per cent change over the natural variations. But the continued
release of CFMs over this period would cause the ozone reduction to rise for a
decade after the detection of its effect, even if termination of CFM release
occurred at that time. The ozone reduction would then still take 65 years or more
to return to half that value.

The N.A.S. have analysed the risks involved in waiting for a period up to two years
for additional information to become available which would reduce the uncertainties

in the predictions. They have concluded that these risks are not overly serious
and have therefore recommended such a course of action. However, it should be
clearly pointed out that much of this additional information should be available
within a shorter period (six months to one year) and that the Academy further
recommended that if such information further substantiates their conclusions
regulatory action should be taken at that time. Furthermore it is unlikely that
the overall uncertainty will ever be reduced to much less than a factor of 2.
Finally we should point out that regulatory procedures in any given country may well
take as much as two to three years. Since the problem is a global one, action on
an international scale may be expected to involve even longer time scales. For
these reasons the authors feel that the current evidence is probably sufficiently
strong to warrant initiation of international decisions at this time.

The Vital Role of Measurement

Reference was made earlier to observations of the CFMs. A brief look at the whole
question of observations is now necessary, because there is a serious monitoring
problem that should vitally concern UNEP and its member states.

There is a drastic imbalance between meteorological and chemical observation systems
in the stratosphere. The former are actively operational, and are internationally
organized by WMO. The basic thermodynamic, dynamical and physical parameters –
temperature, pressure, wind velocity and cloud distribution – are measured routinely
by balloons and satellites to a low but presumably adequate level of precision (for
circulation analysis), over the entire globe, on a daily or even continuous basis.
Nothing of the sort is true for the chemistry of the atmosphere. Even humidity,
fundamental to the meteorologist and climatologist, is measured only crudely and
intermittently above about 15 km (where changes are presumed to be small). It is
vital that adequate chemical monitoring be instituted. The WMO-sponsored proposals
for a research period in this area should be the prelude to an operational system
covering the key constitutents.

A major difficulty has been that the chemistry of the ozone layer is dominated by
trace species present at concentrations down to parts per billion or even per
trillion. Analytical techniques for such low concentrations have advanced greatly
in recent years. They tend to require, however, expensive and bulky equipment.
Hence elaborate means are necessary to lift them into the stratosphere. To get
vertical profiles, and to calibrate satellite observations, very large balloons
are necessary. They have been used for example, at several experimental sites in
Canada, the United States, France and Australia in connection with the studies
described above. Rocket launching is necessary for higher altitude work, and is
cumbersome, costly and hard on the sensors carried. Aircraft can be used in the
lower stratosphere. The R.A.F. Canberra jet and the protype Concordes have both
been used to conduct very sophisticated radiative and "grab-sample" observations
at these levels. For long-range monitoring of total stratospheric burden, however,
there is no doubt that ground-based and satellite technology is the correct answer.

In the past few years we have nevertheless begun to observe many of the trace
species on a scattered basis. The various oxides of nitrogen, and nitric acid,
have been widely measured in the stratosphere. Hydrochloric acid has also been
detected, and its seasonal variations measured.* In view of their key role in
stratospheric chemistry, their measurement is desirable, though not essential to
the argument. OH data are of the highest importance.

The chlorofluoromethanes themselves have been measured extensively by means of
balloon and aircraft carriers. The concentrations have been found reasonably

---

*chlorides have been observed in particles at lower stratospheric altitudes.

uniform in the troposphere (though the northern hemisphere has more than the southern, because of very unequal release). In the stratosphere they have been observed to decrease with height above 15 km at rates reasonably consistent with the model calculations described earlier. The stable man-made tracer sulphur hexafluoride (SF_6), widely used in electrical transformers, has also been detected at these levels, and does not decrease as rapidly with height. Hence the view that chlorine species act in the fashion described above has observational confirmation. If direct measurements of Cl and its oxides were available additional confirmation would be possible.

Therefore there is urgency in the proposals being made on all sides that observation of trace-species in the stratosphere be intensified. Meteorologists will urge this even more strongly than the chemists, because they suspect that frequent synoptic observation of these species will show complexities in distribution and variation akin to those displayed by ozone and water vapour. Until these complexities have been laid bare the assumptions underlying the various model calculations cannot be tested.

ENVIRONMENTAL IMPACT OF STRATOSPHERIC POLLUTION

Concern about the impact of pollution on ozone concentrations in the stratosphere is so recent that there has not been time to take adequate stock of the potential threats to the natural environment, and to human health. The only comprehensive reviews available to the writers were those prepared under the U.S. CIAP and NAS programmes, and the COMESA-COVOS Oxford Symposium Proceedings. Greater emphasis is now being placed on photobiological research in many countries, in response to the perceived threats. In a few years it will be possible to give far better answers to the question "What will the impact be?"

Most of the effects identified so far will result from increased ultraviolet radiation at the earth's surface. These may include:

- impact on human health, especially as regards skin and eyes;

- impact on plants, animals and ecosystems;

- impact on agriculture

Other impacts, not as yet clearly identified, include a potential threat to climatic stability; altered ambient environment for aircraft in the stratosphere; and altered solar irradiation of the boundary layer, where photochemical effects are important.

Ozone Depletion and Ultraviolet Penetration

The various reactions discussed earlier have the effect of depleting from the solar beam all or part of the radiation at very short wavelengths. With present ozone concentrations, the solar irradiance* at the earth's surface contains virtually no power at wavelengths below 295 nm, yet the irradiance at the top of the atmosphere is rich in power down to at least 180 nm. All this radiation is absorbed during its path downwards through the atmosphere. Most of the loss is due to absorption by ozone, with some help from oxygen.

Ozone absorbs solar radiation over a wide range of wavelengths, including weak effects in the visible part of the spectrum (400 to 700 nm) and in the ultraviolet from 310 to 340 nm. But the absorption is very strong below 310 nm, and effectively total below 295 nm. Small reductions in total ozone should have the effect of allowing more ultraviolet to penetrate to the earth's surface. Even today,

---

*i.e., the energy received per unit area and per unit time, or power per unit area.

therefore, there are marked variations in surface ultraviolet irradiance from day
to day, because of rapidly fluctuating total ozone (see Short-Term Variations in
Total Ozone, above). There is also a strong latitudinal gradient. In bright sunshine there is more ultraviolet present in tropical latitudes (with total ozone
well below 0.280 cm) than in high and middle latitudes (with total ozone above
0.350 cm). In all latitudes there is a slight increase of ultraviolet irradiance
with height. Especially in winter the low angle of solar radiation in the middle
and high latitudes is of key importance.

Direct observation of the incident ultraviolet, and especially of the more damaging
wavelengths, has not been carried out routinely in many places. Figure 13.16 shows
how annual ultraviolet incidence varied with latitude over the United States in
1973-74. Annual totals were almost twice as high in 30N as in 45N. Summer
variation was less marked. There is a need for more widespread monitoring of ultraviolet irradiance.

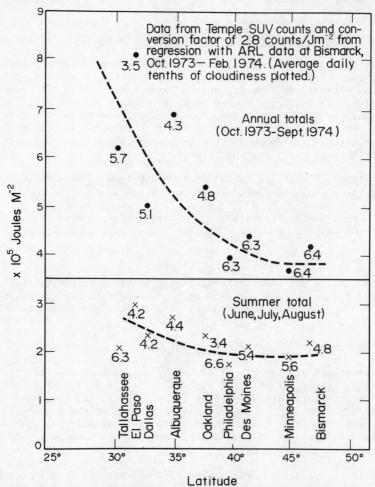

Fig. 13.16. Variation of erythemally-effective ultraviolet radiation
with latitude over the United States, 1973-74. (IMOS)

What effect can be deduced from reduced ozone concentrations? Here we rely on the
CIAP estimate shown in Fig. 13.17.   This gives the percentage change of ultraviolet
(weighted to take account of variable human skin-sensitivity to different wavelengths) corresponding to various percentage decreases of total ozone.   The diagram
indicates that a 1 per cent *decrease* in total ozone should produce a 2 per cent
increase in weighted surface ultraviolet irradiance, up to about a two per cent
decrease in ozone.   More drastic increases should accompany larger ozone reductions.

If this two-to-one ratio is accepted, then the larger day-to-day variations in
total ozone observed in middle and high latitudes - of order 15 to 20 per cent
amplitude (with respect to the local mean) - should produce corresponding day-to-day
variations of order 30 to 40 per cent in ultraviolet irradiance, subject to the
effect of changing cloudiness.

The reduction in total ozone predicted by the chemical modelling discussed above
is of the order of 10 per cent in the long-term steady state, with an uncertainty
factor of two. If the above relations hold good, this reduction should produce a
persistent increase of between 10 and 40 per cent in ultraviolet irradiance of the
earth's surface. The upper bound of this range is similar to the short-term
excesses sometimes observed in the present day middle latitude climate as a result
of transient atmospheric disturbances.

These very general estimates require some adjustment if applied to specific
localities, especially as regards local cloudiness, altitude, day-length and solar

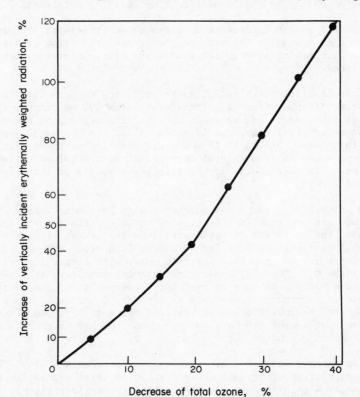

Fig. 13.17. Percentage increase vertically incident erythemally-weighted
ultraviolet radiation corresponding to percentage decrease of total
ozone.   (IMOS, Schulze)

exposure. Thus exceptionally high ultraviolet totals may occur over south-facing, high altitude snowfields. Detailed analysis of such effects has not yet been pushed very far.

The increase in *total* ultraviolet irradiance, disregarding the weighting for human skin-sensitivity, should be significantly smaller than the above estimates. Ozone does not absorb strongly in the longer ultraviolet wavelengths, which penetrate to the earth's surface. Hence a decrease in total ozone should have little impact at these wavelengths, which are not as dangerous as the shorter ultraviolet to the human skin (see below). They may, however, affect plants and animals.

Impact on Human Health

The effects on human health that have been identified include the incidence and prevalence of skin cancers (carcinomas and melanomas), eye effects, and disturbance of vitamin-D synthesis.

It is well established by epidemiologists that there is a connection between sun-burn and skin carcinomas on the one hand, and between sunburn and ultraviolet irradiance on the other. Skin carcinomas are much commoner in low latitudes than in high, and on exposed skin than on concealed. They are also commoner among light-skinned people who sunburn easily than among the dark-skinned. It is very probable, though not conclusively proved, that the actual mechanism whereby carcinomas are produced is the release of carcinogenic substances due to the effect of ultraviolet on DNA – the genetic information carrying molecules.

It follows that a threatened increase in ultraviolet irradiance also implies an increase in skin carcinomas. The same may be true of the far more dangerous melanomas, though the link between the latter and ultraviolet irradiance is less firmly established.

The major impact of ultraviolet radiation in producing sunburn *(erythema)* occurs at wavelengths below 320 nm. Since no ultraviolet below 295 nm reaches the ground (except at high altitudes), the range of ultraviolet wavelengths between 295 and 320 nm is the critical band, and is often called the *erythemally-effective band*, or UV-B, by biologists. Figure 13.18 shows the relative erythemal response of ultraviolet irradiance at ground level corresponding to total ozone levels of 0.200 cm and 0.400 cm. The third curve is the relative response of the skin under laboratory conditions.

The variation of both carcinomas and melanomas with latitude closely resembles that between latitude and erythemally-effective ultraviolet irradiance. Available epidemiological studies in the United States show that among light-skinned people the incidence of carinomas *doubles* for each 8 to 11 degree *decrease* in latitude. A smaller increase is also reported for melanomas, with larger scatter. From these and other data the U.S. National Academy of Sciences concluded that a 10 per cent decrease in ozone would lead to a 20 per cent increase in melanoma mortality and a larger increase in carcinomas – possibly as high as 30 per cent. As a working simplification, the various U.S. investigations agree that *for each one per cent decrease in ozone, there should be a two per cent increase in erythemally-effective ultraviolet irradiance, and at least a two per cent increase in skin cancer.*

This conclusion appears valid for light-skinned persons as a whole. Thus, Queens-land, where a population of predominantly British or Irish origin lives in sub-tropical latitudes with high sunshine totals, has the world's highest incidence of skin cancer. Most peoples of the tropical world have dark skin pigmentation, which gives them protection. Where non-Caucasian light-skinned peoples live in a tropical environment, they, too, have a high skin cancer incidence.

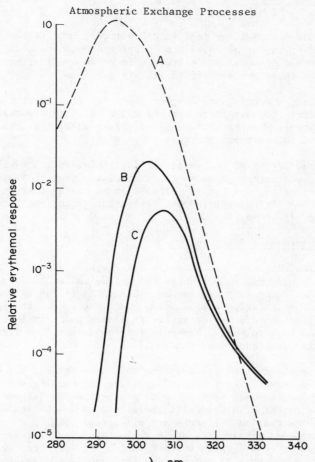

Fig. 13.18. Relative erythemal efficiency at ground level of solar
radiation as a function of wavelength. A is relative response. B.
Erythemal efficiency, total ozone 0.200 cm. C is the same for total
ozone 0.400 cm. (NAS)

The effects of exposure of light skin to erythemally-effective ultraviolet are both
short-term and prolonged. Sunburn follows exposure within an hour or so in those
susceptible (most individuals). But the appearance of carcinomas and melanomas is
usually long-delayed. Typically they develop years after prolonged exposure. Hence
they appear preferentially on those parts of the body most exposed, and long after
the individual is accustomed to strong sunshine.

There remain several incompletely answered questions. If total ozone is indeed
reduced in future by 7.5 per cent, and if the danger to the individual of skin
cancer is hence increased by 15 per cent, this is equivalent to a southward
migration by the individual of only 2-4 degrees of latitude. Hence Caucasian
immigrants to the tropical countries (e.g. U.K. to Queensland, or U.S. to Hawaii)
should already have known a marked proneness to skin cancer. It is not clear
whether adult immigrants are more prone than those of similar racial stock who were
born in the tropics.

These uncertainties have led some environmental authorities to scoff at the increased hazard due to ozone reduction. If the increase is only equivalent to the

effect of moving one's home a few degrees of latitude, this is a risk likely to be willingly accepted by the highly mobile population of western countries. The migrations to Florida, Arizona and California in North America, and to the sunniest parts of Europe, by those who can afford it, are examples.

Mid-latitude peoples, moveover, are subject to quite large short-term variations of ultraviolet exposure. Because these are not enduring, they presumably have less impact than the chronic changes, but it is not clear what relationship exists between chronic and intermittent exposure.

Other aspects of the impact on health are of less importance. There is extensive epidemiological evidence that exposure to artificial ultraviolet sources causes short-term eye irritation, and similar effects among beach-workers and climbers may be due in part to ultraviolet-induced irritation. But long-term effects are uncommon, as far as is known.

Impact on Plants, Animals and Ecosystems

In the case of human skin cancers, there is considerable evidence that ultraviolet radiation in the 295-320 nm range attacks DNA in the cells, and that this is probably the cause of many skin carcinomas. Since DNA is the characteristic genetic material of most living tissues throughout the biota, it might be expected that comparable effects could be identified in plants and animals of all sorts, including crops and farm animals. Moreover many organisms live far more directly exposed to the solar beam than does man.

Much research has been directed at answering this speculation, but not enough to yield comprehensive answers, in the opinion of the photobiologists we consulted. More attention appears to have been given to photosynthesis (depending on visible light) than to the impact of ultraviolet radiation. Nevertheless there is an abundance of evidence that such impacts are widespread, and in most cases negative. Qualitatively it appears likely that an increase in ultraviolet irradiance will have mainly harmful effects on plants and animals. On the other hand a reduction in the downward flux of ozone to the surface (as might accompany a reduction in total ozone) would probably benefit some organisms, though the effect would be small.

Most organisms that are directly exposed to the ultraviolet have in any case evolved protective mechanisms. Their evolution and survival through geological history must have taken account of this necessity. It is not certain how permanent is the present level of total ozone. We do not even know with any precision the composition of the solar beam in the ultraviolet, and we cannot say anything convincing about its variation in time. It may be, therefore, that the present level of ultraviolet irradiance differs substantially from that of the geological past, even the very recent past. Hence there is no certainty that protective mechanisms evolved genetically are in satisfactory equilibrium with the present ultraviolet flux.

Research in this area is so complex that we cannot hope to summarize it adequately. Instead we shall paraphrase and abbreviate conclusions reached by the CIAP panel on biological implications. Among their conclusions we isolate the following:

- The different wavelengths of ultraviolet radiation play quite different roles in many processes, but similar roles in others. In some cases, damage done at one wavelength may actually be offset by repairs at another.

- Ultraviolet in the 295-320 nm band seems to cause stress-enhanced aging in cells. A rapid increase in such radiation would probably do more harm than a slow increase.

- Rapid killing of cells and tissues by ultraviolet - observed in many plant
  and animal species under laboratory conditions - seems to be due to "direct
  photochemical alteration in nucleic acids and/or proteins of the cells".

- Organisms exist in present-day ecosystems that are "directly coupled" to ultraviolet in the 295-320 nm band as a major limiting environmental factor.

In spite of the caution with which this and other research groups approach the
generalisation of their results, it is our experience that most photobiologists
regard the prospect of world-wide increases in the level of ultraviolet irradiance
as a matter of serious concern calling for intensified research, and probably for
public regulation in the near future. Every research photobiologist can produce
naive evidence of the potential impact in the shape of seedlings or fruiting
bodies stunted or misshapen by deliberate exposure to enhanced ultraviolet, or of
insects or microorganisms similarly affected. It is not in dispute that increased
ultraviolet irradiance will have biological effects, some of them harmful. What is
still unclear is the precise nature of the effects for a given level of increased
ultraviolet.

Also unclear in the research literature is the extent to which these effects will
operate on ecosystems, as well as on individual organisms. The function and
dynamics of most ecosystems are still too imperfectly known for the direct impact
of enhanced ultraviolet upon them to be predicted, or even identified, except via
its effect on specific organisms.

Research so far has instead concentrated on the possible effects of climatic change
on the ecosystems. If the prevailing levels of temperature, precipitation,
cloudiness and photosynthetic radiation are altered as a result of stratospheric
pollution, there should be equivalent changes in surface ecosystems. In fact, no
such climatic changes have yet been predicted with any certainty.

A special concern is for aquatic ecosystems, especially oceanic. There is only
a poor understanding of the penetration of waterbodies by ultraviolet radiation.
In the absence of such understanding it has proved hard to arrive at any conclusions
concerning impacts on aquatic systems. Temperature, precipitation, cloudiness and
sunshine regimes, if altered, would clearly have such impacts.

Much of the research conducted by photobiologists has been directed towards crop
species, such as corn, tomato and beans. The comments made earlier, apply with
special force to such species. There has also been a considerable volume of work
on the vulnerability of farm animals to increased ultraviolet irradiance.

Here again it is not yet possible to state categorically what will be the effect
of ultraviolet changes on farm productivity and practice, though agricultural
scientists are as concerned as other biologists that the effects may be adverse.
Again there has been more emphasis on the possible effects of associated climatic
change - i.e., on changes in the physical climate other than in ultraviolet
irradiance.

Impact on Physical Climate
===

So far we have treated the potential direct impact of stratospheric pollution via
its effect on ozone concentrations, and hence on ultraviolet radiation at ground
level. We stressed the possibility, however, of an indirect impact via induced
climatic change.

A reduction in ozone concentrations in the stratosphere should have the effect of
cooling the layers between 30 and 60 km, which are currently warm (see Fig. 13.1).

This warmth is due to the absorption of solar radiation by the ozone. If total ozone is reduced by the order of ten per cent, model calculations indicate a cooling of about 1 C in the lower stratosphere. A small warming - of order 0.5 C - might occur at ground level, though some more recent calculations give a small cooling.

Changes of this sort, which would have unequal geographical effects, would have some impact on atmospheric circulation, changes of which could produce surface effects not foreseen in the above calculations. We do not at present have atmospheric models sophisticated enough to predict these secondary changes. They may well be negligible, but this cannot be taken for granted. In addition one cannot rule out a third tier of effects, due to changes in the exchange time between stratosphere and troposphere. These might alter the details of the chemical balance discussed earlier, and so act as feedbacks to the entire ozone cycle. It is therefore urgently necessary that techniques of climatic modelling be refined so that these secondary and tertiary effects can be assessed.

Still another potential impact on surface physical climate has recently been adduced, involving the chlorofluoromethanes. These gases, and also carbon tetrachloride, strongly absorb infrared terrestrial radiation in the waveband 9,000 to 12,000 nm, the so-called "window" region through which much radiation escapes from the earth's surface towards space. The build-up of chlorofluoro- methanes *in the troposphere* is hence gradually interposing a barrier to the free escape of heat from the surface. Carbon dioxide and water vapour play a similar role, though not in this transparent region of the earth's spectrum.

At the NASA-Langley Research Center in the U.S.A. estimates have recently been made indicating that an increase of present chlorofluoromethane mixing ratios (0.2 to 0.4 ppbv) to about 4 ppbv (equally divided between CF_2Cl_2 and $CFCl_3$) could raise surface temperatures by about 0.9 C. Increases of this magnitude are in fact forecast at present production levels. This heating effect is of the same order of magnitude as the expected heating due to carbon dioxide increase from fossil fuel consumption. Recent calculations in Canada tend to confirm these estimates.

We conclude that decreasing ozone concentrations in the stratosphere, and the build-up of the chlorofluoromethanes in the troposphere, may induce changes in the physical climate of the earth's surface. It is urgent that these possibilities be investigated further.

CONCLUSION

We were not asked to make recommendations for international action in these matters. Nevertheless we found ourselves making such recommendations as we went along. In this formal conclusion to the review we prefer a more general discussion.

It seems to us that a good case has been made out for stratospheric pollution as a serious environmental problem of global scope. Certainty has not been reached at any point, but the weight of scientific evidence points to such a pessimistic conclusion. We believe that the world community will have to tackle the problem. It has been created by advanced technology, and is therefore unequal in its area of origin. But atmospheric dispersion guarantees that its effects will be world-wide.

This opinion is not universally shared. Industry representatives have insisted that valuable technologies should not be unwound unless the case against them is fully proved. Individual scientists also remain sceptical. A distinguished British atmospheric physicist, for example, recently wrote

"An appreciation of the actual complexity of all the mechanisms suggests
that predictions of ozone depletion by diffusion theories are almost
certainly wrong and exaggerated, and the biological effects suggested
on the assumption that they are right are far-fetched and quite
misleading as to the nature, causes and controllability of the symptoms
in question."*

He based this severe criticism mainly on the inapplicability of diffusion models
to the slabby structure so typical of the lower stratosphere, a point we ourselves
raised earlier.

We believe, however, that a substantial case for action has been made, in spite of
the inadequacy of the models. The observations coming in from the stratosphere
increasingly support the arguments put forward by the atmospheric chemists. As
for the biological effects, we agree that uncertainties remain; but biologists
seem to agree that increased ultraviolet irradiance, even if only of a few per
cent, may have serious consequences.

The very striking thing about this newly identified problem - it is barely five
years old as a formulated idea - is that it involves inextricably the human economy,
the soil and biota, the atmosphere, and the spectrum of the sun. It is a fully
integrated problem, incapable of being defined except as a whole, unlikely to be
solved without equally broad understanding. The world ecosystem is at risk, if the
hypotheses are valid.

It has been demonstrated that ozone is created by ultraviolet radiation from the
sun, which is outside our jurisdiction and, so far, outside our range of contamination. But ozone is dissociated mainly by nitrous oxide originating in the
denitrification process that goes on in the soil - which we are disturbing
profoundly. We are also releasing technologically other pollutants that disturb
the equilibrium. If the chemists are right, as we believe, then stratospheric
pollution is preeminently the kind of problem that UNEP was set up to solve - those
where man's technology creates world-wide stresses that are broader than any one
specialized agency can tackle.

Of the issues so far identified, the release of chlorine species looks at this
moment as the most important. Though we became aware of the problem only in 1974,
it has already received a considerable amount of observational confirmation.
Chlorine - containing compounds like the chlorofluoromethanes are among the most
useful and versatile of all chemicals. Refrigerating and air-conditioning systems,
the inflation of plastic "foams," the use of aerosol spraycans, have become worldwide, and they presently depend heavily on these compounds. Hitherto the latter
have seemed harmless. It will not be a light matter to dispense with them, and to
incur the huge capital cost of finding and deploying substitutes. But it may have
to be faced.

Interference in the nitrogen cycle is also something that we must learn to abate.
If aircraft or other devices are to fly in the stratosphere, they will have to be
designed so that their exhaust systems minimize - or preferably eliminate - the
output of nitrogen oxides and, incidentally, sulphur compounds and particles. If
the use of nitrogenous fertilizers continues its present expansion, as presumably
it will, we shall have to examine the consequences closely - as we shall have to do
with such fumigants as methyl bromide.

A full discussion of these questions of control and management is beyond the scope
of the present report. But we believe that it should proceed, and that UNEP should

*R.S. Scorer, 1976: A commentary on ozone depletion theories, Atmospheric
    Environment, 10, 177-180.

sponsor on a world scale an analysis comparable with those attempted nationally or regionally by the U.S., France, the U.K. and the OECD. Within the U.N. family it is clear that WMO, WHO, FAO and UNESCO have a special interest in aspects of this problem. But it transcends them all.

APPENDIX I

List of Major Sources Consulted

WORLD METEOROLOGICAL ORGANIZATION DOCUMENTS

R.J. Murgatroyd, F.K. Hare, B.W. Boville, S. Teweles and A. Kochanski,
1965: The Circulation in the Stratosphere, Mesosphere and lower Thermosphere,
W.M.O. Technical Note No. 70 (WMO-No. 176.TP.87), 206 pp.

Commission for Atmospheric Sciences, Working Group on Stratospheric and Mesospheric
Problems (H.U. Dütsch, Chairman), 1975: Final Report of First Session, Geneva,
Sept. 8-11, 1975, 16 pp. plus ten appendices.

WMO Press Release/No. 315, 1976: Statement on Modification of the Ozone Layer due
to Man's Activities, Jan. 6, 1976, 6 pp.

UNITED STATES PUBLIC DOCUMENTS

(i) Climatic Impact Assessment Program, Department of Transportation.

A.J. Grobecker, S.C. Coroniti and R.H. Cannon, Jr, 1974: Report of Findings,
The Effects of Stratospheric Pollution by Aircraft, DOT-TST-75-50, National
Technical Information Service, Springfield, Virginia 22151, U.S.A., 134 pp.
plus 8 appendices. The latter contain summaries of more extended monographs
available separately as follows:-

CIAP Monograph 1 The National Stratosphere of 1974, DOT-TST-75-51
"   "   2 Propulsion Effluents in the Stratosphere, DOT-TST-75-52
"   "   3 The Stratosphere Perturbed by Propulsion Effluents, DOT-TST-75-53
"   "   4 The Natural and Radiatively Perturbed Troposphere, DOT-TST-75-54
"   "   5 Impacts of Climatic Change on the Biosphere, DOT-TST-75-55
"   "   6 Economic and Social Measures of Biologic and Climatic Change,
DOT-TST-75-56

(ii) National Academy of Sciences Reports.

Environmental Impact of Stratospheric Flight: Biological and Climatic Effects
of Aircraft Emissions in the Stratosphere, Climatic Impact Committee (H.G.
Booker, Chairman), 1975, 348 pp. (A similar report on the fluorocarbons is
also available.)

(iii) Council on Environmental Quality and Federal Council for Science and
Technology Report.

Fluorocarbons and the Environment: Report of the Federal Task Force on
Inadvertent Modification of the Stratosphere, Govt. Printing Office,
Washington, 1975, NSF. 75-403, 109 pp.

UNITED KINGDOM - FRANCE DOCUMENTS

COMESA-COVOS, 1974: Anglo-French Symposium, Oxford, 24-26 September, 1974,
Proceedings, 2 vols., not consecutively paginated. Available from Meteorological Office, Bracknell, Berkshire, England.

COMESA, 1975: Report of the Committee on Meteorological Effects of Stratospheric
Aircraft, R.J. Murgatroyd, Chairman. Meteorological Office, Bracknell, 2 parts,
597 pp.

GENERAL REFERENCES

We consulted large numbers of books and research papers. The following book was
particularly useful:

C.E. Junge, 1963: Air Chemistry and Radioactivity, Academic Press, New York and London, 382 pp. This is still an essential reference work 13 years after publication.

and the following papers:-

M. Ackerman, 1975: NO, NO_2 and HNO_3 below 35 km in the atmosphere, Journal of the Atmospheric Sciences, 32, 1649-1657.

P.J. Crutzen, 1970: The influence of nitrogen oxides on the atmospheric ozone content, Quarterly Journal of the Royal Meteorological Society, 96, 320-325.

H.S. Johnston, D. Kattenhorn and G. Whitten, 1976: Use of excess carbon 14 data to calibrate models of stratospheric ozone depletion by supersonic transports, Journal of Geophysical Research, 81, 368-386.

C. Junge, 1971: Der Stoffkreislauf der Atmosphäre: Probleme und neuere Ergebnisse der luftchemischen Forschung. Jb. Max-Planck-Ges. Förd. Wiss. E.V. 1971, pp. 149-181.

M.B. McElroy, Chemical Processes in the Solar System: A Kinetic Perspective, Harvard University, August 1975, typescript, 233 pp.

R. E. Munn and B. Bolin, 1971: Global air pollution meteorological aspects, Atmospheric Environment, 5, 363-402

R.E. Munn and D.M. Whelpdale, 1975: Global air pollution, paper presented to International Congress of Scientists on the Human Environment, Kyoto, Japan, Nov. 16-26, 1975.

F.S. Rowland, 1976: Possible anthropogenic influences of stratospheric ozone, published in WMO Bulletin.

A.L. Schmeltakopf, P.D. Goldan, W.R. Henderson, W.J. Harrop, T.L. Thompson, F.C. Fehnsenfeld, H.I. Schiff, P.J. Crutzen, I.S.A. Isaksen, and E.E. Ferguson, 1975: Measurements of stratospheric $CFCl_3$, CF_2Cl_2, and N_2O, Geophysical Research Letters, 2, 393-396.

Yuk Ling Yung, Michael B. McElroy and Steven C. Wofsky, 1975: Atmospheric halocarbons: A discussion with emphasis on chloroform, Geophysical Research Letters, 2, 397-399.

PAPER 14

Assessment of the Economic Impact of Restrictions on the Use of Fluorocarbons - An Outline of OECD Activities 1976-1977

Presented by the Organization for Economic Cooperation and Development

OBJECTIVES OF THE STUDY

Following the presentation at the 16th session of the OECD Environment Committee (12th-14th November, 1975) of a review of the potential problems associated with the discharge of fluorocarbons (CFCs) into the environment, the Committee agreed that the possible consequences of regulating the use of fluorocarbons should be studied on an international scale.

A work programme was subsequently approved by the Environment Committee at its 17th session (16th-18th March, 1976). The main tasks of this work programme can be summarized as follows:

(a) identification of the activities of other international organizations in respect of fluorocarbons;

(b) compilation of an inventory of completed, on-going and planned research of relevance in the areas of

    - stratospheric chemistry and dynamics, and

    - biological and non-biological effects of changes in ultraviolet radiation*;

(c) identification of substitutes for fluorocarbons; and

(d) study of policy options (to be carried out in two steps):

    - identification of existing, planned or possible regulatory measures, and

    - study of the economic, technological and social impact of such measures**.

SCOPE OF THE FIRST REPORT

A first report has been prepared, as requested, on the work under items (c) and (d) of the work programme. The report includes:

---

* This aspect was subsequently agreed as being more appropriate for other organizations.

** It was agreed that the analysis should focus primarily on the effects on the *aerosol industry* and that a first report should be presented early in 1977.

- a general description of the structure of the industries producing and using CFCs, including a summary of the information supplied by Member countries regarding CFC production and consumption patterns;

- discussion of a number of possible regulative measures implying varying levels of reduction of CFC emission as well as differing time scales for implementation (i.e. "regulatory scenarios");

- discussion of the problems associated with comprehensive economic impact analyses of the different "regulatory scenarios";

- tentative conclusions to be drawn from an analysis of the presently available data; and

- identification of areas needing further data and/or analysis.

Appendices include additional material on:

- a breakdown by countries and type of usage of 1974 OECD production of F_{11} and F_{12} fluorocarbons;

- estimates of world output and emission of fluorocarbons between 1931 and 1975;

- listing of the fluorocarbon-related activities of other international organizations;

- summaries of existing national legislation relevant to the problems of fluorocarbons in the environment;

- identification of possible substitutes to F_{11} and F_{12} as aerosol propellants.

Since the report has not yet been discussed in the OECD Member countries, it is not possible to give more than a short description of the content of the report.

CONTENT OF THE REPORT

Background Data on the Production and Use of Fluorocarbons

In view of the fact that most of the discussion concerning fluorocarbon emissions to the atmosphere has centered on fluorocarbons F_{11} and F_{12}, the report examines the effects of restriction on the use of these two fluorocarbons.

According to information submitted by OECD Member countries, 1974 OECD production of these fluorocarbons amounted to around 755,000 tons. Net exports from the OECD area were estimated at roughly 55,000 tons, making the consumption within the OECD area some 700,000 tons.

The main uses of F_{11} and F_{12} are as: (1) aerosol propellants; (2) refrigerants; and (3) foaming agents for plastic foams. Personal aerosols accounted for approximately half of the 1974 OECD consumption of these fluorocarbons. In this respect there were no significant differences between the OECD Member countries. As far as the consumption pattern for the other uses of these fluorocarbons is concerned, the differences between countries were more significant.

The production packaging and distribution of fluorocarbons and products using fluorocarbons involve a large number of industrial sectors which would be affected to a varying degree by regulations concerning the use of fluorocarbons F_{11} and F_{12}. For purposes of a more detailed analysis of the posssible economic effects of different regulatory alternatives, the report describes briefly the structure of the industrial sectors supplying fluorocarbons and fluorocarbon-using products.

The following industrial activities are considered in the report:

- production of raw materials and intermediate products for fluorocarbons;
- CFC-synthesis;
- aerosol production;
- refrigeration and air-conditioning systems manufacture; and
- plastic foam production.

Regulatory Options

The number of regulatory options available to the authorities is very large indeed. There are numerous alternatives both with regard to the scope of the regulation and the time scale for its implementation.

The report has selected 5 regulatory options which are discussed in a short-term perspective (i.e. implementation 6 months after a decision) and in a medium-term perspective (i.e. a 3-year implementation lag). The following regulatory alternatives for F_{11} and F_{12} were selected:

- ban on use in personal aerosols;
- ban on use in personal as well as in domestic aerosols;
- ban on all uses as aerosol propellant (with the exemption of medical aerosols);
- ban on all propellant and plastics use; and
- ban on all uses (except replacement uses in refrigeration and A/C systems).

(Note: Ban on use in *new* refrigeration and A/C systems is considered within a 3- and 6-year time-scale).

Economic Impact Analysis

The objective of an economic impact analysis is to identify those industry sectors likely to be affected by the regulations under consideration and to estimate the magnitude of the effects on variables like employment, production, prices, profitability, balance of payments, environment, etc. A detailed analysis of this type has not been possible, but the report gives an indication of the likely short- and long-term economic effects associated with the different regulatory alternatives.

PAPER 15

An Outline of Current Monitoring and Research being Carried out into Stratospheric Ozone and Related Subjects in Australia

Presented by the Government of Australia

This paper summarises the present Australian program of monitoring and research into stratospheric ozone, and associated studies into the manufacture and use of fluorocarbons. It is divided into the following sections:

- A review of the activities of the Australian Bureau of Meteorology to study the thermal structure and circulation of the stratosphere.

- A review of the research currently being carried out by the Commonwealth Scientific and Industrial Research Organisation (CSIRO) into stratospheric ozone, related gaseous pollutants, particulates and ultra-violet radiation.

- Studies into the Australian production and usage of fluorocarbon materials being carried out by the Department of Environment, Housing and Community Development.

THERMAL STRUCTURE AND CIRCULATION OF THE STRATOSPHERE - AUSTRALIAN
BUREAU OF METEOROLOGY

The Australian Bureau of Meteorology has, for the past 20 years, been making routine extended upper air balloon flights to monitor the thermal structure and circulation of the stratosphere. The network of 16 stations for which data is available on a daily basis up to December 1976 is shown in Fig. 15.1. A recent change to smaller balloons will mean that the pressure level reached by 50 per cent of flights will be lowered from 12 mb to 54 mb (30 km to 20 km) approximately. It is proposed, however, that the higher altitude flights will continue indefinitely at Darwin, Alice Springs and Laverton to maintain the climatological record of stratospheric wind and temperature over a range of latitude.

From October 1970 to December 1973, the Bureau of Meteorology carried out a program of rocket probes into the stratosphere from Woomera up to a height of about 70 kilometres to determine the seasonal variation of stratospheric wind patterns. There are no plans to reactivate this program.

The Bureau of Meteorology also operates a total ozone measurement program in Brisbane and Hobart on behalf of the CSIRO Division of Atmospheric Physics.

During the last decade, stratospheric research in the Bureau of Meteorology has been concerned with a diagnostic study of the behaviour of the Southern Hemisphere polar vortex and stratospheric circulation up to 30 millibars (about 23 km) in relation to the Spring time warming. Research has also been undertaken into the

Fig. 15.1. Network of radiosonde stations in Australia

synoptic climatology of winds in the stratosphere with particular emphasis on the transport of radioactive debris.

Figure 15.1 shows the network of radiosonde stations in Australia at which 50 per cent of balloon flights attain an altitude of approximately 30 km (12 mb). Daily observations available for approximately 20 years to December 1976. The high altitude stations at Mawson, Cabey and Macquarie Island were also a part of this network. The program is to continue indefinitely.

STRATOSPHERIC OZONE RELATED GASEOUS POLLUTANTS PARTICULATES AND ULTRA-VIOLET RADIATION – COMMONWEALTH SCIENTIFIC AND INDUSTRIAL RESEARCH ORGANISATION

Two Divisions of CSIRO are conducting research into factors that may affect the ozone layer. The Division of Atmospheric Physics based at Aspendale near Melbourne, Victoria is monitoring ozone concentrations, gaseous pollutants and the incidence of ultra-violet radiation. The Division of Cloud Physics is interested in particles found in both the atmosphere and the stratosphere and has developed methods for collection and analysis.

Division of Atmospheric Physics - Atmospheric Chemistry Program

Ozone. Atmospheric ozone has been monitored for almost 20 years. Six monitoring stations have been established in Australia and they are located at Aspendale (near Melbourne), Cairns, Brisbane, Perth, Macquarie Island and Hobart. Observations of total ozone and of its distribution with height are made by the Umkehr method. It is intended to establish a seventh Dobson ozone station at Alice Springs in the near future. Measurements from the ground stations are made daily.

Chemical ozone sensors of the Mast-Brewer type are flown from Aspendale fortnightly to measure the vertical distribution of ozone. The balloon-borne samplings carried out from Aspendale provide vertical profiles of ozone concentrations from ground level to approximately 27,000 metres.

Since the inception of the ozone monitoring program results have provided no definite evidence of any change in the concentration of stratospheric ozone which can be ascribed to pollutants.

Halocarbons. Measurements have commenced of halocarbons, Freon 11 ($C Cl_3F$) and carbon tetrachloride ($C Cl_4$). Ground level concentrations of Freon 11 are monitored regularly at Aspendale and Cape Grim in Tasmania (the interim site for the Australian Baseline Station). Using aircraft sampling the Freon 11 and carbon tetrachloride content of air at 4,000 metres over Bass Strait and at 9,000 metres over the Great Australian Bight is measured regularly. These gases are also sampled using high altitude balloons.

Oxides of nitrogen. A spectrophotometric method (Dobson spectrophotometer) is being used to establish average background levels of nitrous oxide.

One year's data from both Aspendale and Brisbane suggest that seasonal variations in nitrous oxide concentration closely parallel those for oxide. Work is under way to improve the technique and accuracy of the method used for monitoring this gas.

In the near future it is planned to start measuring nitric oxide in the stratosphere (the first such measurements for the Southern Hemisphere).

Erythemal ultra-violet radiation. A network of six ground level ultra-violet monitoring stations distributed around Australia was established one year ago. However, this program is not sufficiently advanced to determine whether or not there are any systematic trends in changes in the incidence of ultra-violet radiation reaching the land surface in Australia.

Stratospheric water vapour. A balloon-borne infra-red radiometer is used to measure stratospheric water vapour at altitudes of 24 to 27 km. Attempts are being made to use the same balloons to obtain grab-samples of stratospheric air for laboratory measurements of CO_2, halocarbons, and oxides of nitrogen.

NASA flight. The Division recently availed itself of the opportunity to participate in a flight over Tasmania by the NASA laboratory aircraft "Galileo II". On the flight techniques used in Australia to measure Freons were compared, and were found to be consistent with methods used in the U.S.A. Results from the flight confirmed that Freon levels in the atmosphere over Southern Australia differed little from those of the Northern Hemisphere.

Division of cloud physics - particles in the stratosphere. Studies are being made

308 Government of Australia

of the size, composition and concentration of particles found in the lower stratosphere. Direct measurements of these particles began with samplings from U-2
aircraft flights in 1962. From 1967 to 1974 stratospheric particles have been
sampled using an impactor designed by the Division and flown on balloons launched
by the U.S. Atomic Energy Commission. Since 1974 samplings have taken place bi-
monthly from balloons launched by the Department of Science from its Mildura
launching site. The program includes co-operative ventures with the University of
Melbourne and the Max Planck Institute.

Results on the seasonal and longer-term trends in particle concentration are now
becoming available.

Chemical analysis of particles. A technique has been developed for detecting the
common atmospheric ions in particles down to a very small size. The single-particle
analysis technique applies a thin film of conventional chemical reagents to particles
by vacuum deposition. Exposure to water vapour promotes a reaction the products of
which are viewed with an electron microscope.

AUSTRALIAN FLUOROCARBON PRODUCTION AND USAGE - DEPARTMENT OF ENVIRONMENT, HOUSING AND COMMUNITY DEVELOPMENT

Following a 1975 request from the OECD for national figures on fluorocarbon production and usage, the (then) Department of Environment began to collate data from
the aerosol and fluorocarbon manufacturing industries. This data base has
subsequently been maintained, both to provide information for the OECD and to
facilitate assessment of the industrial impacts and likely efficiency of any controls
which may be proposed. Details of information forwarded to the OECD are attached.

Australia produced 14,000 tonnes of CFCs in 1974, the most recent year for which
the OECD has worldwide data. It appears from information becoming available, that
this will rise to some 17,000 tonnes in 1976. Although these tonnages appear large,
they should be viewed in the context of an estimated annual global production of
some 756,000 tonnes, the Australian production representing only 1.96% of the total.
CFCs are produced at two plants in Australia, there being insignificant imports
(768 tonnes) and no exports.

As will be noted from the OECD figures, on a "per capita" basis, Australia has a
very high rate in Tables 15.1 and 15.2 of aerosol usage, typically being the second
or third highest user of most products. On this basis, it is by far the largest
user of insect sprays and repellents. This factor is largely responsible for
Australia's unique consumption of roughly equal percentages of fluorocarbon usage
in domestic and personal products (approx. 35% total CFC production each); most
other OECD countries use a far higher proportion in personal products (typically
55% personal, 15% domestic).

There has been an increasing trend for some other aerosols to be propelled by
other gases, most commonly butane, which has a cost advantage. This has especially
been the case for insect sprays, which present relatively few formulation problems.
It is unlikely the trend will extend to personal products such as hairsprays as the
product is too flammable and CFCs have good flame retardant properties.

No firm regulatory measures are being planned for the present pending the outcome
of joint international studies into the dimensions of this problem, and establishment of the need for such controls.

TABLE 15.1 Aerosol Use and Production[a] in Selected Countries, 1970 and 1974. (Country Survey by U.S. and Canada: IMOS report, p. 91; Chemical Specialties Manufacturers Association, Inc.)

1970	Total Use[b] (units per 1000 people)	Insecticides	Paints	Household	Hair Spray	Deodorants	Production (Million Units)
Australia	NA	NA	NA	NA	NA	NA	NA
Canada	6240	330	520	1640	1450	1070	133
Finland	NA	NA	NA	NA	NA	NA	NA
France	4400	318	99	596	1690	318	277
Italy	3100	847	19	394	1050	414	173
Japan	1600	453	120	253	519	213	194
Netherlands	4400	430	78	1000	1290	650	90
Sweden	3600	253	101	406	1320	885	20
Switzerland	8200	305	168	1540	3020	1930	60
United Kingdom	5500	432	234	1470	1820	576	361
United States	13000	545	1045	2864	2230	2840	2554

1974							
Australia	9256	1834	856	2732	1637	2149	125
Canada	8020	400	620	2000	1200	1740	180
Finland	3546	375	4	750	417	2000	NA
France	NA	NA	NA	NA	NA	NA	NA
Italy	NA	NA	NA	NA	NA	NA	NA
Japan	2090	622	154	362	507	244	230
Netherlands	5808	815	41	1190	1705	1502	155
Sweden	NA	NA	NA	NA	NA	NA	NA
Switzerland	NA	NA	NA	NA	NA	NA	NA
United Kingdom	NA	NA	NA	NA	NA	NA	NA
United States	13500	600	1185	3410	2090	3360	2849

[a] All aerosol propelled units (or containers) are included, not just those using fluorocarbons as a propellant.

[b] "Total" figures exceed sum of individual specialty uses since later does not cover food-related uses.

TABLE 15.2   Summary of Replies in Answer to two OECD Questionaires on the Australian
             Fluorocarbon Industry

AUSTRALIAN FII and F12 PRODUCERS

Australian Fluorine Chemicals P/L

Pacific Chemical Industries P/L

TOTAL AUSTRALIAN PRODUCTION AND USAGE (tonnes, 1970-1975)

	1970	1971	1972	1973	1974	1975 (est)
Aerosols	6,834	7,461	8,390	9,647	10,700	11,840
Refrigeration	1,570	1,714	1,927	2,216	2,000	2,700
Foam	831	907	1,020	1,173	1,300	1,410
Total	9,235	10,082	11,337	13,036	14,000	approx 16,000

Approximate total production capacity 25,000 tonnes

ANALYSIS OF FLUOROCARBON CONTENT IN AEROSOLS, AND TOTAL USE PER 1000 POPULATION

	Total Aerosol Production (X1000)	Units Per 1000 Population	Fluorocarbon Propelled Aerosols per 1000 Population	% Containing Fluorocarbons	% of Total Fluorocarbons Production
Insect sprays	24,471	1,834	917	50)	
Household	24,516	1,838	92	5)	
Air Fresheners	11,923	894	447	50)	35
Paints	11,419	856	541	60	
Hair Sprays	21,833	1,637	1,555	95)	4
Other Personal (deodorants, etc)	28,665	2,149	2,041	95)	36
Pharmaceutical	645	48	34	70	<1
TOTAL	126.5 million	9,256	5,417	approx 58	approx 76

AUSTRALIAN AEROSOL INDUSTRY (PRODUCTION AND DISTRIBUTION)

Volume of domestic wholesale sales $90,000,000
   "   "   "   retail   " $126,000,000
   "   "   imports $75,000

Number employed in production and distribution of aerosols approx 3000

AUSTRALIAN REFRIGERATION INDUSTRY

Value of Manufacturers' sales	$86,878,000
Value of Exports	$13,793,000
Value of Imports	$92,422,000
Capital Employed in Refrigeration and air conditioning	$ 1,771,000
value added	$39,552,000
number employed	3,704

PAPER 16

*Stratospheric and Ozone Research Activities in Canada**

Presented by the Government of Canada

The recent work that has been done in Canada in response to concern about anthropogenic modification to the stratospheric ozone content has developed from an extensive prior Canadian involvement in ozone research. Ozone monitoring, related stratospheric measurements and laboratory photochemical studies, photochemical modelling and atmospheric data analysis had all been in progress for many years before 1970 when the supersonic aircraft/ozone question was first raised.

Ozone monitoring at the five stations of the Canadian Dobson network has been in continuous operation since 1960 and the ozonesonde network since 1965. Also the Canadian government has run the World Ozone Data Centre (on behalf of WMO) since its inauguration in 1964. Development of a modern standard replacement for the Dobson spectrophotometer was started at the University of Toronto in 1968 by Brewer et al.

In the early sixties the aircraft and balloon measurement programme run by the Canadian Armament Research and Development Establishment (CARDE) accumulated a large data bank of stratospheric infrared spectral measurements which continues to yield useful data on several trace constituents (i.e. Lowe and McKinnon- 1972). Key measurements of the ozone-producing solar ultraviolet flux were made by the University of Toronto group of Brewer and Wilson in 1965 and high altitude ozone concentrations were measured by Evans and Llewellyn in 1970 at the University of Saskatchewan.

Laboratory studies relating to atmospheric ozone photochemistry were pursued by several groups including those at York University, the National Research Council and at CARDE.

Photochemistry models were developed by Hampson in 1966 at CARDE and also, primarily for the upper atmosphere, by Gattinger in 1968 at the National Research Council. Models which incorporated the combined effects of photochemistry and atmospheric transport (two and three dimensional) were constructed by the McGill Arctic Meteorological Research Group which had done extensive analysis of stratospheric motions in relation to ozone (Boville and Hare in cooperation with Godson - 1961).

*Prepared by Dr B.W. Boville and Dr W.F.J. Evans of the Atmospheric Environment Service, Toronto, Canada.

During the early seventies the Atmospheric Environment Service (AES), which is the government's lead agency in stratospheric research, had greatly strengthened its expertise in the area largely by recruitment from the Canadian university community. As a result it was then ready to mount an accelerated research programme, which includes in-house research and support of university and industry research projects, and to move in collaboration with other national and international agencies to meet the challenge of evaluating the new threats to the ozone layer.

General guidance to the AES is provided by an "Advisory Committee on Stratospheric Pollution" which has met twice yearly since 1973. The committee comprises a wide cross-section of government, university and industry scientists who are involved in various aspects of stratospheric research. The committee also, in conjunction with the AES, serves to coordinate overall research in this area.

SURVEY OF CURRENT RESEARCH ACTIVITIES

Progress in stratoshperic research must be based on the combination of laboratory investigations of relevant atmospheric photochemistry, atmospheric modelling of the laboratory data and the testing of models against atmospheric measurements. Only then can the models be used with any confidence to project the future effects of man's activities on the ozone layer.

Atmospheric Measurements

The verification of models against measurements of stratosphere composition and transport is both extremely important and at the same time, very difficult due to the advanced instrumentation required to detect very small amounts of the important trace gases.

The AES Stratospheric Balloon Measurements Projects is a multiple experiment balloon payload designed to make simultaneous measurements of key stratospheric trace gases important in the photochemical balance of the ozone layer. The 1974 experiment was flown at 35 km to study the nitrogen chemistry of the stratosphere by making simultaneous measurements HNO_3, NO_2, NO, O_3 and temperature in conjunction with solar ultraviolet flux measurements. In the program, half of the twelve experiments were conducted by AES scientists and half by Canadian university scientists. Payload engineering and field support of Project STRATOPROBE I (Table 16.1) were provided by Canadian industry (SED Systems Ltd.).

In the 1975 program, experiments to measure freons, HCl, HF and ClO were incorporated to supplement the measurements of nitrogen chemistry constituents. These new experiments are designed to evaluate the presence of chlorine compounds of natural and anthropogenic origin. In the STRATOPROBE II series, separate flights of the nitrogen chemistry (Table 16.2) and chlorine chemistry (Table 16.3) payloads were made.

In 1976, two flights of a combined payload to measure both chlorine and nitrogen chemistry were flown from Yorkton in August. The experiment complement, the experimenters and affiliations involved in the STRATOPROBE III Project are described in Table 16.4. The AES balloon project is coordinated by W.F.J. Evans, J.K. Kerr and D.I. Wardle of the Atmospheric Environment Service.

In addition to the STRATOPROBE Project, there are a number of other related stratospheric measurement activities, most of them at universities. These include other balloon experiments as well as ground based aircraft and rocket measurements. A listing of these experimental activities is given in Table 16.5 which also indicates the location and personnel involved.

TABLE 16.1   The 1974 AES Balloon Project: Stratoprobe I – The Nitrogen Chemistry
             Experiment.

Experiment	Constituents	Experimenters	Organization
SOLAR ABSORPTION			
Scanning Solar Radio-meter	H_2O, O_3, N_2O	W. Evans, S. Bain	AES
NO_2 Spectrophotometer	HO_2, O_3	A. Brewer, J. Kerr, T. McElroy	Univ. of Toronto AES
NO_2 Scanning Photo-meter	NO_2	D. Wardle	AES
U.V. Flux Spectro-meter	U.V. Flux O_3	D. McEwen	Univ. of Sask.
EMISSION			
HNO_3	HNO_3	W. Evans, C. Lin	AES
Scanning Emission Radiometer	H_2O, O_3	W. Evans, C. Mid-winter	AES
Far I.R. Michelson	O_3, H_2O, HNO_3	T. Clark, D. Kendall	Univ. of Calgary
I.R. Spectrometer	OH, O_2 ($^1\Delta$)	E. Llewellyn	Univ. of Sask.
SAMPLING			
NO Snooper	NO	H. Schiff, B. Ridley	York Univ.
O_3 Sampler	O_3	R. Olafson, J. Kerr	AES
Ozonesonde	P, T, O_3, H_2O		AES

Table 16.2   Stratoprobe II – 1975 Chlorine Payload Configuration Experiments.

Experiment	Constituents	Experimenters	Organization
SOLAR ABSORPTION			
Gaspec	HCl, NO	H. Zwick, R. Wiens	Barringer
Michelson Interferometer	HCl, NO, NO_2, CH_4	H Buijs	Bomem
NO_2 Spectrophotometer	NO_2	T. McElroy	AES
ClO Spectrophotometer	ClO	J. Kerr	AES
U.V. Scanning Spectrometer	O_3	D. McEwen	Univ. of Sask.
SAMPLING			
Nitric Oxide Snooper	NO	B. Ridley, H. Schiff	York Univ.
Ozone Chemiluminescent Sampler	O_3	R. Olafson, T. McElroy	AES
Modified Ozonesonde	O_3	R. Olafson	AES
Whole Air Sampler Spheres	$F11$, $F12$, CCl_4	H. Schiff, T. Thompson	York Univ. NOAA
SKY RADIATION			
Scattering Spectrometer	ClO, OH	D. Wardle	AES
6 to 14µ Scanning Radiometer	HNO_3, O_3, N_2O_5	W. Evans	AES
11µ Nitric Acid Radiometer	HNO_3	W. Evans, R. O'Brien	AES

TABLE 16.3   Stratoprobe II - 1975 Nitrogen Payload Configuration Experiments.

Experiment	Constituents	Experimenters	Organization
SOLAR ABSORPTION			
Gaspec	HCl, NO	H. Zwick, R. Wiens	Barringer
Michelson Interferometer	HCl, NO, NO_2, CH_4	H. Buijs, G. Vail	Bomem
NO_2 Spectrophotometer	NO_2	T. McElroy	AES
NO_2 Scanning Photometer	NO_2	D. Wardle	AES
U.V. Scanning Spectrometer	O_3	D. McEwen	Univ. of Sask.
SAMPLING			
Nitric Oxide Snooper	NO	B. Ridley, H. Schiff	York Univ.
Ozone Chemiluminescent Sampler	O_3	R. Olafson	AES
Modified Ozonesonde	O_3	R. Olafson, T. McElroy	AES
SKY RADIATION			
IR Spectrometer	OH^*, $O_2(^1\Delta)$	E. LLewellyn	Univ. of Sask.
6 to 14μ Scanning Radiometer	HNO_3, O_3, N_2O_5	W. Evans, R. O'Brien	AES
Nitric Acid Radiometer	HNO_3	W. Evans, R. O'Brien	AES
Far IR Michelson Interferometer	O_3, H_2O, HNO_3, N_2O	T. Clark	Univ. of Calgary

TABLE 16.4 Stratoprobe III - 1976 Chlorine-Nitrogen Payload Experiments.

Experiment	Constituents	Experimenters	Organization
SOLAR ABSORPTION			
NO_2 Scanning Photometer	NO_2	D. Wardle	AES
Michelson Interferometer	HCl, HF, NO_2, CH_4	H. Fast, H. Buijs	AES; Bomem
NO_2 Spectrophotometer	NO_2	T. McElroy	AES
ClO Spectrophotometer	ClO	J. Kerr	AES
U.V. Scanning Spectrophotometer	O_3, U.V. spectra	D. McEwen	Univ. of Sask.
U.V. Fluxmeter	U.V. Flux	J. Williamson	AES
SAMPLING			
Nitric Oxide Snooper	NO	B. Ridley, H. Schiff	York Univ.
Ozone Chemiluminescent Sampler	O_3	R. Olafson	AES
Modified Ozonesonde	O_3, T, n(M)	R. Olafson, T. McElroy	AES
Whole Air Sampling Spheres	F11, F12, CCl_4, N_2O	B. Ridley, H. Schiff	York Univ.
SKY RADIATION			
Scattering Spectrometer	ClO, OH	D. Wardle	AES
6 to 14μ Scanning Radiometer	HNO_3, O_3, N_2O_5, CH_4	W. Evans, R. O'Brien	AES
11μ Nitric Acid Radiometer	HNO_3	W. Evans, R. O'Brien	AES
Far I.R. Michelson Interferometer	O_3, H_2O, HCl, N_2O	T. Clark, D. Kendall	Univ. of Calgary

TABLE 16.5

Measurement	Experimenters	Organization
Solar Absorption Measurements of HCl, HF with a balloon borne Michelson Interferometer.	H. Buijs	Bomem
Ground based monitoring of stratospheric gases with a high resulution Michelson Interferometer.	H. Buijs	Laval Univ.
Analysis of stratospheric trace gases from CARDE balloon and aircraft. I.R. Data.	R.P. Lowe	Univ. of Western Ontario
Spectrophotometric Measurements of NO_2 and O_3 from ground and aircraft.	A.W. Brewer, C.T. McElroy, J.B. Kerr	Univ. of Toronto AES
Balloon flight measurements of nitric oxide in the stratosphere.	B.A. Ridley, H.I. Schiff	York Univ.
Grab sampling of stratospheric air and subsequent laboratory analysis for freons and N_2O.	B.A. Ridley, H.I. Schiff	York Univ.
High resolution ground based solar spectrum measurements for detection of ClO.	R.W. Nicholls, W. Fabian	York Univ.
Balloon measurements of stratospheric constituents by submillimeter interferometer in emission and absorption.	T.A. Clark, et al.	Univ. of Calgary
Rocket measurements of water vapour in the stratosphere.	W.F.J. Evans, R. O'Brien	AES
Rocket measurements of high altitude ozone from O_2 ($^1\Delta$). Ground based monitoring of stratospheric NO_2.	E.J. Llewellyn	Univ. of Sask.

Stratospheric Monitoring

Ozone monitoring is important to establish the climatology of ozone and to verify predictions of ozone depletion which should ultimately be detectable as trends in the world total ozone. Current activities in this direction include maintaining and improving the precision of the Dobson network and the ozonesonde network as well as continuing evaluation of satellite BUV ozone data. In order to overcome instrumental and operational deficiencies in the Dobson instrument, a new total ozone spectrometer has been developed and is now in initial stages of commerical production. This is a grating instrument with digital electronics and capable of being highly automated, a feature which should significantly reduce the amount of operator time required for measurements.

TABLE 16.6

Canadian ozone monitoring network. Ozonesondes, Dobson ozone instruments.	R.A. Olafson, A. Asbridge	AES
Publication of World Ozone Data.	L. Morrison	AES
Satellite BUV Ozone Measurements.	C.L. Mateer	AES
Solar Ultraviolet Flux Monitoring in Saskatchewan.	D.J. McEwen	Univ. of Sask.
Development of new network ozone spectrophotometer.	J.B. Kerr, D.I. Wardle, C.T. McElroy (contractor - SED Systems Ltd.)	AES

Laboratory Studies

These comprise studies of chemical reaction rates and of photo absorption cross-sections; more accurate data is required in both areas for stratospheric modelling. As well, spectroscopic studies are used to parameterize certain detailed absorption processes and to aid spectroscopic measurement of trace constituents in the stratosphere.

Modelling Studies

Models of the natural stratosphere including photochemistry and transport are required to simulate the ozone layer and to project ozone depletions due to anthropogenic increases in NO_x by aircraft and fertilizer usage as well as freon release into the troposphere. Fortunately, modelling work of this type is being carried out in Canada. Models to predict climatic effects of ozone depletion as well as the direct greenhouse effect of freons on the climate are also required for an evaluation of stratospheric pollution problems.

TABLE 16.7

Photochemical reactions of atmospheric interest: quenching of $O(^1D)$ by N_2O, O_2, N_2, etc.	H.I. Schiff	York Univ.
Spectroscopic studies of stratospheric molecules, spectroscopy of ClO, O_2.	R.W. Nicholls, et al.	York Univ.
Single particle detection of atmospheric species, photodissociation studies.	J.W. McConkey	Univ. of Windsor
Studies of absorption of infrared radiation by ozone and NO_2.	C. Young	Univ. of New Brunswick
Laboratory reactions of O and $O(^1D)$ with Species of Atmospheric Interest: Photolysis of ozone, hydroxyl reactions with freons.	R.J. Cvetanovic, et al.	NRC
Ultraviolet Absorption Properties of Fluorocarbons.	C. Sandory	Univ. of Montreal
Absorption spectroscopy of ClO.	J.A. Coxon	Dalhousie Univ.

TABLE 16.8

A 2-dimensional photochemical meteorologically interactive steady state model of the stratosphere. A time dependent version is being tested.	R.K.R. Vupputuri	AES
One and 2-dimensional time-dependent photochemical models of the ozone layer with parameterized transport.	J.C.McConnell	York Univ.
Calculation of stratospheric ultraviolet fluxes.	R.W. Nicholls, M.W. Cann	York Univ.
General Circulation Models and Climatic Variation; Effects of Freons on Climate.	G.J. Boer, et al.	AES

Related Research Projects

The precursor of the natural destruction process of ozone is nitrous oxide produced by microbiological organisms at the earth's surface. Studies of microbiological production in the soil are complemented by measurements of tropospheric nitrous oxide and of nitrous oxide in the oceans. The flux from the oceans into the atmosphere is at present crucial to the evaluation of the nitrogen fertilizer/ozone problem and is highly uncertain.

Tropospheric measurements of halocarbons bear directly on freon ozone destruction and continue to be necessary. Report studies of the effects of stratospheric ozone reduction are also in progress.

TABLE 16.9

Tropospheric measurements of Freons.	L. Elias	NRC
Tropospheric measurements of N_2O.	J.W. McKenny	Univ. of Windsor
Microbiological Studies of Nitrous Oxide and nitrogen fertilizers.	R. Knowles	Macdonald Coll. of McGill Univ.
Measurements of N_2O content in oceans.	T. Yoshinari	Macdonald Coll.
Environmental Studies of Stratospheric Pollution.	F.K. Hare	Univ. of Toronto
Survey Report Studies of Freon Ozone Depletion.	R.E. Munn	AES

FUTURE PLANS IN STRATOSPHERIC OZONE LAYER RESEARCH

- To continue a broad program of research and development on the ozone layer in government, industry and universities.

- To continue to operate the Canadian ozone network. Edmonton, Resolute, Churchill, Goose Bay and Toronto have Dobson instruments. Ozonesondes are flown each Wednesday (except Toronto).

- To continue publication of World Ozone Data.

- To test the new ozone spectrophotometer under operating conditions and carry out intercomparison tests with view to replacing the Dobson instruments in the Canadian network.

- To continue one and two dimensional modelling of the ozone layer and the effects of stratospheric pollution.

- To conduct flights of Project Stratoprobe in 1977 to measure the chlorine-nitrogen chemistry of the ozone layer.

- To participate in international programs of research and monitoring and in coordinated intercomparisons of experimental measurements of trace stratospheric constituents.

PAPER 17

Sur l'Etude des Problemes de l'Ozone Atmospherique en Belgique

Presented by the Government of Belgium

ABSTRACT

This paper outlines the observational, experimental and theoretical work on the stratospheric ozone layer carried out in Belgium.

It also reviews the ultraviolet spectrum and the photochemistry of ozone together with current knowledge on the catalytic hydrogen, nitrogen and chlorine systems.

Future activities in Belgium will include regular observations of total ozone, as well as its vertical distribution, by the Institut Royal Meteorologiques. These will be complemented by observations conducted in the stratosphere and on mountain tops by the Institut d'Astrophysique of the University of Liège and by information on minor stratospheric constituents obtained through measurements of solar infra red radiation.

Finally, at the Institute d'Aéronomie spatiale and at the University of Brussels laboratory, experimental studies and stratospheric observations will be undertaken with balloons, rockets and satellites (Spacelab) as well as theoretical studies involving simulation of stratospheric conditions through one and two dimensional mathematical models.

In this last case, the experimental, observational and theoretical aspects will be related so as to exploit new developments in the hydrogen, nitrogen and chlorine chemistry of the ozone layer.

An integrated research programme on all aspects of stratospheric phenomena must be continued in order to increase the accuracy of quantitative predictions of ozone layer variations.

INTRODUCTION

Dès les premières observations chimiques de l'ozone atmosphérique au cours du dix-neuvième siècle, on constate déjà un intérêt pour ce problème en Belgique, particulièrement dans le cadre de l'hygiène générale. Plus de 150 stations d'observations élémentaires, qui avaient été disséminées dans les villes et les campagnes, avaient fourni, pendant plus d'une douzaine d'années, des données que Van Bastelaer[1], en 1892, avait analysées, mais auxquelles on ne peut plus

321

accorder de valeur aujourd'hui.

Dès que les sondages stratosphériques deviennent possibles, on retrouve un intérêt qui se manifeste à l'Institut Royal Météorologique de Belgique; par example, dans les travaux successifs de Jaumotte[2] et Nicolet[3][4] consacrés à la fois aux aspects aéronomiques et météorologiques liés à la distribution verticale de l'ozone et aux variations dues à des phénomènes dynamiques.

Aujourd'hui, des sondages réguliers d'ozone sont effectués par l'Institut Royal Météorologique de Belgique à Uccle-Bruxelles. Les données sur l'ozone qui comprennent la quantité totale, la pression partielle, le rapport de mélange, la quantité par couches, les moyennes et également les "ozonograms" sont régulièrement publiées dans le Bulletin Trimestriel de l'Institut Royal Météorologique de Belgique.

Les mesures de la quantité totale d'ozone sont effectuées à l'aide du spectrophotomètre de Dobson et sa distribution verticale est obtenue à l'aide de la sonde du type Brewer. Les analyses de De Muer[5][6][7] fournissent toutes les indications sur les observations effectuées régulièrement depuis 1969, en particulier sur les conditions de transport de l'ozone.

D'autre part, des recherches théoriques, expérimentales et observationnelles ont été développées à l'Institut d'Aéronomie Spatiale de Belgique à Uccle-Bruxelles, et également en collaboration avec diverses institutions étrangères en France et aux USA. L'Université de Bruxelles et l'Institut d'Aéronomie peuvent collaborer grâce à l'aide du Fonds National de la Recherche Scientifique. Enfin, l'Institut d'Astrophysique de l'Université de Liège, par ses observations à grande dispersion du spectre infrarouge solaire, participe à l'observation des constituants minoritaires dans la stratosphère.

REVUE DES TRAVAUX EFFECTUES EN BELGIQUE DANS LE CADRE DES CONNAISSANCES ACTUELLES DE L'OZONE STRATOSPHERIQUE

Photochimie de l'Ozone et la Radiation Ultraviolette

Le rayonnement ultraviolet du soleil de longueurs d'onde inférieures à 242 nm (2420 Angströms) est à l'origine de la formation de l'ozone suivant le processus $O + O_2 + M \rightarrow O_3$ puisqu'il provoque la photodissociation de l'oxygène moléculaire suivant le processus $O_2 + hv \rightarrow O + O$.

Il convient de signaler que la connaissance de cet ultraviolet est requise avec une certaine précision. C'est pourquoi à l'Institut d'Aéronomie de Belgique, on s'est attaché à l'étude de sa détermination exacte. On retrouve dans les publications d'Ackerman[8], d'Ackerman et Simon[9], de Simon[10][11][12] et de Samain et Simon [13] les résultats obtenus par des mesures en fusée au-delà de 100 km ou en ballons stratosphériques à 40 km qui constituent des données spectrales pour l'étude de la photodissociation dans la stratosphère.

L'objet actuel de cette recherche à l'Institut d'Aéronomie est de déterminer avec plus de précision et de détail le spectre solaire dans le domaine d'absorption des bandes de Schumann-Runge de l'oxygène (200 - 175 nm) qui joue un rôle important dans la photodissociation de nombreux constituants minoritaires de la stratosphère comme, par exemple, les chlorofluorométhanes.

D'ailleurs, l'Institut d'Aéronomie, par une série de travaux successifs (Ackerman, Biaumé et Nicolet[14]; Ackerman et Biaumé[15]; Ackerman, Biaumé et Kockarts[16]; Kockarts[17][18][19]; Biaumé[20][21]; Cieslik et Nicolet[22]) a déterminé les conditions d'absorption de la radiation solaire au sein de la structure de rotation des bandes de O_2 couvrant le domaine spectral de longueurs d'onde inférieures à 200 nm.   La détection de l'ultraviolet solaire dans les régions spectrales requises sera poursuivie tout comme les recherches expérimentales et théoriques liées aux problèmes de photodissociation dans la stratosphère.

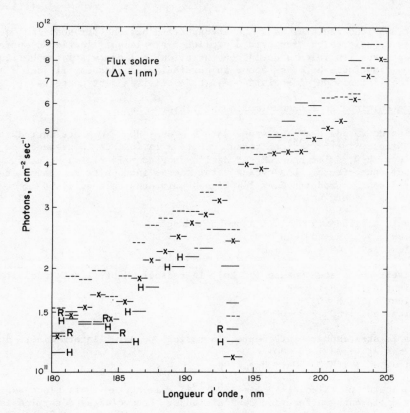

Fig. 1. Flux solaires observés dans le domaine spectral compris entre 180 et 205 nm.

Afin de préciser les conditions de formation et de destruction de l'ozone stratosphérique, nous les présentons, d'abord dans une atmosphère d'oxygène pur, par le schéma suivant :

Photodissociation de l'oxygène,
(Coefficient J_2) ; $O_2 + h\nu \rightarrow O + O$.
Formation de l'ozone,
(Coefficient k_2) ; $O + O_2 + M \rightarrow O_3 + M$.
Photodissociation de l'ozone,
(Coefficient J_3) ; $O_3 + h\nu \rightarrow O_2 + O$.
Destruction de l'ozone et reformation de l'oxygène,
(Coefficient k_3) ; $O + O_3 \rightarrow O_2 + O_2$.

En équilibre photochimique, on écrit (Chapman, 1930) :

$$n^2(O_3) = \frac{J_2}{J_3} \, n \, (M) \, n^2 \, (O_2) \, \frac{k_2}{k_3}$$
\hfill (I)

ou, pour toutes conditions, si v est la vitesse de transport,

$$\frac{\delta n(O_3)}{\delta t} + \text{div}\left[n(O_3)\, v_{O_3}\right] + 2\,\frac{k_3\, J_3}{k_2\, n(M)\, n(O_2)}\, n^2(O_3) = 2\, n(O_2)\, J_2$$

(II)

Ces équations fondamentales dans une atmosphère d'oxygène pur sont les équations auxquelles on doit se référer afin d'expliquer simplement l'action d'autres constituants sur les valeurs de la concentration de l'ozone stratosphérique. Les constantes cinétiques k_2 et k_3 étant aujourd'hui bien connues, il faut s'assurer que les coefficients de dissociation J_2 et J_3 utilisés sont corrects.

Le Système Chimique de l'Ozone Impliquant l'Hydrogène

Il y a plus de 25 ans que le système O_3 – H – OH – HO_2 – O a été introduit par Bates et Nicolet[23][24][25] en tenant compte à la fois de la présence de la vapeur d'eau, H_2O, du méthane CH_4 et de l'hydrogène moléculaire H_2 dans l'atmosphère supérieure. Les cycles catalytiques introduits par Bates et Nicolet[25] sont reproduits dans les lignes suivantes tels qu'ils doivent encore être utilisés aujourd'hui :

(a) à haute altitude (50 km), l'atome d'hydrogène
(Coefficient a_2) ; $H + O_3 \rightarrow OH + O_2$
(Coefficient a_5) ; $OH + O \rightarrow H + O_2$

(b) à partir de la stratopause (50 km), le radical OH par l'intermédiaire de O
(Coefficient a_5) ; $OH + O \rightarrow H + O_2$
(Coefficient a_1) ; $H + O_2 + M \rightarrow HO_2 + M$
(Coefficient a_7) : $HO_2 + O \rightarrow OH + O_2$

(c) dans la stratosphère inférieure, le radical OH par l'intermédiaire de O_3
(Coefficient a_6) ; $OH + O_3 \rightarrow HO_2 + O_2$
(Coefficient a_{6c}) : $HO_2 + O_3 \rightarrow OH + 2\, O_2$

On sait aujourd'hui (Nicolet[26][27][28][29]) comment ces trois processus jouent des rôles différents en fonction de l'altitude. Le cycle (a) détermine la distribution de l'ozone dans la mésosphère, le cycle (b) ne peut être effectif qu'au dessus de 50 km, tandis que le cycle (c) intervenant plus bas est loin d'être suffisant dans la basse stratosphère. C'est pourquoi l'introduction des oxydes d'azote a été requis pour expliquer le comportement de l'ozone.

Néanmoins, comme l'ozone stratosphérique dépend des conditions de transport dans la stratosphère inférieure, en tout cas au dessous de sou maximum de concentration (25 km), des modèles stratosphériques ont été introduits dès 1971 à l'Institut d'Aéronomie en vue de détecter les effets des ruptures d'équilibre chimique ou photochimique (Nicolet et Vergison[30]; Nicolet et Peetermans [31][32]).

L'application à CH_4 et à CO par Nicolet et Peetermans[32] indique combien la distribution de CH_4 (Ackerman et Muller[41]) est liée aux conditions de transport et, en conséquence, influence la production de CO et de H_2O stratosphériques.

C'est pourquoi au simple schéma de la figure 2, il faut substituer celui de la figure 3 où interviennent NO et CO dans le rapport des concentrations OH et HO_2, en particulier dans la stratosphère inférieure. C'est pourquoi les conditions aéronomiques de CH_4 et de CO sont à l'étude à l'Institut d'Aéronomie sous diverses formes comme, par exemple, par une étude des variations des conditions de photodissociation de CO_2 en fonction de la température et par de nouvelles observations stratosphériques de CH_4.

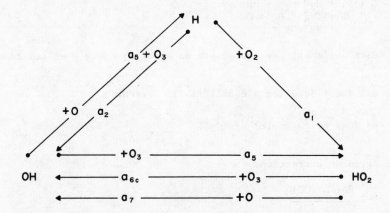

Fig. 2. Schéma des réactions de H, OH et HO_2 dans une atmosphère d'oxygène et d'hydrogène.

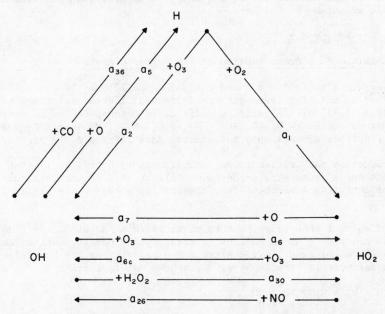

Fig. 3. Schéma des réactions de H, OH et HO_2 en présence de tous les constituants minoritaires stratosphériques.

En fin de compte, il faut souligner que les solutions à ce problème des constituants hydrogénés agissant directement ou indirectement sur l'ozone stratosphérique ne seront définitives que lorsqu'on possèdera une vue correcte de la distribution verticale de H_2O, CH_4 et H_2 dans toute la stratosphère jusqu'à 50 km, en fonction de la latitude et des saisons.

Ainsi, on peut indiquer (Nicolet[27]) que l'action des constituants de l'hydrogène issus de la vapeur d'eau, du méthane et de la molécule d'hydrogène sous l'action de la réaction

$$O\ (^1D)\ +\ H_2O \rightarrow OH\ +\ OH$$
$$+\ CH_4 \rightarrow CH_3\ +\ OH$$
$$+\ H_2 \rightarrow OH\ +\ H$$

se traduit essentiellement par les réactions a_5 et a_7 entre O et les radicaux OH et HO_2.

Le résultat est que l'équation d'équilibre (1) devient :

$$n^2\ (O_3) = \frac{J_2}{J_3}\ n\ (M)\ n^2\ (O_2)\ \frac{k_2}{k_3\ (I\ +\ A)} \qquad (A)$$

où A est le terme de correction

$$A = \frac{2a_5\ n\ (OH)}{2k_3\ n\ (O_3)}$$

qui joue un rôle seulement à partir de 50 km. Il sera surtout fonction de la variation des concentrations de H_2O, CH_4 et H_2 et du processus de reformation de H_2O suivant le mécanisme

$$OH\ +\ HO_2 \rightarrow H_2O\ +\ O_2\ .$$

Le Système Chimique de l'Ozone Impliquant les Oxydes d'Azote

Le rôle des oxydes d'azote fut d'abord envisagé dans l'ionosphère dès 1945 (Nicolet[33][34]), mais fut également considéré pour l'étude des problèmes du ciel nocturne (Nicolet[35][36][37]) dans le cadre de la chemosphère où on retrouve l'ensemble des constituants N. NO, NO_2, NO_3, N_2O_4, N_2O_5, HNO_2 et HNO_3 qui est à la base des réactions aéronomiques introduites dans l'étude de la stratosphère.

Après l'introduction par Crutzen et par Johnston en 1970-1971 de l'effet des oxydes d'azote sur l'ozone stratosphérique, l'Institut d'Aéronomie a consacré également ses activités expérimentales, observationnelles et théoriques à ces problèmes.

On retrouve d'abord l'idée émise pour la première fois par Nicolet[27] de la production du monoxyde d'azote NO dans la stratosphère par la réaction de l'atome excité d'oxygène $O(^1D)$ avec l'oxyde nitreux N_2O produit par les bactéries dans le sol et diffusant vers la stratosphère. Le schéma est le suivant :

$$O\ (^1D)\ +\ N_2O \rightarrow NO\ +\ NO$$
$$\rightarrow N_2\ +\ O_2$$

où la proportion des deux processus est de l'ordre de 50 . La réaction catalytique dans la stratosphère, qui peut s'exprimer par

(Coefficient b_4) ; $O_3\ +\ NO \rightarrow NO_2\ +\ O_3$
et
(Coefficient b_3) ; $NO_2\ +\ O \rightarrow NO\ +\ O_2$
est tempéré par l'action de la photodissociation rapide
(Coefficient J_{NO_2}) ; $NO_2\ +\ hv\ (\lambda > 300\text{-}400\ nm) \rightarrow NO\ +\ O$

La disparition des oxydes d'azote s'effectue certainement (Nicolet[37][38]) par la formation dans la stratosphère inférieure de l'acide nitrique suivant le mécanisme $OH\ +\ NO_2\ +\ M \rightarrow HNO_3\ +\ M$.

Une étude détaillée de Brasseur et Nicolet[39] montre comment les transports s'effectuent (Fig. 4). Alors que NO a tendance à monter dans la stratosphère, l'acide nitrique pénètre dans la troposphère où il est soluble dans l'eau.

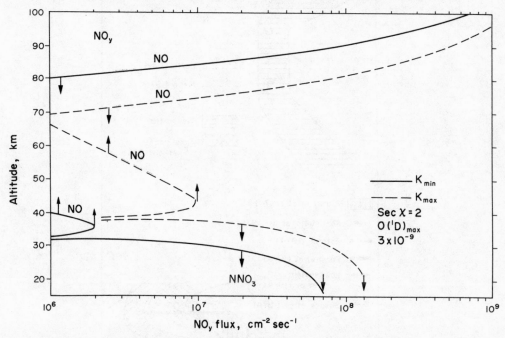

Fig. 4. Transport vertical du monoxyde d'azote NO et de
l'acide nitrique dans deux sens opposés.

Les multiples observations effectuées dans la stratosphère par l'Institut d'Aéronomie sur NO_2 (par exemple, Ackerman et Muller[40][41]; Ackerman, Fontanella, Frimout, Girard, Louisnard et Muller [42]; Ackerman[43] et Muller [44]) ont apporté après les premières vues de la distribution verticlae de ces constituants, des indications sur l'extrême variabilité de NO et de NO_2 (Fig. 5 et Fig. 6).

Ceci indique combien de nouvelles observations sont encore requises et que des recherches expérimentales et théoriques sont nécessaires. En particulier, si le problème de HNO_3 dans sa photodissociation (Biaumé[45]) est bien clarifié et permet de démontrer (Brasseur et Nicolet[39]) que HNO_3 n'est plus en équilibre avec NO_2 et NO; il faut toutefois noter qu'un autre processus suggéré au laboratoire par Simonaitis et Heicklen à Pennsylvania State University, Ionosphere Research Laboratory, peut apparaître dans la stratosphère (Nicolet[46]). Il s'agit de la réaction

$$HO_2 + NO_2 + M \rightarrow HO_2 NO_2 + M$$

conduisant à l'acide pernitrique qui serait très stable à des températures stratosphériques de l'ordre de 200 K, en tout cas en l'absence de radiation solaire. Il faudrait approfondir l'étude de ce composé avant de tirer des conclusions définitives sur l'effet quantitatif précis des oxydes d'azote sur l'ozone stratosphérique, d'autant plus que les conditions aéronomiques de NO_3 et de N_2O_5, appelés souvent à la rescousse pour expliquer les anomalies observées dans le comportement des oxydes d'azote, sont loin d'être déterminées. On se préoccupe d'ailleurs de cette question à l'Institut d'Aéronomie.

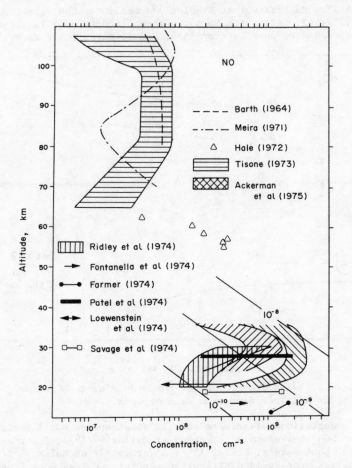

Fig. 5. Distribution verticale du monoxyde d'azote observée
dans l'atmosphère. A noter l'extrême variété des données
d'observation.

D'ailleurs, si au processus généralement reconnu comme essentiel dans la production
de NO, c'est-à-dire l'oxydation de N_2O, on ajoute l'effet du rayonnement cosmique,
comme l'a déterminé Nicolet[47], ou l'effet des irruptions de protons solaires
comme le suggère Crutzen, on ne peut négliger a priori les injections artificielles
de NO dues aux bombes thermonucléaires atteignant la stratosphère ou les avions
volant dans la stratosphère.

Dans le cas des bombes, les analyses ne concordent pas. Suivant celles effectuées à
l'Institut d'Aéronomie par Brasseur[48], il apparaît difficile de prouver
l'existence d'un effet. Des irrégularités se manifestent dans le contenu total
d'ozone dont l'origine reste inexpliquée. L'application à des modèles mathématiques
avec conditions aux limites fixées a été effectuée également à l'Institut
d'Aéronomie par des modèles à une dimension (Nicolet et Vergison[30]; Nicolet et
Peetermans[31]; Brasseur et Nicolet[39]) et permet de conclure à la sensibilité
des résultats aux conditions introduites dans les modèles. Un modèle à deux
dimensions développé par Brasseur[48] permet de déterminer sous quelles formes on
peut représenter les observations de l'ozone et ensuite de les appliquer à des
prédictions de perturbation de la couche d'ozone par des effets artificiels après

avoir considéré les effets des constituants minoritaires que sont les composés
hydrogénés et les oxydes d'azote (Brasseur et Bertin[49]).

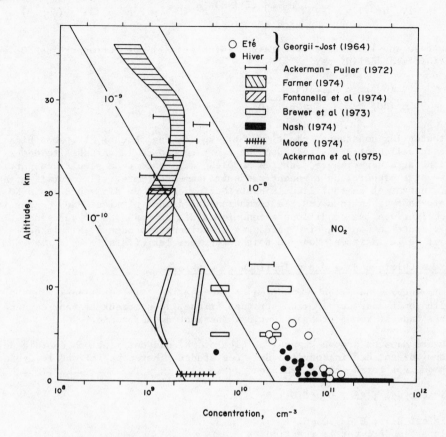

Fig. 6. Distribution verticale du dioxyde d'azote observée
dans l'atmosphère.

Ces divers modèles seront repris au cours des mois prochains afin de tenir compte
de paramètres plus précis dans le calcul de l'évolution des conditions aéronomiques
de la stratosphère. Avant de s'engager définitivement, il est certainement
nécessaire de connaître mieux tous les aspects du problème aéronomique de N_2O et
d'évaluer avec plus de certitude à la fois les sources terrestres et marines de
ce constituant naturel produit par les bactéries, sources pour lesquelles il y a
actuellement des contestations.

Quoi qu'il en soit, l'action des oxydes d'azote se manifeste sur l'ozone sous la
forme de l'addition d'un terme de correction (comme dans l'équation A)

$$B = \frac{2\, b_3\, n\, (NO_2)}{2\, k_3\, n\, (O_3)}$$

de telle sorte que l'on écrit, d'abord pour les conditions d'équilibre,

$$n^2(O_3) = \frac{J_2}{J_3}\ n(M)\ n^2(O_2)\ \frac{k_2}{k_3\ (I + A + B)} \qquad\qquad (A + B)_1$$

Government of Belgium

et ensuite pour les conditions générales

$$\frac{\delta n(O_3)}{\delta t} + \text{div}\left[n(O_3)\ v_{O_3}\right] + 2\ \frac{J_3\ k_3\ (I+A+B)}{k_2 n\ (M)\ n(O_2)}\ n^2\ (O_3) = 2\ n(O_2)\ J_2 \qquad (A + B)_2$$

Ceci suppose que l'on admet avec raison (Nicolet[37]) que le rapport de NO_2 et NO est parfaitement défini par

$$\frac{n\ (NO_2)}{n\ (NO)} = \frac{b_4\ n(O_3)}{J_{NO_2} + b_3\ n(O)}$$

Comme toutes les constantes cinétiques k_2, k_3, a_5, a_7, b_3, b_4, ... sont bien connues actuellement; il s'agit donc de tenir compte de la variabilité des conditions stratosphériques dont les modèles, qui tentent de simuler en particulier la basse stratosphère, ne tiennent pas suffisamment compte. Il faudrait connaître en tout instant et en tout lieu la distribution exacte du dioxyde d'azote NO_2 dans la stratosphère. En tous cas, la formation des acides nitrique HNO_3 et pernitrique HNO_4 souligne que les conditions aéronomiques des composés de l'hydrogène et des oxydes d'azote ne sont pas indépendantes et doivent dépendre à la fois de la saison et de la latitude, dont on doit déterminer les effets.

Le Système Chimique de l'Ozone Impliquant de Chlore

Depuis la suggestion émise par Molina et Rowland, en 1974, de l'action des chlorofluorométhanes sur l'ozone stratosphérique, de nombreux travaux ont été effectués dans le monde et également à l'Institut de Aéronomie.

On trouvera dans un exposé général de Nicolet[46] une description générale des réactions qui ont été introduites dans ces études. Retenons ici que le cycle catalytique est formé par la réaction double :

(Coefficient d_2) ; $Cl + O_3 \rightarrow ClO + O_2$
et
(COefficient d_3) ; $ClO + O \rightarrow Cl + O_2$
Ce cycle est modéré par la réaction
(Coefficient d_4) ; $ClO + NO \rightarrow Cl + NO_2 + hv \rightarrow Cl + NO + O$
et également par
(Coefficient J_{ClO}) : $ClO + hv \rightarrow Cl + O$
qui rétablit l'atome d'oxygène et dès lors l'ozone.

On peut voir aisément (Nicolet[46]), lorsqu'on considère l'ensemble des valeurs numériques des coefficients de réaction qui sont connues par les recherches de laboratoires effectuées aux USA et en Grande Bretagne, que l'on peut toujours écrire dans le champ de radiation du soleil, l'équation d'équilibre basée sur les réactions que nous venons d'écrire

$$d_2\ n(Cl)\ n(O_3) = n(ClO)\ \left[J_{ClO} + d_3 n(O) + d_4\ n(NO)\right]$$

avec

$$\frac{n(NO_2)}{n\ (NO)} = \frac{b_4\ n(O_3) + d_4\ n(ClO)}{J_{NO_2} + b_3\ n(O)}$$

De là, le terme de correction comme dans le cas du radical OH et de la molécule NO_2 s'écrit

$$C = \frac{2 \ d_3 \ n(ClO)}{2 \ k_3 \ n(O_3)}$$

et l'équation de l'ozone en équilibre photochimique dans la stratosphère supérieure s'écrit, [cf. équation (A) et (A + B)],

$$n^2 \ (O_3) = \frac{J_2}{J_3} \ n(M) \ n^2 \ (O_2) \quad . \quad \frac{k_2}{k_3 \ (I + A + B + C)} \qquad\qquad (A + B + C)_1$$

et, dans le cas général

$$\frac{\delta n(O_3)}{\delta t} + \text{div} \left[n(O_3) \ v_{O_3} \right] + 2 \ \frac{J_3 \ k_3 \ (I + A + B + C)}{k_2 \ n(M) \ n(O_2)} \ n^2 \ (O_3) = 2 \ n \ (O_2) \ J_2$$

$$(A + B + C)_2$$

On voit ainsi simplement comment les constituants minoritaires provenant de l'hydrogène, de l'azote et du chlore sous la forme de OH, NO_2 et ClO influencent l'ozone stratosphérique. Cependant, il y a interprétation des effets des différents constituants tant dans les processus de formation que de destruction de chacun d'entre eux. Ainsi, il faut ajouter que l'action de CH_4 suivant le processus

$$Cl + CH_4 \rightarrow CH_3 + HCl$$

produit l'acide chlorhydrique qui peut passer dans la troposphère où il est soluble dans l'eau. Toutefois, la présence de HCl est modérée par la réaction de

$$OH + HCl \rightarrow Cl + H_2O$$

reformant l'atome de chlore. Mais le problème est loin d'être résolu, même dans l'analyse des réactions chimiques où intervient le chlore comme celle, par exemple, de ClO avec ClO qui peut conduire à des composés complexes. Ainsi, la présence de $ClONO_2$ a suscité de nombreux travaux aux Etats-Unis. A l'Institut d'Aéronomie, on s'est d'abord consacré à l'observation stratosphérique et la détection de HCl dans la stratosphère jusqu'à 35 km (Fig. 7) a été une des premières observations de sa distribution verticale (Ackerman, Frimout, Girard, Gottignies et Muller[50]).

D'autre part, des calculs inédits de Brasseur indiquent quelles seraient les distributions des principaux constituants du chlore et leur action sur l'ozone dans un modèle stratosphérique à deux dimensions.

Cependant, il faut noter que, plus encore que pour les autres constituants minoritaires, le chlore manifeste son action en fonction de la nature de l'intrusion dans la stratosphère des différents chlorofluorométhanes et des actions ultérieures du rayonnement ultraviolet solaire dans le domaine de longueur d'onde de l'ordre de 200 nm. C'est pourquoi une série de nouvelles analyses des coefficients d'absorption des produits naturels comme le chlorure de méthyle CH_3Cl ou d'autres produits industriels a commencé au laboratoire (Wisemberg 1976).

On retiendra combien il est difficile d'avoir une vue exacte de la distribution (ou de la production) du chlorure de méthyle comme produit de combustion (par exemple, 1 milligramme de CH_3Cl par gramme de tabac consumé dans une cigarette). Est-ce que son intrusion dans la stratosphère s'effectue sous l'effet de la puissante convection verticale de la troposphère dans les régions tropicales où l'agriculture (fumerons) est à l'origine d'une production considérable comme d'ailleurs tout produit naturel fumant qui se consume. Il s'agit, en effet, de

332   Government of Belgium

déterminer la part exacte du produit naturel CH_3Cl qui pénètre quotidiennement dans
la stratosphère par rapport aux produits industriels comme CF_2Cl_2 et $CFCl_3$ et par
rapport à d'autres produits comme le tétrachlorure de carbone CCl_4 ou tout composé
halogéné (Nicolet[51]).

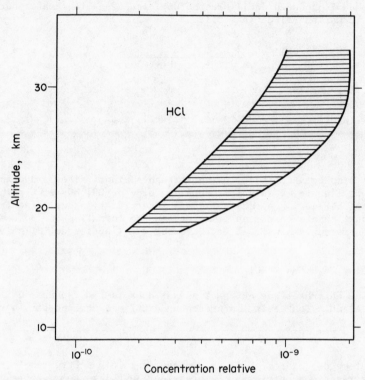

Fig. 7. Distribution verticale de HCl dans la stratosphère.

D'ailleurs, il convient dès que l'on parle des chlorofluorométhanes de considérer
la production du fluor dans la stratosphère résultant de la dissociation de CF_2Cl_2
et de $CFCl_3$. En effet, le fluor agit également sur l'ozone atmosphérique par un
cycle catalytique identique à Cl

$$F + O_3 \rightarrow FO + O_2$$

et $$O + FO \rightarrow F + O_2$$

tout en pouvant être sans doute également tempéré par

$$FO + NO \rightarrow F + NO_2$$

L'action de CH_4 se manifeste également comme pour le chlore

$$F + CH_4 \rightarrow HF + CH_3$$

avec toutefois une différence due à la grande stabilité de l'acide fluorhydrique.
Cette production de HF doit conduire à la présence de cette molécule dans la
stratosphère. Elle a été détectée par Zander[52] de l'Institut d'Astrophysique
de l'Université de Liège par des observations solaires effectuées par ballon de
27,5 km d'altitude. La raie d'absorption à 4038,97 cm^{-1} est actuellement observée
par l'Institut d'Astrophysique de l'Université de Liège (Zander, Roland, Delbouille,

dans le journal "Le Soir" du 11 décembre 1976) en haute montagne (Jungfraujoch) en Suisse.

De telles observations devraient permettre d'étudier la variation de HF en fonction des saisons et, éventuellement, de suivre son accroissement car il apparaît que le fluor dans la stratosphère ne peut résulter que des produits industriels introduit dans celle-ci sous forme de chlorofluorométhanes. L'acide fluohydrique doit sans doute passer dans la troposphère où il est soluble dans l'eau afin d'éliminer le fluor de la stratosphère. Il conviendrait donc que les modèles simulant les propriétés de l'ozone stratosphérique considèrent également le comportement du fluor au même titre que les autres constituants minoritaires. Néanmoins, il convient que l'étude expérimentale des réactions du fluor, tout comme celles comportant le brome (Fig. 8), soit développée avec l'idée d'une adaptation aux conditions aéronomiques dans la stratosphère.

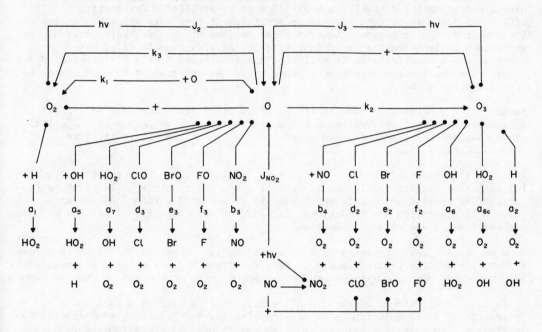

Fig. 8. Réactions intervenant dans la formation et la destruction de l'ozone

En tous cas, il convient comme pour les autres constituants minoritaires, d'introduire l'action sur l'ozone par le terme de correction

$$D = 2 \, f_3 \, n(FO)/2k_3 \, n(O_3)$$

Encore une fois, un monoxyde FO intervient comme l'élément agissant. Il en serait d'ailleurs de même pour BrO dans le cas du brome.

En fin de compte, en considérant les valeurs numériques des concentrations requises de OH, NO_2, ClO, FO et BrO pour pouvoir jouer un rôle dans l'équilibre de l'ozone stratosphérique, on voit que pour des valeurs A = B = C = D = I, la concentration de OH devrait être supérieure à 5×10^7 cm^{-3} à 50 km, que celle de NO_2 doit atteindre au moins 2×10^9 cm^{-3} à 30 km et devrait dépasser 3×10^8 cm^{-3} à 50 km. Quant à celle de ClO, elle ne peut être inférieure à 5×10^8 cm^{-3} à 30 km et à 5×10^7 cm^{-3} à 50 km.

Les monoxydes FO et BrO devraient probablement atteindre des concentrations identiques à celles de ClO, si on admet que les constantes cinétiques de FO avec O sont du même ordre de grandeur.

CONCLUSIONS

Il apparaît de ce résumé des activités passées et actuelles de recherche en Belgique sur la stratosphère et, en particulier, de celles qui sont associées à l'ozone atmosphérique, que les mêmes types de travaux continuerant. Ce sont d'abord l'observation régulière de l'ozone atmosphérique par sa quantité totale et sa distribution verticale à l'Institut Royal de Météorologie de Belgique.

Dans un autre ordre d'idées, l'Institut d'Hygiène et d'Epidémiologie de Belgique a marqué son intérêt (cf. Dekleermaker[53]) pour ce problème. Cet Institut effectuera des observations régulières pratiquement continues au niveau du sol de l'ozone et de ses précurseurs anthropogènes (oxydes d'azote et hydrocarbures) dans une région semi-résidentielle et industrielle (Anvers) dans le cadre des observations automatiques de la pollution atmosphérique; à cela s'ajouteront des observations par intermittence de l'ozone et de ses précurseurs lors de campagnes sur le territoire belge.

Ensuite, les observations dans la stratosphère et en haute montagne de l'Institut d'Astrophysique de l'Université de Liège sur le rayonnement infrarouge du soleil conduisant à l'enregistrement des absorptions de constituants minoritaires stratosphériques.

Enfin, à l'Institut d'Aéronomie Spatiale de Belgique et à l'Université de Bruxelles, les recherches expérimentales au laboratoire, les observations de la stratosphère par ballons, fusées et satellites (Spacelab) et les études théoriques allant jusqu'à la simulation des conditions aéronomiques stratosphériques par des modèles mathématiques à une et deux dimensions.

Dans ce dernier cas, les trois aspects expérimentaux, observationnels et théoriques seront liés afin de tenir compte à la fois des progrès que l'on peut accomplir dans les études des composés hydrogénés, des oxydes d'azote et des constituants halogénés.

Il est, en effet, encore nécessaire d'apporter plus de précision dans les "prédictions" de l'évolution de la stratosphère surtout lorsqu'on veut apporter une appréciation quantitative exacte. En d'autres termes, les recherches doivent être poursuivies avec des vues synthétiques de tout l'ensemble des phénomènes stratosphériques.

TABLE 1   Inventaire des Travaux Actuels Belge sur les Problèmes de
            l'Ozone Atmosphérique

(Janvier 1977)

Institution	Observations atmosphériques	Mesures de laboratoire	Modèles
I.R.M.	ozone total (sol) profil vertical O_3 (sondes) rayonnement solaire et terrestre turbidité atmosphérique	--	--
U.Lg.	spectre IR solaire haute résolution (haute montagne, ballon) constituants mineurs	--	--
U.L.B.	--	analyse des données	Modèles à une et deux dimensions
I.A.S.	constituants mineurs par spectre IR en émission et en absorption du sol à 40 km (ballons, avions), en orbite (préparation Spacelab) U.V. solaire, ballons, satellite (1978), (préparation Spacelab) mesures O_3, NO, mobilité ionique par fusée (mésosphère, haute stratosphère)	calibration sources et détecteurs UV-IR sections efficaces d'absorption	Modèles à une et à deux dimensions. Analyses de données de réactions chimiques
I.H.E.	Monitoring de O_3, NO_x, C_nH_n, CO au niveau du sol dans le cadre du réseau automatique de la pollution atmosphérique dans la région anversoise. Observations de O_3, NO_x, C_nH_n, CO par des laboratoires mobiles.		

I.R.M.   Institut Royal Météorologique, Avenue Circulaire 3, B-1180 Bruxelles
U.Lg.   Institut d'Astrophysique de l'Université de Liège, Cointe-Sclessin
         (Belgique)
U.L.B.   Université Libre de Bruxelles (Prof. Nicolet), Avenue Den Doorn 30,
         B-1180 Bruxelles
I.A.S.   Institut d'Aéronomie Spatiale de Belgique, Avenue Circulaire 3, B-1180
         Bruxelles
I.H.E.   Institut d'Hygiène et d'Epidémiologie, Section Air, Rue Juliette Wytsman
         14, B-1,050 Bruxelles   G.VERDUYN, Chef de Service.

REFERENCES AUX TRAVAUX BELGES

1. D.A. VAN BASTELAER, L'ozone atmosphérique et l'ozométrie en Belgique, Ciel et
    Terre, 13, 485, 509 et 533, 1892.

2. J. JAUMOTTE, La diffusion des gaz dans le champ de la pesanteur et l'ozone
    atmosphérique, Inst. Roy. Mét., Mémoires VI, 1936.

3. M. NICOLET, Le problème atomique dans l'atmosphère supérieure, Inst. Roy.
    Mét., Mémoires XI, 1939.

4. M. NICOLET, L'ozone et ses relations avec la situation atmosphérique, Inst.
    Roy. Mét., Miscellanées XIX, 1945.

5. D. DE MUER, Ozone and total oxydant level in the surface layer at Uccle, Inst.
    Roy. Mét., Publ., Série A, No87, 1975.

6. D. DE MUER, Vertical ozone distribution over Uccle (Belgium) from six years of
    soundings, Beit. Phys. Atm., 49, 1, 1976.

7. D. DE MUER, The vertical ozone distribution over Uccle (Belgium) in relation
    to simultaneous observations of wind and temperature, Dresden Ozone Symposium,
    to be published in the Proceedings 1977.

8. M. ACKERMAN, Ultraviolet solar radiation related to mesopheric processes, in
    mesospheric models, Reidel Publ. Dordrecht (Holland), pp. 149-159, 1971.

9. M. ACKERMAN and P. SIMON, Rocket measurement of solar fluxes at 1216 Å, 1450 Å
    and 1710 Å, Solar Physics, 30, 345, 1973.

10. P. SIMON, Balloon measurements of solar fluxes between 1960 Å and 2300 Å,
    Proc. 3e CIAP Conference, Broderick and Hast editors, pp. 137-142, 1974.

11. P. SIMON, Observation de l'absorption du rayonnement solaire par ballons
    stratosphériques, Bull. Acad. Roy. de Belgique, 60, 617, 1974.

12. P. SIMON, Nouvelles mesures de l'ultraviolet solaire dans la stratosphère,
    Bull. Acad. Roy. de Belgique, Cl. Sc. 61, 399, 1975.

13. D. SAMAIN and P. SIMON, Solar flux determination in the spectral range 150-210
    nm, Solar Physics, in press, 1976.

14. M. ACKERMAN, F. BIAUME and M. NICOLET, Absorption in the spectral range of the
    Schumann-Runge bands, Canad. J. Chem., 47, 1834, 1969.

15. M. ACKERMAN and F. BIAUME, Structure of the Schumann-Runge bands from the 0-0
    to the 13-0 band, J. Mol. Spectrosc. 35, 73, 1970.

16. M. ACKERMAN, F. BIAUME adn G. KOCKARTS, Absorption cross sections of the
    Schumann-Runge bands of molecular oxygen, Planet. Space Sc. 18, 1639, 1970.

17. G. KOCKARTS, Penetration of solar radiation in the Schumann-Runge bands of
    molecular oxygen in *Mesospheric Models and Related Experiments*, p. 160,
    Reidel, Dordrecht (Holland), 1971.

18. G. KOCKARTS, Absorption par l'oxygène moléculaire dans les bandes de Schumann-
    Runge, Aeronomica Acta, Brussels, No107 (3 volumes), 1972.

19. G. KOCKARTS, Absorption and photodissociation in the Schumann-Runge bands of molecular oxygen in the terrestrial atmosphere, Planet. Space Sc., 24, 589, 1976.

20. F. BIAUME, Structure de rotation des bandes 0-0 à 13-0 du système de Schumann-Runge de la molécule d'oxygène, Acad. Roy. Sc. Belgique, Mémoires, 40, N°2, 1972.

21. F. BIAUME, Détermination de la valeur absolue de l'absorption dans les bandes du système de Schumann-Runge de l'oxygène moléculaire, Thèse de Doctorat, Université de Bruxelles, Aeronomica Acta, Bruxelles, N°100, 1972.

22. S. CIESLIK et M. NICOLET, The aeronomic dissociation of nitric oxide, Planet. Space Sc., 21, 925, 1973.

23. D.R. BATES and M. NICOLET, Théorie de l'émission du spectre de la molécule OH dans le ciel nocturne, C.R. Acad. Sc. Paris, 230, 1943, 1950.

24. D.R. BATES and M. NICOLET, Atmospheric hydrogen, Publ. Astr. Soc. Pacific, 62, 106, 1950.

25. D.R. BATES and M. NICOLET, The photochemistry of water vapor, J. Geophys. Res., 55, 301, 1950.

26. M. NICOLET, Ozone and hydrogen reactions, Ann. Géophys., 26, 531, 1970.

27. M. NICOLET, Aeronomic reactions of hydrogen and ozone, in *Mesospheric Models and Related Experiments*, pp. 1-51, Reidel Comp. Dordrecht (Holland), 1971.

28. M. NICOLET, Aeronomic chemistry of the stratosphere, Planet. Space Sc., 20, 1671, 1972.

29. M. NICOLET, An overview of aeronomic processes in the stratosphere and mesosphere, Canad. J. Chem., 52, 1381, 1974.

30. M. NICOLET et E. VERGISON, L'oxyde azoteux dans la stratosphère, Aeronomica Acta, Bruxelles, A-91, 1971.

31. M. NICOLET and W. PEETERMANS, The production of nitric oxide in the stratosphere by oxidation of nitrous oxide, Ann. Géophys., 28, 751, 1972.

32. M. NICOLET and W. PEETERMANS, On the vertical distribution of carbon monoxide and methane in the stratosphere, Pure Appl. Geophys., 106, 1400, 1973.

33. M. NICOLET, Contribution à l'étude de la structure de l'ionosphère, Inst. Roy. Mét. de Belgique, Mémoires, 19, pp. 1-162, 1945.

34. M. NICOLET, Ionospheric processes and nitric oxide, J. Geophys. Res., 70, 691, 1965.

35. M. NICOLET, Nitrogen oxides and the airglow, J. Atm. Terr. Phys., 7, 297, 1955.

36. M. NICOLET, The aeronomic problem of nitrogen oxides, J. Atm. Terr. Phys., 7, 152, 1955.

37. M. NICOLET, Nitrogen oxides in the chemosphere, J. Geophys. Res., 70, 679, 1965.

38. M. NICOLET, Photochimie de l'ozone dans la stratosphère sous l'action des oxydes d'azote et des composés de l'hydrogène, Bull. Acad. Roy. Sc. Belgique, Cl. Sci., 57, 935, 1971.

39. G. BRASSEUR et M. NICOLET, Chemospheric processes of nitric oxide in the mesosphere and stratosphere, Planet. Space Sc., 21, 939, 1973.

40. M. ACKERMAN et C. MULLER, Stratospheric nitrogen dioxide from infrared absorption spectra, Nature, 240, 300, 1972.

41. M. ACKERMAN and C. MULLER, Stratospheric methane and nitrogen dioxide from infrared spectra, PAGEOPH, 106, 1325, 1973.

42. M. ACKERMAN, J.C. FONTANELLA, D. FRIMOUT, A. GIRARD, N. LOUISNARD et C. MULLER, Simultaneous measurements of NO and NO_2 in the stratosphere, Planet. Space Sc., 23, 651, 1975.

43. M. ACKERMAN, NO, NO_2 and HNO_3 below 35 km in the atmosphere, J. Atm. Sc., 32, 1649, 1975.

44. C. MULLER, Détermination de l'abondance de constituants minoritaires de la stratosphère par spectrométrie d'absorption infrarouge, Thèse de Doctorat, Université de Bruxelles, Aeronomica Acta, Bruxelles, A-N°174, 1976.

45. F. BIAUME, Nitric acid vapor absorption cross-se-tion spectrum and its dissociation in the stratosphere, J. Photoch., 2, 139, 1973.

46. M. NICOLET, Stratospheric Ozone : An introduction to its study, Rev. Geophys. Space Phys., 13, 593, 1975.

47. M. NICOLET, On the production of nitric oxide by cosmic rays in the mesosphere and stratosphere, Planet. Space Sc., 23, 637, 1975.

48. G. BRASSEUR, L'action des oxydes d'azote sur l'ozone dans la stratosphère, Thèse de Doctorat, Université de Bruxelles, Aeronomica Acta, Bruxelles, A-N°173, 1976.

49. G. BRASSEUR et M. BERTIN, Un modèle bi-dimensionnel de la stratosphère, Aeronomica Acta, Bruxelles, A-N°146, 1975.

50. M. ACKERMAN, D. FRIMOUT, A. GIRARD, M. GOTTIGNIES and C. MULLER, Stratospheric HCl from infrared spectra, Geophys. Res. Letters, 3, 81, 1976.

51. M. NICOLET, Conscience scientifique face à l'environnement atmosphérique, Bull. Acad. Roy. de Belgique, Cl. Sc., 61, 1039, 1975.

52. R. ZANDER, Présence de HF dans la stratosphère supérieure, C.R. Acad. Sc. Paris, 281, 213, 1975.

53. L. DEKLEERMAKER, In welke mate kan een 3,5 millimeter ozonlaag de toekomst van ons leefmilieu medebepalen, Instituut voor Hygiene en Epidemiologie, November 1975.

PAPER 18

Research in the Field of the Stratospheric Ozone Layer

Presented by the Government of Federal Republic of Germany

Mr Chairman,

distinguished delegates,

First of all, on behalf of the delegation of the Federal Republic of Germany, I would like to thank the United States Government for providing the opportunity of meeting here in Washington, D.C. to discuss scientific matters concerning the stratospheric ozone layer. We also express our gratitude for the initiative taken by UNEP and the support provided by it.

The Government of the Federal Republic of Germany takes a serious view of the danger of a modification of the ozone layer as a result of the effects of certain halo-carbons and other air pollutants. It is therefore considering an intensification of its research effort in order to contribute towards finding a solution to the scientific issues that have emerged. At present we can only confirm the opinion of the United States National Academy of Science that a great deal of uncertainty still prevails in this field, and it therefore requires a great deal of additional research.

The bulk of present research work in the Federal Republic of Germany involves the investigation of the presence, behaviour and effects of halocarbons. On the government side, a research program covering the years 1976 to 1979 is currently being executed with funds amounting to DM 4.7 million. Research projects requiring a further DM 2 million are in preparation. Other research activities are being carried out by industry to the order of approximately DM 1 million annually. The projects concern the improvement of analytical measuring methods, the measuring of vertical and horizontal profiles, as well as the North-South distribution of halocarbons with the aid of aircraft probes or registering balloons, and also laboratory tests of the photochemical reactions taking place in the atmosphere. Although priority is given to halocarbons, nitric oxides and other trace gases in the atmosphere are also detected at the same time so that scientific conclusions can be based on the evaluation of as comprehensive a range of data as possible.

The following institutions are participating in the program - the universities of Freiburg, Wuppertal, Bonn, Kiel, Bochum, the Max Planck Institutes in Lindau and Mainz, the Nuclear Research Establishment Jülich, the Battelle Institute and the German Weather and Meteorological Service.

Bonn University is also carrying out a project involving the measurement of short-wave solar radiation, which is greatly influenced by variations in the ozone layer. This research, which uses actinometric photometry, is to be carried out close to the earth in order to be able to correlate fluctuations in ultra-violet radiation with possible biological effects over a long period. Another project being carried out at Heidelberg University aims at detecting natural sources of halocarbons by examining datable samples of ground water. The findings are to be combined to provide a trend analysis. The Max Planck Institute in Mainz is preparing a study on the metabolization of halocarbons in the biosphere.

The halocarbons program was launched in the Federal Republic of Germany in 1976. A symposium to be held in Mainz in mid-1977 is planned. During the course of the symposium the program will be presented and the first results will be put up for discussion.

At present an emphasis is being placed on elaboration of the basis for data measurement of the actual presence and behaviour of halocarbons. In the later phase of the program, which is to begin this autumn, we plan to include studies of the possible effects on the climate caused by halocarbons and other air pollutants. Climate models, in particular, are to be developed in order to provide some clues about the effects of emissions.

Moveover, studies of the biological consequences, which are of interest to us all, are being considered. We believe that it is necessary to clear up the scientific uncertainty that exists in this subject, so that further speculation may be eliminated.

Investigation of these issues within a reasonable period will require considerable funding and effort. We are interested in an efficient international division of efforts, and are prepared to make the results of our work readily available and to concentrate our effort within this international framework in order to close the gaps in knowledge.

In our view, the United Nations Environmental Programme is the appropriate forum to coordinate the international research effort on these global environmental problems. The Federal Republic of Germany is willing to cooperate in the coordination of the scientific endeavours undertaken by the various governments and non-governmental organizations.

PAPER 19

Programme Française de Recherche sur l'Ozone Stratospherique

Presented by the Government of France

ABSTRACT

Research on the stratospheric ozone layer was intensified in France beginning in 1972 when a committee was created for the study of the consequences of stratospheric flights (COVOS).

The research conducted by COVOS concerned the spectometric measurement from balloons or airplanes, of minor nitrogen oxyde compounds (NO, NO_2, NO_3, H) in the stratosphere, the study of small-scale turbulence, the study of the biologic effects of ultra violet rays, the measurement of the nitrogen oxyde content of airplane engine exhausts and, lastly, models, particularly two-dimensional models. The results obtained do not indicate that any sizable reduction of the ozone layer can be expected from the use of planes of the Concorde type in the foreseeable future.

Additional research was started in 1976 under the DGRST, the CNSF, and the Weather Bureau. The main purpose of that research was to check the ozone layer to study the minor chloride and oxyde compounds in the atmosphere and in the laboratory, and the development of new methods of exploration of the atmosphere.

In general, France considers that the many unknown factors affecting knowledge of the stratosphere require a long-term research program which should be coordinated on the international level. Accordingly, France is delighted with the step taken by UNEP.

PROGRAMME FRANÇAIS DE RECHERCHE

L'intérêt des chercheurs français pour la stratosphère et pour l'ozone ne date pas d'aujourd'hui puisqu'on doit notamment à Fabry Buisson la première détermination absolue de la quantité d'ozone contenue dans l'atmosphère. Ceci se passait en 1921.

Cependant nous nous limiterons dans cet exposé aux recherches plus récentes. Comme au Royaume-Uni le développement du programme "Concorde" a amené le Gouvernement français à se préoccuper des conséquences possibles de l'introduction directe dans la stratosphère d'effluents gazeux issus des moteurs des avions supersoniques. Il a donc été créé en 1972 un comité d'études pour les conséquences des vols stratosphériques (COVOS). Ce comité, placé auprès du Ministère des Transports et du Ministère de la Qualité de la Vie, a été chargé de piloter les recherches en France sur ce problème et de conseiller le Gouvernement.

Le COVOS a travaillé jusqu'en 1976 et son rapport a été remis au Gouvernement il y a quelques semaines.

Ainsi durant cette période un effort important de recherches a été accompli en France. Nous donnons maintenant un rapide aperçu des principaux résultats obtenus.

Dans le domaine de la mesure des constituants, les profils verticaux de NO, NO2 et NO3H ont été mesurés à plusieurs reprises au moyen d'un spectromètre infrarouge embarqué soit en avion (Caravelle, Concorde) soit en ballon. Les résultats obtenus ont montré la grande variabilité spatiale et temporelle des concentrations de ces composés.

Ce travail est le fruit d'une coopération entre une équipe français de l'Office National d'Etudes et de Recherches Aéronautiques (ONERA) et de l'Institut Aéronomie Spatiale de Belgique. Nous tenons à souligner au passage combien la coopération franco-belge tout au cours de ce programme et encore actuellement a été active et fructueuse.

D'importantes recherches ont également été menées sur les phénomènes dynamiques et notamment sur l'étude des turbulences à petite échelle. Des mesures ont été faites soit à partir du sol en utilisant la scintillation des étoiles, soit à partir de ballons en utilisant un anémomètre différentiel. Ces recherches ont permis de mettre en évidence l'apparition fréquente dans la stratosphère de couches turbulentes de faible épaisseur (deux cents mètres en moyenne) s'étendant sur quelques centaines de kilomètres. Ces résultats montrent que finalement le transport turbulent joue un faible rôle dans la stratosphère et conduisent à adopter dans les modèles à deux dimensions un coefficient de transport vertical, Kzz, beaucoup plus faible que celui qui était considéré auparavant.

Le COVOS a encouragé les recherches sur les modèles; notamment un modèle photochimique bi-dimensionnel a été établi par le groupe du Professeur NICOLET en collaboration avec une équipe française. Ce modèle bi-dimensionnel présente bien évidemment des incertitudes dues à l'absence d'un nombre suffisant de données expérimentales. Il a été néanmoins appliqué au cas hypothétique d'une flotte de cent "Concorde" volant sept heures par jour. La réduction d'ozone est alors de 0.25 %, ce qui n'est pas contradictoire avec les résultats du CIAP qui prévoit pour le même nombre d'avions une réduction de 0.7 % avec un facteur 3 d'incertitude.

Le COVOS a également suscité des recherches sur les effets biologiques des rayonnements ultra-violets. Le Docteur LATARJET, en France, est le premier à avoir mesurer l'effet nocif du rayonnement UVB en 1935. De nouvelles études ont été menées en 1972, 1973 sur l'effet érythemal et cancérigène et conduisent à des résultats qui sont analogues a ceux obtenus par d'autres auteurs : 7 à 9 % d'augmentation de l'effet pour 5 % de réduction de l'ozone. De même l'effet abiotique et mutagène sur les micro-organismes a été étudié et là aussi des résultats analogues à ceux obtenus ailleurs ont été trouvés.

Enfin le COVOS a suscité des recherches sur la mesure des quantités d'oxyde d'azote rejeté par les réacteurs d'avion, ainsi que sur les moyens de diminuer cette quantité.

Ce tableau des études menées par le COVOS montre ce qu'il a été possible de faire avec les moyens français. Il est bien certain que la France ne dispose ni des compétences ni des moyens financiers qui lui permettrait de traiter elle-même tous les problèmes. Tout ce que nous souhaitons est d'apporter une contribution significative à un effort de recherches qui ne peut être qu'international.

Nous passons maintenant aux recherches plus récentes commencées en 1976.

En 1976 plusieurs événements se sont produits. Tout d'abord il a été décidé que les recherches commencées par le COVOS devraient se poursuivre avec un objectif plus large que celui du COVOS, celui de s'intéresser à l'ensemble des problèmes et des risques induits par les pollutions de la stratosphère. C'est la Délégation Générale à la Recherche Scientifique et Technique qui a été chargée de prendre ainsi le relai du COVOS. La DGRST est un organisme placé auprès du Ministère de l'Industrie et de la Recherche et a à la fois pour tâche de coordonner l'action des différents organismes de recherches de tous les Ministères et de piloter des recherches sur des crédits propres.

D'autre part, un accord tripartite a été signé en mai 1976 entre les Etats-Unis, le Royaume Uni et la France pour coopérer sur les recherches concernant la stratosphère en notamment la surveillance de l'ozone.

Il y a maintenant trois organismes pilotes en France pour la recherche stratosphérique : la DGRST, le Centre National d'Etudes Spatiales (CNES) et la Direction de la Météorologie.

La DGRST a financé en 1976 une douzaine de projets de recherches dont nous citerons ici les principaux.

Il y a d'abord des études de laboratoires :
1) les vitesses de réaction des constituants chimiques sont mesurées par un laboratoire du CNRS à Orléans, qui utilise des techniques dérivées de la chimie des flammes : un réacteur à décharges micro-ondes permet d'obtenir à partir des molécules, des atomes ou des radicaux libres. Les premiers résultats sur les réactions $CL + H_2O_2$ et $CL + HNO_3$ seront disponibles dans un mois. D'autres réactions de l'atome de Chlore de la molécule CLO et du radical HO_2 seront ensuite étudiées.

2) D'autres mesures de laboratoires concernent les spectres à haute résolution, notamment dans l'infra-rouge et dans le domaine des micro-ondes. Il est indispensable de disposer de ces spectres pour pouvoir faire des mesures par spectrométrie des constituants mineurs de l'atmosphère. Des mesures sont en cours pour O_3, HNO_3, ClO, ClO_2 et H_2CO.

Pour ce qui est des mesures dans l'atmosphère, le spectromètre à grille déjà cité et utilisé pour la mesure des composants azotés a été perfectionné et automatisé. Il peut être utilisé pour d'autres composés notamment les composés chlorés. En particulier des mesures de HCl ont déjà été faites en 1976 à partir d'avions. En octobre 1976 ce spectromètre a été embarqué sur le CONVAIR 990 de la NASA pour mesurer à différentes altitudes et simultanément NO, NO_2, HCl, HNO_3, O_3, CH_4 et H_2. Ces mesures sont en cours de dépouillement.

En 1977 on cherchera à utiliser le spectromètre pour mesurer $CFCl_3$ et CF_2Cl_2, CCl_4, NH_3, HF et ClO à partir de ballons et d'avions.

D'autres méthodes de mesures sont en cours de développement : en particulier la mesure continue de NO_2 à partir du sol par laser, fonctionnant soit en fluorescence soit par absorbtion.

Nous avons également en projet d'équiper la station de mesure de l'ozone de Montlouis qui fait partie du réseau international et qui fonctionne actuellement avec un spectromètre DOBSON d'un interféromètre à transformer de Fourier qui permettrait de mesurer de façon continue et plus précise l'ozone ainsi que certains constituants mineurs.

Dans le domaine de la dynamique on étudie les échanges troposhpère-stratosphère par
une méthode originale consistant à mesurer le rapport des activités du Plomb 210
et du Polonium 210 dans un échantillon. Ceci nécessite un prélèvement et un
filtrage des aérosols dans la stratosphère.

Enfin la DGRST soutient l'amélioration des modèles et notamment du modèle
physicochimique à deux dimensions qui a été mentionné plus haut. Son application
à l'étude de l'effet des chluoro fluoro méthanes sera faite dans quelques mois.

Le Centre National d'Etudes Spatiales consacre la plus grande partie de ses efforts
au développement et à l'utilisation des ballons ainsi que des satellites ou des
laboratoires spatiaux habités.

Il a mis au point des ballons ouverts de volume allant jusqu'à 300 000 m³ qui
peuvent être lancés depuis la base d'Aire-sur-Adour ou la base de Gap en France
ainsi parfois que de l'étranger. Ces ballons emportent des charges lourdes à
35 Kms où ils peuvent rester un jour entier.

Le CNES est engagé actuellement dans le développement de ballons surpressurisés de
25 mètres de diamètre pouvant emporter des charges de 10 Kg à 23 Kms pendant
plusieurs semaines ce qui permettra d'explorer la stratosphère de façon continue
et de bien mesurer les variations journalières.

Ces ballons doivent utiliser des matériaux nouveaux et posent de difficiles
problèmes de contrôle. On pense qu'ils seront opérationnels en 1979.

Plusieurs programmes d'utilisation sont déjà prévus, notamment les études de
turbulences et de circulation atmosphériques en coopération avec l'Université du
Wisconsin.

Parmi les autres programmes soutenus par le CNES il faut mentionner la mesure de
la vapeur d'eau stratosphérique avec un nouveau capteur qui sera essayé au
printemps en ballon, des expériences en satellites : la France participe à OSO 8
à l'expérience LIMS de NIMBUS G et le spectromètre à grille infra-rouge sera
embarqué sur le Spacelab.

Enfin la Direction de la Météorologie a entrepris des études qui s'orientent dans
deux directions :
1) la surveillance de l'ozone. La Météorologie a établi une station à Biscarrosse
sur la côte Atlantique française qu'elle a équipée d'un spectromètre DOBSON et
d'où elle fait régulièrement des sondages, ainsi que des tirs de fusées
météorologiques. De cette façon la corrélation entre l'ozone et les conditions
météorologiques y compris dans la mésosphère pourra être bien observée.

D'autre part la Météorologie met en oeuvre un appareil de contrôle simultané de
l'ozone, de NO_2 et de NO_3H qui peut être embarqué sur des avions de lignes
commerciales. Cet appareil a été développé sous l'égide du COVOS. Il est
actuellement en cours d'essais.

2) La seconde orientation des travaux de la Météorologie est l'étude par méthode
statistique des données d'ozone et des données météorologiques recueillies, aussi
bien à Biscarrosse que dans les autres stations du réseau international en vue de
déterminer l'influence des paramètres météorologiques et de contribuer à définir
quel serait le réseau idéal qui permettrait d'assurer réellement une surveillance
de l'ozone et de déceler des variations qui ne seraient pas dues à des causes
naturelles. C'est là évidemment une question très importante.

CONCLUSION

Nous considérons en France que la recherche sur les perturbations de la stratosphère et notamment de l'ozone est très importante pour l'avenir de l'humanité et doit être considérée comme une oeuvre de longue haleine. Elle nécessite un programme à long terme s'échelonnant sur plusieurs années. Il est manifeste que de nombreuses inconnues et incertitudes pèsent encore sur le fonctionnement physico-chimique de cette machinerie extrêmement complexe qu'est la stratosphère. Il est donc indispensable de développer les méthodes et les outils, notamment les capteurs et les véhicules qui permettront de mesurer un grand nombre de constituants à des moments et en des lieux différents. Ceci représente un très gros effort auquel la France est diposée à participer dans une mesure en rapport avec son niveau scientifique et son potentiel économique.

Un tel effort doit être coordonné au niveau international et c'est pourquoi la France se réjouit de l'initiative prise par le programme d'environnement des Nations-Unies. Cette initiative devrait permettre la coordination et l'entente entre les nations pour aboutir en commun à des résultats véritablement significatifs. La réunion de Washington constitue nous l'espérons le premier pas dans cette direction.

PAPER 20

A Report on the Atmospheric Research Activities in Italy

Presented by the Government of Italy

The Italian National Research Council is carrying out research related to environmental problems. Certain aspects of the impact of human activities on the ozone layer are receiving particular attention.

The research effort is taking place in goverment as well as industrial laboratories. Some research is being undertaken in collaboration with other countries.

Italy is in favour of international collaboration and of participation with international scientific organizations.

The main fields of activity in which Italy is at present involved and in which it intends to develop research efforts in the near future are:

- modelling of the atmosphere;
- atmospheric measurements, both tropospheric and stratospheric;
- laboratory atmospheric chemistry.

In the modelling field, work is being carried out at the University of Rome and Aquila on 1-D ozone photochemical models and their extension to 3-D models. A fairly elaborate radiative transfer model is now available, and sensitivity tests for the effect of multiple scattering are being carried out with the photochemical model.

A 1-D mathematical model is also being developed at the Polytechnic of Milan. This will provide a quick and reliable tool to examine and screen kinetic and photochemical reactions of concern in the stratospheric ozone controversy. Emphasis is being placed on the overall kinetic scheme as well as on the method of acquiring and processing it by computer program. In this way different proposed reactions and/or kinetic constants may be compared.

In the field of stratospheric measurements, the University of Florence Space Physics Laboratory is preparing, in collaboration with Queen Mary College, London, and the National Physical Laboratory, Teddington, a submillimeter infrared balloon experiment. The proposed measurements of mixed ratios of nitrogen and chlorine compounds include ClO and $ClONO_2$.

At the Universities of Rome and Aquila stratospheric aerosols are being studied using two optical radars operated from the ground. An experiment has been prepared to excite the fluorescence of NO_2 with a dye laser. OH will also be studied by means of this technique.

Research activities are also planned for tropospheric photochemistry and sink at the University of Ferrara and at the Atmospheric Physics Institute of Bologna.

In the field of atmospheric chemistry, work is being carried out on the evaluation of nitrogen oxides and chlorine oxides photochemistry in the laboratories of the Universities of Rome and Camerino.

Industry in Italy has shown a willingness to collaborate by undertaking its own research program and also by agreeing to participate in any effort contributing to a better understanding of the impact of human activities on the ozone layer.

PAPER 21

Review of Research Activities Related to the Fluorocarbon Threat to the Ozone Layer

Presented by the Government of the Netherlands

In the Netherlands we have no comprehensive scientific research on the threat to the ozone layer from fluorocarbons. We have concentrated on socio-economic research. However, Dutch scientists participate in research undertaken by the World Meteorological Organization. The Netherlands Organization for Applied Scientific Research is undertaking a laboratory investigation of the tropospheric residence times of chlorocarbons in relation to stratospheric ozone depletion.

Owing to their concern about a possible depletion of the ozone layer, the Netherlands authorities have decided to take international research as a basis for its policy towards fluorocarbons.

In particular, great importance is being given to the report of the U.S. interagency task force on inadvertant modification of the stratosphere of June, 1975, the study of the National Academy of Sciences of September, 1976, and its evaluation by the IMOS task force.

As national research has been concentrated on socio-economic aspects, it seems relevant to say a few words on the socio-economic consequences of restricting the release into the atmosphere over the Netherlands of F-11, F-12 and other perhalogenated fluorocarbons. This study has been executed by an independent consulting group in collaboration with the Netherlands chemical industry. In December, 1976, the group published a report. To give a general idea of the comparative situation, the Netherlands are - according to provisional OECD figures - producing 4% and consuming 2% of the whole of F-11 and F-12 in OECD countries. 70% of the national consumption is in the application area of aerosols. Refrigeration and plastic foams consume respectively 14% and 12%.

As far as F-11 and F-12 alone are concerned aerosols are believed to account for 75-80% of emissions from all sources. Hence the concentration of the study on aerosols.

If a ban on existing fluorocarbon aerosol propellants was imposed in the Netherlands in about 18 months time, the immediate result would be a loss of some 80 million guilders on a total added value of 300 million guilders per year and a loss of some 750-950 jobs on a total employment of about 2300 people. The losses would be of a similar magnitude if the ban was imposed in about four years time and a satisfactory replacement fluorocarbon propellant had not been produced. If this possibility were to become a reality, the loss of added value would be reduced to some 12-16 million guilders, but up to 450 jobs would still be at risk. The study goes on to state that the emergence within four years of a new environmentally safer fluoro-carbon propellant to replace F-11 and F-12 is not improbable provided that govern-ments soon give firm guidance as to which fluorocarbons may eventually be banned.

Mr Chairman, I have tried to give a brief review of our present socio-economic research. We are willing to cooperate in the coordination of research through UNEP. Thank you.

PAPER 22

U.K. Studies of the Stratosphere and the Possible Effects of Stratospheric Pollution

Presented by the Government of the United Kingdom

INTRODUCTION

In the United Kingdom there has long been a tradition of stratospheric research starting with the work of Dobson and his colleagues at Oxford and continuing steadily up to the present with close co-operation between the Meteorological Office and the universities. In addition to the earlier work, surface observations of total ozone have been made regularly using Dobson instruments at a number of British stations both in the U.K. and abroad since the early 1950s.

The Brewer instrument for measuring ozone amounts electrochemically was developed for use on balloons, in aircraft and for research purposes generally. Measurements of winds, temperatures, radiation and humidity in the upper atmosphere have also been made by a variety of methods using radio sondes, satellites and aircraft. A great deal of information was gained from studies carried out by the U.K. Atomic Energy Authority and others of the injection of radioactive isotopes into the upper atmosphere during the various nuclear weapon testing programmes. Laboratory and theoretical studies of ozone were also made in the universities, particularly at Oxford, and the Meteorological Office was well advanced in the development of tropospheric general circulation models and had made studies using low resolution models of the radiative and dynamical aspects of the stratosphere and mesosphere by the late 1960s. Consequently there was considerable background and expertise available in the U.K. on stratospheric studies when in the early 1970s as a result of concern over the possible effects of the exhaust fumes of supersonic aircraft in the stratosphere, general attention was attracted internationally to the possibility that human activities might produce serious effects on the environment by disturbing the natural balance.

This subject was then studied in the U.K. in conjunction with projected Concorde aircraft operations and a major research programme was carried out between 1972 and 1975 on possible effects of stratospheric aircraft under the direction of an expert committee (COMESA) as described in the next section. The overall results of this work were broadly in agreement with those of similar programmes in the U.S.A. and in France, but clearly further work is necessary and other possible pollutants have to be considered. Subsequently a formal Tripartite Agreement was made in May 1976 between France, U.S.A. and U.K. to maintain close collaboration in stratospheric monitoring and research. This present report is an extension of the U.K. first six-monthly report (Nov. 1976) made under the Tripartite Agreement.

COMMITTEE ON METEOROLOGICAL EFFECTS OF STRATOSPHERIC AIRCRAFT (COMESA)

In 1972 COMESA was set up to direct and co-ordinate a programme of research, initially of 3 years' duration, aimed at:

- determining the natural composition of the stratosphere and assessing the changes which might arise from the operation of aircraft (supersonic aircraft in particular);

- assessing the meteorological effects, if any, consequent on these changes.

The terms of reference of the Committee restricted its work to the assessment of meteorological effects and consequently studies of possible biological, social and economic consequences were not attempted. The end point of the work was regarded as the prediction of the likely effects on the ozone in the stratosphere (which determines the amount of solar ultra-violet radiation penetrating to the earth's surface) and the magnitude of possible changes in climate, particularly surface temperature, which might arise as a result of increased aircraft operations in the stratosphere. The studies by COMESA produced a substantial increase of knowledge of the natural composition of the stratosphere and its governing processes and this enabled an assessment to be made of the general magnitude of the changes which might arise from the operation of stratospheric aircraft and the consequential meteorological effects. The COMESA report concluded that the possible effect on the ozone layer from comparatively small numbers of supersonic aircraft will not be detectable. However, the possibility remained of some effects if very large numbers of SSTs with present-day emission characteristics flew at higher operational altitudes and the effects of further expansion of stratospheric flight by subsonic aircraft of improved performance also required further study. The Committee therefore advised the continuance of research into the basic stratospheric processes particularly those relevant to the estimation of effects due to aircraft emissions. Further reports have since been produced by the COMESA workers in such areas as studies of the penetration of solar ultra-violet radiation to the surface, radiation and general circulation models, 1-D modelling studies, the effects of nuclear tests on the ozone layer and the natural sources of NO_x in the stratosphere. Work has also been continued to improve the representation of both photochemical and dynamical aspects in 1-D and 2-D models and a general programme of measurements of minor atmospheric constituents of importance in pollution studies has been started in the Meteorological Office.

EFFECTS OTHER THAN AIRCRAFT EMISSIONS

The possible effect of various chemicals released at the surface, particularly chlorofluorocarbons, on stratospheric ozone has since about 1974 become a subject of even more international concern. In view of this the U.K. Department of the Environment started a programme of work under contract at Atomic Energy Research Establishment (AERE) Harwell to study the effects of chlorofluorocarbons on stratospheric ozone. The programme includes laboratory studies of reaction rates and kinetics (e.g. hydroxyl reactions with various chlorocarbons), measurements of ground level concentrations of chlorofluorocarbons 11 and 12 and carbon tetrachloride, and aircraft sampling. In addition, Harwell has undertaken comprehensive one-dimensional modelling calculations using results obtained by their measurement programme and also assessed the results obtained in modelling studies elsewhere. The continuing work includes further studies of the role of chlorine nitrate in the ozone chemistry and also two-dimensional modelling of tropospheric composition.

Using results from this programme and other available scientific data the Central Unit on Environmental Pollution, Department of the Environment, published in April 1976 a report: Pollution Paper No 5 "Chlorofluorocarbons and their Effect on Stratospheric Ozone". The report concluded that, while scientifically many of the hypotheses put forward were plausible and warranted further attention, *there was no*

immediate cause for precipitate action. It emphasized the need for a better understanding of the problem and in an attempt to reduce some of the uncertainties surrounding the matter it recommended that further research be undertaken within the
next two years to help resolve some of the outstanding issues. In the meantime, it
felt prudence demanded that attempts should be made by industry to seek alternatives
which will meet the needs of the public and industry in general but which will not
pose the same potential threat to the stratosphere. The results given in the U.S.
National Academy of Sciences report "Halocarbons: Environmental Effects of
Chlorofluoromethane Release" published 13 September 1976 were broadly the same and
would appear to lead to rather similar conclusions.

STRATOSPHERIC RESEARCH ADVISORY COMMITTEE

In view of the range of stratospheric pollutants being discussed (which also may
include other chlorine and bromine compounds, additional nitrogen oxide releases
arising from the use of agricultural fertilisers and others) and their possible
inter-reactions, the U.K. decided in May 1975 that interests in stratospheric pollution research could best be served by co-ordinating the various activities through
a single governmental organisation. It was agreed the Central Unit on Environmental
Pollution (CUEP) in the Department of the Environment should assume this role and in
May 1976 a Stratospheric Research Advisory Committee was set up to advise the U.K.
Government on co-ordinating stratospheric pollution research in the U.K. It com-
prised representatives from the U.K. Meteorological Office, AERE Harwell, Universities, Industry, Government Research Laboratories and Government Departments.

U.K. STRATOSPHERIC RESEARCH PROGRAMME

In addition to continuing basic studies of the natural stratosphere by the Meteorological Office, other Government agencies and the universities, the U.K.
Government is now supporting a specially organised programme of research into the
effects of pollutants on stratospheric ozone. The programme can be divided into
three main groups:-

- laboratory studies of the chemical kinetics of stratospheric species;

- atmospheric measurements of trace species;

- computer modelling techniques to predict the effects of stratospheric pollutants.

This research is intended to provide a better understanding of the complex interaction of trace species both in the troposphere and stratosphere. Studies being,
or soon to be supported include:

Chemical Kinetics

- Laboratory studies on the detection of peroxyl (HO_2) radicals by laser magnetic
  resonance and extension of the technique to ClO and HCO radicals. The work will
  give information on absorption spectra, absorption coefficients and reaction
  rates of these species.

- Studies of the reaction rates of ClO and Cl with NO_2 and their variations with
  temperature and pressure leading to an elucidation of the ClO, NO_2 reaction as a
  possible sink for chlorofluorocarbons.

- Studies of the reactions of free radical species remaining after the initial
  chlorine radical is removed from chlorofluorocarbons 11 and 12, including studies
  of the behaviour of $COCl_2$, COF_2 and COFCl.

Atmospheric Measurements

- Continued measurement of chlorofluorocarbons and other halocarbons at ground level

in the U.K. to establish background concentrations, long term trends and atmospheric residence times.

- Measurement of tropospheric concentrations of nitrous oxide on an automatic basis
  to establish the magnitude of any N_2O source at ground surface.

- Shipboard measurements of oxides of nitrogen, methyl chloride and methyl bromide
  in the atmosphere and coastal waters around the U.K. to estimate the role of the
  ocean as a source/sink of atmospheric halocarbons and nitrous oxide.

- A programme of upper air sample collection and measurement using commercial
  flights of Concorde. This will concentrate on the measurement of stable species
  including halocarbons, nitrous oxide and carbon monoxide. Emphasis will be on
  the collection of well standardised data at frequent intervals for the determining of vertical profiles up to 55,000 feet including an investigation of the
  statistical variations and interrelationships between individual chemical species
  and with stratospheric ozone.

Computer Modelling Studies

- Maintenance of a one-dimensional computer modelling capability to permit rapid
  assessment of the implications of new kinetic data on ozone depletion calculations
  and to provide some input parameters for two-dimensional models.

- A two-dimensional modelling programme to study the effects of meridional and
  vertical mean and eddy motions and their interaction with the chemical kinetics.
  A simplified model with 3 vertical columns is also contemplated, which will
  allow insertion of surface releases into the simulated atmosphere at realistic
  latitudes, incorporating some meridional as well as vertical transport together
  with a detailed photochemical scheme of the type used in 1-D models. This
  simplified model will allow more experimental determinations to be made than are
  possible with the full 2-D model.

- A two-dimensional modelling programme of the transport features of chlorofluorocarbons in the stratosphere including an evaluation of the treatment of chlorofluorocarbons both as inert tracers and with parameterised destruction by photochemical reactions for residence time studies.

ON-GOING U.K. CONTRIBUTIONS TO STRATOSPHERIC MONITORING

In addition to its regular measurements at radio-sonde stations of wind and temperature up to about 25 km in the global network primarily for weather forecasting and
climatological purposes the U.K. Meteorological Office is continuing to maintain
four stations making total ozone measurements using the Dobson spectrophotometer.
These are located at Bracknell and Lerwick in the U.K., St. Helena, and Mahe
(Seychelles). At present it appears likely that the long term series of regular
ozone observations at Oxford will have to be discontinued and those at Bracknell
used to continue the record for central England. In addition measurements are also
being made by the British Antarctic Survey (BAS) at their Antarctic stations Halley
Bay, Argentine Island and S. Georgia. All the regular U.K. observations are transmitted to the World Weather Centre in Toronto for inclusion in the global record
and general distribution. The U.K. also co-operates internationally in the World
Meteorological Organisation's studies relating to the ozone layer and the standard-
isation of observations.

Apart from the planned use of Concorde aircraft to obtain data on other stratospheric
constituents all the other U.K. measurements at present being made or planned are
primarily for research purposes. The U.K. Meteorological Office operates a
Meteorological Research Flight at Farnborough which carries out limited flight
programmes to measure stratospheric humidity and ozone, and possibly other constituents; a rocket launching station at South Uist where measurements of wind and

temperature are made for special studies such as stratospheric sudden warming
phenomena and calibration of satellite measurements; and also has a laser installa-
tion which is useful for aerosol studies. Teams from Oxford University and the
National Physical Laboratory also collaborate in International Projects studying the
stratosphere using large balloons and satellites. The Oxford measurements of global
stratospheric temperature fields using the Nimbus D and E satellites have been
followed by the higher altitude measurements from Nimbus F. In addition Oxford and
NPL will be co-operating in the use of Nimbus 6 and also Space Lab. in the future
to obtain new global measurements of several stratospheric constituents. Studies
are also being made by the Meteorological Office to extend the use of satellite
observations to be obtained from Tiros N, Meteosat and others for operational
purposes.

CONCLUSIONS

The U.K. has contributed very considerably to present understanding of the stratosphere and intends to continue actively its studies both of natural processes and
the likely effects of pollutants in this region of the atmosphere. At present it is
devoting considerable resources both to national studies and international collaboration through its Tripartite Agreement with the U.S.A. and France, work within the
European Economic Community and co-operation in the World Meteorological Organisation. However, there is considerable scope for expansion of these activities and
increased collaboration with scientists of other nations, particularly in the area
of stratospheric measurements, if adequate funding were made available.

PAPER 23

U.S. Investigations to Evaluate the Potential Threat of Stratospheric Ozone Diminution

Presented by the Government of the United States

HISTORY OF UNITED STATES CONCERN RELATING TO THE EARTH'S OZONE
CONCENTRATION

Currently the United States is engaged, together with a number of other nations, in
a large research effort related to the problem of assessing the impact of human
activities on the earth's stratospheric ozone concentration and the resulting
effects of ozone diminution on the earth. This research effort involves many hundreds of participants in governmental, industrial, academic and other non-profit
institutions and includes many activities undertaken internationally, either in
collaboration with national science organizations in individual countries or in the
framework of international scientific organizations.

U.S. investigations have focused in succession on several specific areas of human
activities which tend to increase the supply of ozone-destroying substances in the
stratosphere over and above the natural processes of ozone creation and destruction.
These include:

- Aircraft engine emissions in the stratosphere or upper troposphere.

- Release of chlorofluoromethanes in the atmosphere.

- Release of nitrous oxide in the atmosphere.

U.S. studies to a large extent have been directed toward specific issues of potential adverse affects in the stratosphere. However, there is increasing recognition
of the importance of stratospheric modification as a long-term problem whose implications and dimensions are not very well known.

As many of the impacts of human activities on the stratosphere involve long incu-
bation periods by human time scales, measured in decades or longer, and even longer
recovery periods, nations individually and collectively must learn how to anticipate,
to study, to prevent where possible, and to cope with when necessary, the full
dimensions of this problem. This will require a long-term program of continued
research and cooperative international effort, not only into atmospheric physics
and chemistry, but also biological, ecological, agricultural and climatic research
and critical examination of the many difficult economic, social, political and
legal questions.

The U.S. Congress has devoted particularly close attention to the full dimensions

of the scientific concern that the scale of aerospace, industrial and agricultural
operations was potentially threatening to modify the radiative properties of the
atmosphere. In 1971 Congress directed that a Climatic Impact Assessment Program
(CIAP) be undertaken to study and provide within four years a report to Congress on
the possible effects of high-altitude aircraft on the ozone layer. The CIAP Report,
completed in 1975, concluded that there is cause for concern from the operation of
large fleets of civilian high-altitude aircraft in or near the stratosphere using
currently available aircraft engine technology.

In 1975, the U.S. National Academy of Sciences, in a report prepared with governmental funding, reviewed the conclusions of the CIAP study, and reaffirmed that
there was cause for national and international concern from the reduction of the
amount of ozone in the stratosphere. The National Academy's Climatic Impact
Committee, recognizing the stratosphere is a world resource that knows no boundaries, recommended that "international organizations concerned with atmospheric
science, biological science, medical science and aerospace science devote increased
attention to studies concerned with the stratosphere and with possible inadvertent
modification thereof."

Interest in chlorine chemistry in the stratosphere accelerated rapidly in 1974
with regard to the effects of chlorofluoromethanes in the atmosphere, following
the work of Rowland and Molina, Cicerone and Stolarski, McElroy and Wofsy and
others. At that time the fluorocarbon research program of the Manufacturing
Chemists Association was redirected to emphasize stratospheric processes.

The U.S. strategy for a long-range, comprehensive program of research into the
phenomena of the stratosphere and the potentially adverse effects of inadvertent
modification of the stratosphere has evolved from a June 1975 report of the
Interagency Task Force on Inadvertent Modification of the Stratosphere (IMOS), a
February 1976 report of its Subcommittee on Biological and Climatic Effects
Research (BACER), a second report by the U.S. National Academy of Sciences first
issued in 1976, and the expressed policy of the U.S. Congress, which in 1976
further authorized and directed that a long-term, comprehensive program of research
be undertaken of the phenomena of the upper atmosphere "so as to provide for an
understanding of and to maintain the chemical and physical integrity of the
Earth's upper atmosphere."

The U.S. long-term research plan, now under preparation, has as its goal the
development of a better understanding of the physics and chemistry of the stratosphere, the extent to which a resulting increase in the amount of solar ultraviolet radiation may adversely modify the biological balance of the planet, the
incidence of radiation-caused skin cancer, and the earth's climate, and study of
associated economic, political, social and legal problems in the national and international sphere. The research plan will define a research program that builds on
and extends the current assessments of the impacts of aircraft engine emissions,
chlorofluoromethanes and other chlorinated compounds, and nitrogen fertilizers;
that extends the detailed analyses of the Climatic Impact Assessment Program; that
coordinates the research techniques of the space age, using satellites and the
U.S. Space Shuttle as well as airborne, rocket and balloon platforms, and that
applies modern medical and biological sciences using advanced statistical methods
and probing at the cellular and molecular levels for an understanding of effects
mechanisms.

INVESTIGATIONS OF ATMOSPHERIC PROPERTIES AND PROCESSES

Field Measurements and Monitoring

U.S. investigations of the fragile nature of the earth's ozone shield have included
the exploration of the principal source and sink terms and linking atomic and

molecular species of the major photochemical systems; monitoring ozone on a global
scale; monitoring of solar radiation; determination of stratospheric circulation;
aerosol size distribution, composition and number density; and atomic oxygen concentration in the stratosphere. Only a few of the necessary measurements have been
made to date. Simultaneous measurements are required of one or more of the
radicals in each of the nitrogen, hydrogen, and chlorine systems, and of hydroxyl
radical, atomic oxygen, and excited oxygen atoms, which play major roles in each
system. Unfortunately, all the necessary instrumentation and measuring techniques,
especially for simultaneous measurements, essential in order to fully understand
the physics and chemistry of the atmosphere, have not been adequately developed at
the present time.

Platforms for field measurements include balloons, sounding rockets, aircraft and
satellites, each of which has its own unique capabilities. Ground-base instruments
also play an important role in stratospheric measurements. The United States
operates an extensive ground-based network for measurement of some of the stratospheric species, including ozone, carbon dioxide, aerosols, and at some stations,
oxides of nitrogen, sulfur dioxide, and solar ultraviolet radiation.

Measurements were made under the CIAP program of the following stratospheric characteristics between 1971 and 1975: ozone, vertical temperature profiles, water
vapor, oxides of nitrogen, methane (CH_4), nitric acid (HNO_3), carbon dioxide,
oxides of sulfur, hydroxyl radical (OH), solar UV flux, and aerosols. In the
troposphere, measurements were made of the vertical temperature profile, ozone,
aerosols, CH_4, NH_4, chlorofluorobarbons, oxides of nitrogen, SO_4, total solar radiation and UV flux from 280 to 320 nm. Data from these measurements were employed
in the development and use of several models of one, two, or three dimensions
(1-D, 2-D, or 3-D).

The current program of field measurements continues those initiated during CIAP,
with increased emphasis on the chlorine system (CI, ClO_x) and its sources,
particularly trichlorofluoromethane (F-11) ($CFCl_3$) and dichlorodifluoromethane
(F-12) (CF_2Cl_2), and its sinks (hydrogen chloride, HCl), methane, carbon tetrachloride, radicals of the hydrogen system, especially OH, the nitrogen system
(NO, NO_2, N_2O, N_2O_5) and the compounds intermediate between systems, such as
chlorine nitrate, nitric acid.

Satellite instruments now include some to measure ozone on Nimbus 4 and Atmospheric
Explorer E, vertical temperature profile on the NOAA operational environmental
satellites, and solar flux, temperature, ozone and water vapor on Nimbus 6.
Future satellite instruments on the Nimbus G will measure ozone, nitric acid,
nitrogen dioxide, nitrous and nitric oxide, water vapor, methane, carbon monoxide,
aerosols and solar flux.

Monitoring of the total ozone column in the United States was introduced on a
systematic basis in 1957 as part of the International Geophysical Year activities.
Currently the United States operates eleven Dobson spectrophotometer stations as
part of the World Meteorological Organization's network of ground-based ozone
stations. Data from the U.S. Dobson stations are checked for quality and then forwarded monthly to the World Ozone Data Center in Canada.

Direct measurements of the vertical profile of ozone were conducted in the United
States through two major balloon sounding networks totalling 25 stations. Soundings were taken weekly from 1963 through to 1966. Today routine weekly ozone
balloon soundings are made at only a single station (Wallops Island, Virginia).

Rocket ozonesondes were developed during the 1960s to study the photochemical
region from 25 to 60 km. Rocketsondes have been used primarily for special purposes

such as short-term latitudinal surveys, satellite calibration, and the development
of an ozone climatology. Currently, monthly soundings are conducted at only a
single station (Wallops Island, Virginia). It is planned to add another rocket
ozonesonde station at Fort Churchill in February 1977.

Satellite measurements of ozone, which offer the unique capability of global measurements from the same instrument, have been conducted systematically by two
methods, backscattered solar ultraviolet (BUV) radiation and thermal emission in
the infrared band of ozone. The BUV instrument was flown on the Nimbus 4 satellite
from 1970 to date, and on the Atmospheric Explorer E from late 1975 to the present.
The Limb Radiance Infrared Radiometer (LRIR) was flown on Nimbus 6, providing data
for only the six months lifetime of this cryostatted system.

The largest set of satellite ozone data is therefore that obtained by the BUV
instrument. These data are currently being reduced and it is planned that the
entire body of data will be made available to the scientific community by the end
of 1978. Both high level profiles and total ozone column can be retrieved from
these data. The data were acquired at approximately local noon time.

Recent work in statistical time series modeling by Hill, Sheldon and Tiede appears
to provide a promising technique for the detection and early warning of significant
trend changes from a globally representative sample of ozone stations.

Laboratory Measurements

Laboratory experiments have provided significant contributions to the understanding
of stratospheric properties and processes in photochemistry, reaction rates, formation and properties of aerosols, heterogeneous reactions, and temperature and
pressure dependence studies.

In 1975, the U.S. National Bureau of Standards published *Chemical Kinetic and
Photochemical Data for Modeling Atmospheric Chemistry*. Since then, some of the
uncertainties in reaction rates are being further refined. However, much remains
to be done with respect to very reactive species under stratospheric conditions,
such as measuring reaction rates for OH and HO_2 and several other significant
reactive species. New measurement techniques are needed, which should lead to the
development of instruments with the sensitivity needed for stratospheric measurements in the chlorine, nitrogen, and hydrogen systems. For example, there are at
present very limited methods for measurement of several stratospheric constituents,
including OH, HCl, ClO, $ClONO_2$, O, and HF. Instrument improvements and new developments are required for the accurate measurement of many species. Recent laboratory measurements have studied atom or radical-molecule reactions over wide ranges
of temperature. The best experimental results may show reproducibility of 15 percent and accuracy of ± 10 to 20 percent within one study as well as agreement to
± 20 to 30 percent between different studies which may employ substantially
different techniques.

Modeling and Analysis

Much of the current knowledge about the chemical structure and the coupling processes of the stratosphere has been derived from a specific type of mathematical
model, the one-dimensional (1-D) model. By explicitly treating only vertical transport and vertical concentration distribution of the trace species in the model
atmosphere, the 1-D model is able to include explicit treatment of a large number
(50 to 100) of chemical reactions. The vertical transport coefficient employed -
the so-called "eddy diffusion coefficient" - is an empirical prescription derived
from observations of the net vertical transport of a long-lived species such as
methane or carbon-14. As such, it can be argued that the effects of the two

horizontal axes have crudely been accounted for. In fact, the major justification
for the applicability of such models is that they appear to represent atmospheric
behavior reasonably well, though actual "validation" of the models' accuracy is
rendered impossible at present owing to a lack of accurate data with which to make
a precise experimental comparison.

General-circulation models (GCM) of the atmosphere represent the opposite extreme;
they explicitly treat all three dimensions of motion, as nearly as possible, on the
basis of physical principles of thermodynamics and fluid mechanics. The detailed
representation of transport in these models makes it impractical at present, within
existing limits or computer capacity, to treat explicitly a very large number of
chemical reactions therein. The complexity and cost of development and use of such
models also explains why perhaps only a handful of research groups in the world are
actively pursuing their development.

An attractive compromise is, conceptually, a two-dimensional (2-D) model, which
treats transport less completely than a GCM, but more realistically than a 1-D
model, and treats chemistry more completely than a GCM but contains fewer chemical
reactions than the 1-D models. "Validation" of the accuracy of the chemical system's
representation by a 2-D model may be attained by comparison and test with results
from a 1-D model. Validation of the 2-D model's transport parameterization – motion
along longitude is treated explicitly while zonal symmetry is assumed – is more
difficult, however. At present, the 2-D models have relied upon comparison of
transport results with data obtained on the transport of radioactive debris, or
comparison of observed and calculated trace species distributions. The lack of
a good data base presents the same difficulties for this type as for the 1-D model.
In particular, the small amount of data available at altitudes in excess of about
20 km makes meaningful validation of transport representative of 2-D models above
that altitude difficult.

In the study of various possible anthropogenic effects on ozone, the choice of
model must be considered in light of the problem under investigation. Thus, 1-D
models prove quite satisfactory for assessments of the effects of CFMs upon ozone.
Diurnal effects are difficult to treat accurately, it is true, as are spatial
inhomogeneities in species concentrations for a 1-D model. It appears, however,
that such effects are relatively small, and can be satisfactorily approximated so
that the over-all reliability of the 1-D model is quite high for the CFM problem.
For the aviation problem on the other hand large uncertainties are injected, even
on the conceptual level, for studies which employ 1-D models. This situation
arises because aircraft exhaust contaminants are injected directly into the 10 to
20 km region of the atmosphere, where transport processes are most poorly represented by the 1-D parameterization. Accordingly, uncertainties up to a factor of
three arise from the fact that different fractional amounts of aircraft emission
may be diffused upward to the vicinity of the ozone layer.

Studies in the United States, then, emphasize the importance of transport
parameterizations much more strongly for studies involving aircraft and space
shuttle emissions than those involving CFMs, and nitrogen fertilizers. As a result,
considerable effort is being expended on more realistic representation of transport
for aircraft emissions, employing 2-D and 3-D models.

Emission Inventory

Information on the production of halocarbons in the United States is available from
U.S. Tariff Commission Reports. World production figures are harder to obtain in
general, and are not available at all for the countries of eastern Europe. An
attempt to assemble worldwide data was made first by the DuPont Company in view of
the intense interest concerning chlorofluoromethanes.

In December 1975 results of three further studies and a joint U.S.-Canadian report to the OECD became available and provided a more accurate and detailed picture. Production estimates for the principal halocarbons in 1973 have been summarized from these sources. For the two chlorofluoromethanes of most concern, they show a 1973 world production of $CFCl_3$ (F-11) as 368,000 metric tons/year, with the United States producing 45 percent of the world total; and for CF_2Cl_2 (F-12) as 441,000 metric tons per year, with the United States producing 55 percent of the world total.

Figures for the total worldwide production and release into the stratosphere of F-11 ($CFCl_3$) and F-12 (CF_2Cl_2) to the present have been gathered by the Manufacturing Chemists Association for the period of manufacture up to 1958 and year by year from 1958 through 1975. The total produced in the world for F-11 comes to 3.44 million metric tons, and for F-12, to 5.09 million metric tons, almost all of this in the Northern Hemisphere. Of the total produced, it is estimated that 2.93 million metric tons of F-11 have been released to the atmosphere (85 percent) and 4.41 million metric tons of F-12 (87 percent).

For F-11, the data correspond to a worldwide growth rate of 13.9 percent per year for the period 1965-74, with a doubling time of 5.3 years. The corresponding data for F-12 are a worldwide growth rate of 9.9 percent per year, with a doubling time of 7.4 years. Production of both F-11 and F-12 apparently peaked in 1973-74, with a reported decline of 15 percent in 1975.

Studies of the global nitrogen system have just begun, and estimates of nitrogen and oxides of nitrogen sources by different investigators are still too divergent to indicate the direction of any conclusions.

Present Status of Stratospheric Investigations

In summary, one may say that to date true globally average profiles for the significant stratospheric parameters involved in the creation and destruction of ozone, as well as that for ozone, are not available. Average vertical profiles of various trace gases have been obtained at a few latitudes, and in some cases at a single latitude. It requires some judgement to estimate how well these profiles are representative of a true global average. Probably for the longer-lived trace gases the global mean profiles will not deviate significantly from the ones measured so far. Concentrations of the shorter-lived constituents strongly depend on a number of parameters and vary considerably in the stratosphere, both in time and space. So little is known about reactive species such as atomic oxygen, HO and HO_2 radicals, and ClO that even single isolated measurements must be regarded as important.

Thus, the data available from field measurements and experimental laboratory studies define certain guidelines or broad bandwidth data from which the modeled profiles cannot deviate too much. Nevertheless, since the 1-D models which have been the primary tools for arriving at overall estimates of ozone depletion incorporate a vertical diffusion coefficient which has been empirically adjusted to a best fit with an assumed global mean profile for specific trace constituents, such as methane, interpretation of the results requires an understanding of both possible observational variabilities and model limitations.

With an understanding of the limitations of current 1-D models, it is possible to quantitatively estimate the steady-state reduction of ozone, based on explicit consideration of from 50 to 100 chemical reactions, including sources, sinks and interaction among the major systems, the "self-healing" feature of the stratosphere in face of ozone reduction, and various other considerations. The tentative conclusion of the National Academy of Sciences, based on data available and current understanding, is that the release of chlorofluoromethanes at the 1973 production rates will reduce the ozone balance at steady state by about 7 percent, with an uncertainty range of from 2 to 20 percent. Considerably more research on strato- spheric properties and processes is needed over the next decade.

Aircraft engine emissions are primarily found in the transport-dominated lower stratosphere. A fuller consideration of the influence of transport on the distri- bution and concentration of oxides of nitrogen and other engine emissions will require 2-D and 3-D models. Multidimensional models also provide better under- standing of many stratospheric processes and their development is a priority requirement.

Further research is needed to arrive at verifiable conclusions concerning the perturbation of the ozone balance to be expected from the increase of N_2O produc- tion from denitrification of nitrogen fertilizers. However, it has been assumed that each intervention in the stratosphere tending to increase the concentration of ozone-destroying radicals is linearly additive in contributing to eventual ozone reduction.

POSSIBLE EFFECTS OF OZONE DIMINUTION ON THE BIOSPHERE

Stratospheric ozone prevents most of the solar UV radiation at wavelengths below 310 nm from reaching the earth's surface. Decreasing amounts of stratospheric ozone will permit increasing amounts of harmful UV-B radiation (280-320 nm) to reach the earth's surface. Therefore, it is extremely important to understand, in a quantitative manner, the direct effects of increased UV-B flux to life on earth and indirectly through climate change which in turn affects life. The United States is conducting a multi-agency Biological and Climatic Effects Research Program (BACER) to produce necessary knowledge in order to arrive at proper policy and regulatory decisions for the protection of human health and the biosphere, which in this case means protecting the stratosphere from human activities which may modify the stratosphere.

The program is designed to accomplish the following:

- Greatly expand the data base for more accurate estimates of known effects (e.g., skin cancer incidence rates and correlations).

- Determine whether a number of suggested biological effects are of significance and to what degree.

- Significantly improve estimates of the quantitative effects on key biological species (e.g., agricultural crops, selected native species and aquatic organisms) and provide estimates of how near such species are to their limits of tolerance for increased UV radiation or other related environmental changes.

- Initiate a number of basic investigations on cellular/molecular mechanisms in order to understand the interactions of UV-B radiation with living matter and to identify physiological "indicators" or "predictors" of UV sensitivity in humans, other animals or plants.

The principal objective for the first phase of the program is to decrease uncer- tainties in existing scientific knowledge and investigate new potential problem areas by December 1977.

The long-term BACER program has plans for conducting supporting studies with plants, animals, and other organisms as well as studies at the molecular and cellular levels to understand UV damage and repair mechanisms and determine indicators of effects.

Topical areas of BACER programs are health effects, biological (nonhuman) effects, climatic effects, instrumentation and monitoring, and economic assessment.

Possible Effects of Increased UV-B on Human Health

Studies are being conducted to more accurately quantify the relationship between skin cancer incidence and UV-B radiation dose which has been demonstrated by available evidence from epidemiological studies and some laboratory studies. A worldwide increase in UV-B radiation flux can be expected to increase skin cancer incidence. The biological amplification factor (ratio of change in skin cancer incidence to change in UV-B flux) is judged by various investigators to be in the range from approximately unity to about three or more. The physical amplification factor (ratio of change in UV-B flux to change in atmospheric ozone concentration) is calculated to be (negative) two. The ozone depletion is presently estimated to reach an equilibrium level ranging from 2 to 20 percent. The overall amplification factor can be in the range from 2 to greater than 6. The resulting effect to skin cancer rates can range from 4 to greater than 120 percent. Therefore, it is important to more accurately quantify this cause-effect or dose-response relationship for both melanoma and nonmelanoma skin cancer types.

The U.S. basic cancer incidence reporting program, which includes melanoma skin cancer, will be augmented by a special one-year effort beginning in April 1977. This effort will investigate nonmelanoma skin cancer at different latitudes, and therefore different UV fluxes, with reporting from selected U.S. cities or regions. This reporting will include skin types and lifestyle parameters. Discussions with some other nations are in progress to extend this reporting to the necessary international scope.

Possible Effects of Increased UV-B on Biological Systems Other Than Human

The objective of the nonhuman biological effects research program is to assess the impact of UV-B radiation on terrestrial and aquatic organisms, especially those used for the production of food, and to determine the ecological significance of increased UV-B radiation.

Terrestrial studies include assessing the effects of increased UV-B exposure on:

- plant growth and development, and key physiological processes such as photosynthesis and respiration;

- production of selected economically important agricultural and forage crops, and forest tree seedlings;

- stability of agricultural chemicals (fertilizers and pesticides);

- the agricultural ecosystem level (livestock); and

- insects, pathogens and their interaction with plants.

Terrestrial studies are being conducted in the laboratory in controlled plant growth chambers, in greenhouses, and in the field utilizing enhanced UV-B radiation from artificial sources.

Aquatic studies include the UV-B tolerance of aquatic organisms and ecosystems; for example, determining the sensitivity of eggs and larvae of commercially important marine species and the sensitivity and activities of surface living organisms (neuston). An important study concerns the influence of UV-B irradiation on N_2-fixation by blue-green algae, which are an essential link in the aquatic food chain and known to be highly dependent on solar radiation for growth.

Instrumentation Research

Meaningful biological research in the UV-B region requires controllable radiation sources specific to this region (280-320 nm) and instrumentation for measuring the radiation flux over narrow bands in this region. Owing to the almost total unavailability of such essential instrumentation specific for the UV-B spectral region, research is being conducted to fulfill those requirements. Specifically, present research and development aims at improving and standardizing radiation sources in the UV-B region for use in biological effects experiments, constructing a portable spectroradiometer for field use, and developing a laboratory precision spectroradiometer to serve as a standard for calibrating detectors and sources. Workshops have been held to bring biologists and instrumentation experts together to discuss research methodologies and techniques.

Possible Effects of Ozone Diminution on Climate

Ozone diminution may threaten changes in the climate of the stratosphere and the earth's radiation characteristics. These changes may result in alteration in the earth's tropospheric climate. The potential adverse effects on man and the biosphere by inadvertent modification of the stratosphere is a global problem that deserves serious attention on an international scale.

Investigations are being organized to approach the questions of the direction of possible resulting climate change, the extent, the timing, and the geographic distribution of change. One perspective-setting question concerns the relative magnitude of possible climate change from stratospheric modification and from radiation balance change associated with tropospheric accumulation of constituents such as CO_2 and CFMs.

An earlier section of this paper addressed the U.S. activities for obtaining new information on the climate of the stratosphere, its physical, chemical, and optical properties. As one unique aspect of the developing U.S. Climate Program Plan, plans are being developed to increase the capability to observe a broad variety of parameters on many scales from satellite altitudes to provide much of the upper atmosphere data required for climate research, particularly with regard to ozone and aerosols.

Because of the interconnections and complex feedback effects of the climatic system (sun, stratosphere, troposphere, hydrosphere, cryosphere, and biosphere), the stratospheric climate cannot be studied in isolation. The effects of stratospheric perturbations must be examined as they appear throughout each of the climate elements. The key to understanding the climatic system, and thus answering these questions, lies in the development of models that simulate the dynamics of the real system.

Numerical general circulation models are currently being studied at several laboratories. Less detailed models, including highly parameterized numerical models and statistically and empirically based models are also under study in several U.S. institutions and agency laboratories.

Existing modeling efforts benefit from empirical studies of the extent and causes of climate variability, which often yield the first estimate of the relative influences of the elements and processes affecting climate change. These diagnostic approaches to climate research are carried out on a small scale throughout university and government laboratories. Development of these empirical studies is seen as a necessary first step in improving our understanding of how the climate system works and how it is evolving. In recognition of the importance of these studies, new support has been directed to diagnostic techniques, coordination of ongoing related research, and to implementation of plans for expansion.

An adequate data base is needed in all modeling attempts - to test sophisticated numerical models, to provide a basis for empirical studies, and to make actuarial assessments of climatic change and its impact on food production. Paleoclimate information is assembled and analyzed. The largest of these efforts, the Climate/ Long Range Investigation Mapping and Prediction (CLIMAP) program provides information on global temperature through analysis of deep-sea sediment cores. Instrumented records of climate since the mid-16th century are available and archived in World Data Centers. Other major sources of climatic data in the United States are maintained by several agencies. A U.S. Climate Program Plan calls for a major expansion of data management capabilities to accommodate increased research activities in climate.

Possible Economic and Social Effects of Ozone Diminution and its Control

The United States is examining the economic and social impacts of the biological and physical response mechanisms associated with perturbations in the stratosphere. Any change in man's climate and environment is likely to lead to a change in man's activities since many of these activities depend on climate and environment. The social costs accompanying these changes is the difference in value of production before and after the environmental changes plus any costs incurred during this transition.

This research will attempt to address the socioeconomic considerations for evaluating the terrestrial impacts of stratospheric ozone reduction. Principal emphasis will be the impact of environmental change on agriculture and ecology, human behavior, i.e., changes in clothing, fuel consumption, etc., materials weathering and marine resources. In order to evaluate these impacts the research results on climate variables, temperature changes and ultraviolet radiation will be used as inputs into this assessment.

The ultimate measurement of socioeconomic impacts is difficult to conceptualize and then convert to common units of measure. However, the results of this research will give environmental decisionmakers order of magnitude estimates on the costs associated with changes in climate and ultraviolet radiation from stratospheric ozone reductions.

In view of the present concern in the United States over the impacts of releases of CFMs, it is necessary to examine the economic impacts of regulation on the chemical manufacturers and the aerosol industry. In this regard the United States has undertaken an extensive study to examine the impact of possible aerosol regulations. We have been cooperating with the industry trade association, the chemical firms and the aerosol fillers and marketers to assure that we are using the most up-to-date financial, engineering and product substitution data. If and as consideration of regulation of other ozone-destroying substances becomes advisable, similar assessments will be necessary in relation to the affected industry.

CONCLUDING REMARKS

In this paper we have indicated the investigations that are now ongoing or planned in the United States to understand and evaluate the potential threat of stratospheric ozone diminution. Unfortunately, the investigations which have already been provided for do not include all the investigations which are needed for the above purpose. A full picture of all the investigations required is emerging as this meeting proceeds. As the meeting continues, we look forward to further discussion to identify a consensus evaluation concerning the priority of investigational needs which are not yet provided for, and effective mechanisms for concerned nations and international bodies to join in providing for meeting those needs.

Report On The Meeting

Report on the Meeting

DRAFT REPORT OF THE UNITED NATIONS ENVIRONMENT PROGRAMME MEETING OF
EXPERTS DESIGNATED BY GOVERNMENTS, INTERGOVERNMENTAL AND NON-
GOVERNMENTAL ORGANIZATIONS ON THE OZONE LAYER, WASHINGTON, D.C.,
1-9 MARCH 1977

1. Opening of the Meeting

The meeting was opened by the Chairman, Mr David A. Munro, Special Advisor to the
Executive Director of the United Nations Environment Programme, at 10:45 am on
1 March 1977 in the Loy Henderson Conference Room at the Department of State,
Washington, D.C.

On behalf of the Department of State of the United States of America Mr Lindsay
Grant, Deputy Assistant Secretary for Environmental and Population Affairs, Bureau
of Oceans and International Environmental and Scientific Affairs, offered a warm
welcome to Dr Mostafa K. Tolba, the Executive Director of the United Nations Environment Programme.

In his opening address Dr Tolba welcomed the delegates to the meeting, thanked the
Government of the United States for acting as host to the meeting and acknowledged
the continuing support given to UNEP by the United States.

Dr Tolba recalled the decision of the Governing Council which called for the meeting
and went on to review the significance of the natural ozone layer and the possibility
of its modification by man.

UNEP's concern with outer limits, especially as it related to dangers to the ozone
layer, was explained and Dr Tolba noted the high priority given to this global
pollution problem within the outer limits programme. He said that in close
cooperation with the specialized agencies, a thorough state of the art review
covering the natural and perturbed stratosphere and the possible consequences for
mankind, should be undertaken. This would enable identification of necessary action
by governments individually or jointly.

In reply to Dr Tolba's opening address, Dr Robert M. White of the National Oceanic
and Atmospheric Administration, speaking on behalf of the Government of the United
States of America, expressed his pleasure with the UNEP decision to convene the
meeting on the ozone layer in the United States and extended a welcome to the
participants on behalf of the United States Government. He cited the complexity,
pervasiveness, and global scope of environmental issues raised by contamination of
the ozone layer as factors that make meetings such as this so vital. Only through
such meetings can the international scientific community form an assessment of our

understanding of the problem and what must be done to help solve the problem. He indicated that he views UNEP's role as catalyst and coordinator in dealing with this global environmental problem as not only appropriate and significant but perhaps critical.

Dr White stressed that only through science can we understand and predict the future consequences of human actions on the protective ozone layer. The hazards involved are universal, and the causes are widely dispersed throughout the world. Clearly, the problem and its resolution are of concern to all nations. He expressed his hope that this attempt by UNEP to further scientific understanding through international exchange of ideas would help to determine the relative importance of various projects and more clearly define priority actions, would provide guidance on what should be done by specific nations and international agencies, and would indicate where coordination of the work of several participating agencies and nations is most necessary.

2. Participants of the Meeting

  (a) Representatives of Governments

Belgium,	Bolivia,	Canada,
China,	Denmark,	Egypt,
Finland,	France,	Federal Republic of Germany,
Guatemala,	India,	Italy,
Ivory Coast,	Democratic Peoples Republic of Korea,	The Republic of Korea,
Libyan Arab Republic,	Netherlands,	Nicaragua,
Norway,	Philippines,	Sweden,
Switzerland,	Thailand,	Togo,
Uganda,	USSR,	United Kingdom,
United States of America	Uruguay,	and Venezuela

  (b) United Nations Bodies

  Department of Economic and Social Affairs of the United Nations

  United Nations Institute for Training and Research

  European Economic Commission

  (c) Representatives of Specialized Agencies

  The Food and Agricultural Organization

  International Civil Aviation Organization

  The World Health Organization

United Nations Educational Scientific and Cultural Organization

The World Meteorological Organization

(d)   Representatives of Non-governmental Organizations

International Chamber of Commerce

International Council of Scientific Unions/Scientific Committee
on Problems of the Environment ICSU/SCOPE

3. Agenda

The meeting approved the agenda.

4. Documentation

A number of background papers relating to various aspects of the ozone layer were
placed before the meeting by experts designated by Governments, United Nations
bodies, Specialized Agencies and Non-governmental Organizations.

The reports of research activities in various countries and the views expressed
by the specialized agencies and other organizations were considered by the meeting and used to prepare an overview of environmental aspects of stratospheric
ozone depletion.

Agenda Item 3: Review of the Risk to the Ozone Layer and Possible Effects on
                Mankind.

Agenda Item 4: Ongoing and Planned Activities (Contributions from Agencies,
                Governments and Other Bodies)

The existence of the Earth's ozone layer and its vital importance to life on this
planet have been known for many years. It is also known that man evolved and
adapted over millions of years, to a particular level of ozone in the atmosphere.
But only recently has it been realized, that some of man's activities could lead
to stratospheric pollution and to a significant reduction in ozone amounts, in a
few decades.

It is now generally accepted that a wide variety of naturally occurring and man-made
substances can lead to changes in the ozone layer, although most of these effects
are small and difficult to identify.

At the meeting, the World Meteorological Organization reported on its actions and
proposals, some undertaken in cooperation with UNEP, to determine the state of the
natural ozone layer and its modification by man's activities. Among the possible
ozone depleting substances, particular attention has been given to the nitrogen
oxide emission of aircraft and to the emissions of chlorofluoromethanes (CFMs).

There was a large measure of agreement from reports by individual nations and by
the WMO and ICAO with the UNEP review that depletion of the ozone layer by current
aircraft emissions is probably negligible, but that CFM emissions are a matter of
concern. The estimates of long-term (100-200 years) ozone depletions from

continued CFM emissions at the 1973 level are near 8% with an uncertainty range from 2% to 20%.

The depletion role of natural events such as volcanoes and solar activity and other of man's activities such as nuclear explosions and the use of nitrogen fertilizer are not sufficiently well understood and will have to be studied within the proposed research and monitoring programme.

Concern over depletions in the ozone layer arises from the danger of increased penetration of the atmosphere by damaging wavelengths of the sun's ultraviolet radiation (UV-B) which could have deleterious effects on man and the biosphere and possible climatic effects.

The World Health Organization reported that the effects of chronic exposure to increased solar UV would be insidious and could result in accelerated skin aging and in increased incidence of skin cancer, but emphasized that all presently available quantitative estimates of the effects are subject to great uncertainty.

The Food and Agriculture Organization reported that any threat to plant productivity, no matter how subtle, is of concern in view of the existing global demands for food and fiber production, but that it is presently impossible to predict the magnitude of the effect on plant life due to any particular stratospheric ozone reduction.

The presentations of nations and organizations emphasized that considerable progress has been made in our understanding of the ozone depletion mechanism and the consequences for man, but stressed the need for overall coordination and for support of the necessary monitoring and research activities to increase man's knowledge of this vital part of his global environment.

The Department of Economic and Social Affairs of the United Nations reviewed available information on production of CFMs and indicated how, with more data and some elaboration of methodology, it might be possible to develop a useful cost/benefit analysis of the various options for regulating the use of ozone-damaging substances.

The representative of the International Chamber of Commerce reported on the relevant research in universities being supported by the Manufacturing Chemists Association.

The reports of countries reflected the fact that much of the research, measurement and monitoring related to the ozone layer is being undertaken at the national level. Research related to atmospheric physics and chemistry, numerical modelling, and the measurements of trace species was outlined in some detail, and reviews of a variety of efforts in the human and biological areas were given.

The review enabled the meeting to define gaps and make recommendations for a World Plan of Action on the ozone layer.

Agenda Item 5:   Recommendations

The recommendations of the meeting are included in the World Plan of Action under the following sections:

1.   The Natural Ozone Layer and its Modification by Man's Activities.

2.   The Impact of Changes in the Ozone Layer on Man and the Biosphere.

3.   Socio-economic Aspects.

Priorities

The actions recommended under each of the above mentioned sections and sub-sections are arranged in a descending order of priority.

Distribution of Work

The World Plan of Action indicates, by means of a parenthetical addition to each particular action, that organization which will assume the leading role in carrying out that action and those organizations which will offer a supporting role.

Coordination Mechanism

The recommendations of the meeting for the coordination mechanism are included in Section 4 of the World Plan of Action under the title, "Institutional Arrangements".

Others

The meeting did not wish to discuss actions to control emissions into the atmosphere of ozone-damaging substances. However, it considered that it would be desirable to investigate all possible modalities for such control including the socio-economic impacts of their application.

AGENDA FOR THE MEETING

1. Opening of the meeting.

2. Adoption of the agenda

3. Review of the risk to the ozone layer and possible effects on mankind.

4. Ongoing and planned activities (contributions for agencies, Governments and other bodies).

5. Recommendations:

    (a) Priorities;

    (b) Distribution of work;

    (c) Coordination mechanisms;

    (d) Others

6. Report of the meeting.

7. Closure of the meeting.

World Plan of Action

World Plan of Action

THE NATURAL OZONE LAYER AND ITS MODIFICATION BY MAN'S ACTIVITIES

The natural stratosphere and its ozone layer have been the subject of extensive research over the last fifty years. The need for jet aircraft to fly at high altitudes has long spurred the interest of the world's meteorological community. The curiosity of atmospheric physicists and chemists has provided the main thrust to illuminate the dynamics and photochemistry of the ozone layer.

In recent years research, associated with man's accelerating activities in the stratosphere and aerospace, has greatly modified our perception of ozone photochemistry. It is widely accepted that natural ozone photochemistry is controlled to a large extent by the action of nitric oxide derived from nitrous oxide which emanates from the earth's surface as a result of denitrification processes. This has added a new dimension to the ozone problem and new interest in studies of the earth's nitrogen cycle. It is now thought that stratospheric ozone is influenced by a variety of naturally occurring substances such as methane and methyl chloride. A continuing long-term programme is essential to our further understanding of these natural phenomena and as a basis for evaluating new effects.

Quite recently the impact of other substances such as the nitrogen oxides from aircraft exhausts and the man-made chlorofluoromethanes (CFMs) has been receiving intense scrutiny. As these are all trace substances with atmospheric concentrations as low as a fraction of a part per billion, the problems of measurement analysis and modelling have changed by an order of magnitude from those encountered when our understanding was based on the assumption of the simple Chapman (pure oxygen) system. Our knowledge and understanding have been greatly increased by a number of intensive national programmes, and there is a large measure of agreement on the model predictions that current aircraft emissions have minimal effects on the ozone layer but that the effects of continuing emissions of CFMs at the 1973 or higher levels are a matter of concern.

There are many known gaps in our knowledge of factors affecting the ozone layer, and there may be factors that are as yet unrecognized. An intensive and well coordinated monitoring and research programme related to the occurrence of trace substances in the atmosphere, to test the model predictions and narrow their range of uncertainty, is particularly important.

Recommendations

The coordinated research and monitoring programme already initiated by the World Meteorological Organization*, to clarify the basic dynamical photochemical and radiative aspects of the ozone layer and to evaluate the impact of man's activities on the ozone balance should be encouraged and supported.

Specifically this should include action to:

Monitor Ozone (WMO). Design, develop and operate an improved system (including appropriate ground-based, airborne and satellite subsystems) for monitoring and prompt reporting of global ozone, its vertical, spatial and temporal variations, with sufficient accuracy to detect small but statistically significant long-term trends.

---

*See UNEP/WG. 7/5 Atmospheric Ozone, a paper submitted by WMO.

<u>Monitor solar radiation (WMO)</u>. Design, develop, and implement a system (including
appropriate ground-based, airborne and satellite subsystems) for monitoring the
spectral distribution of solar radiation extraterrestrially to better understand
the formation of ozone, to determine variations of solar flux and to clarify the
influence of such variations on the earth's radiation budget and on the accuracy of
ground-based measurements.

<u>Simultaneous species measurements (WMO)</u>. Design, develop, and implement an inten-
sive short-term programme of simultaneous measurements of selected species in the
odd-nitrogen, odd-hydrogen and odd-chlorine families to provide a better understanding of the ozone balance and to test the model predictions. This should be
supported by a longer-term programme on all reactive species including their
temporal and spatial variations and base line inventories.

<u>Chemical reactions (WMO)</u>. Design, develop, and implement a broad programme to
accurately determine chemical and photochemical reaction rates and quantum yields
at appropriate temperatures and pressures; to provide accurate atomic and molecular line strength and location data for use in laboratory and remote sensing
measurements of trace species; and to assess the sensitivity of chemical reaction
schemes used in atmospheric models.

<u>Development of computational modelling (WMO)</u>. Develop a hierarchy of improved
(1-D, 2-D, 3-D) computational models by which the interrelationships of chemical,
radiative, hydrodynamic and thermodynamic processes controlling the troposphere,
and the stratosphere, may be established.

<u>Large-scale atmospheric transport (WMO)</u>. Undertake studies to better describe and
quantify large-scale atmospheric circulation and transport. Particular emphasis
should be given to representation of vertical transport and transfer between the
troposphere and the stratosphere.

<u>Global constituent budgets (WMO)</u>. Develop a much improved understanding of the
budgets of atmospheric constituents which affect the ozone balance, including their
sources and sinks. At present this should give particular emphasis to the chlorine
and nitrogen systems.

THE IMPACT OF CHANGES IN THE OZONE LAYER ON MAN AND THE BIOSPHERE

A prediction of depletion of ozone due to man's activities would have little meaning unless it could be shown, at least qualitatively, that it would have significant effects on man and his environment.

It has been demonstrated that excessive exposure to ultraviolet radiation at 254 nm
(radiation of a short wavelength than the UV-B which reaches the ground) causes
tumors in laboratory animals and has deleterious effects on certain plants.
Extensions of the studies to the UV-B band (from 290-320 nm) have been made by
theoretical and experimental work and by a series of epidemiological studies. There
is some evidence that increased UV-B would be associated with an increase in skin
cancer and possibly in eye damage in susceptible sections of the human population.
It is also likely that large increases in UV-B would damage nucleic acids and
proteins and thereby have deleterious effects on plant and aquatic communities, but
the effect of smaller changes in UV-B is highly uncertain. Although it is clear
that significant progress has been made in a wide range of related research activities in the biological and medical sciences there are still many gaps in our knowledge of the effects of increased UV-B.

Recommendations

A wide variety of investigations of the impact of ozone depletion and increased

ultraviolet radiation (UV-B) on man and the biosphere should be encouraged, supported and coordinated. Specifically it is proposed that action be undertaken on:

UV radiation. (i) *Monitor UV-B radiation* (WMO/WHO, FAO): monitor as far as possible with currently available instrumentation the spectral distribution of the
UV-B radiation and the erythemally weighted integrated radiation intensities at the
earth's surface. This should be done for at least a complete solar cycle at
globally distributed sites (and where possible at stations where ozone is being
measured and/or skin cancer data being collected and/or plant effects being studied.
(ii) *Develop UV-B instrumentation* (WMO/UNEP, WHO, FAO): develop improved
instrumentation and methods for measurements and providing precise levels of UV-B
irradiance (both broad-band and narrow-band) including satellite techniques. (iii)
Promote UV-B research (WMO, WHO, FAO): develop standardized methodologies for
conducting UV-B research. Promote investigations to enable a better understanding
of the spectral distribution of UV radiation, effects of flux, and effects of
factors other than ozone such as atmospheric conditions, ground albedo, etc., in
determining the amount and wavelength of UV reaching the ground.

Human health. (i) *Statistics on skin cancer* (WHO): Obtain improved worldwide
statistics of skin cancer incidence and analyze such data in relation to latitude
(with particular emphasis on locations where UV and total ozone can be measured
simultaneously), type of cancer, age of onset, morbidity rates, location of cancer
on body, sex, skin type, genetic and ethnic background, occupation and lifestyle.
Develop internationally agreed upon protocols for design, collection and analysis
of skin cancer data and develop improved data storage, retrieval and dissemination
mechanisms. (ii) *Research on induction mechanisms* (WHO): conduct experimental
and theoretical studies at the molecular, cellular, and tissue levels of the
mechanisms of induction of skin cancer, skin aging and eye damage by UV-B, as
functions of dose, dose rate, and wavelength, with attention to possible synergistic
factors including other wavelengths and possible effects of UV-B on DNA. (iii)
Other health aspects (WHO): study potential impact of increased UV-B on aspects
of human health other than skin cancer; develop a model for predicting human photobiological response for various skin types, genetic and ethnic background and
environmental factors; and conduct research and modelling on the mechanisms of
Vitamin D production and effects.

Other biological effects. (i) *Responses to UV-B* (FAO): conduct studies on the
physiological, biochemical, and structural responses to increased UV-B radiation of
selected wild and agricultural plants, animal species, and microorganisms. These
should emphasize dose-response, reciprocity, synergisms, antogonisms, action spectra
and interactions of stress factors on these organisms. (ii) *Plant communities*
(FAO): study the effect of UV-B radiation on terrestrial plant communities (both
agricultural and natural) *inter alia* by means of modelling studies of the impacts
of UV-B radiation. (iii) *Aquatic systems* (FAO): study the effect of UV-B
radiation on aquatic plants and animals emphasizing those related to primary
productivity in the oceans; develop improved measurements of UV-B penetration into
aquatic environment and study its effects on plankton.

Effects on climate. (i) *development of computational modelling* (WMO): further
develop modelling capability to consider the effects of ozone changes and also
those of CFM and similar compounds on the earth's radiation balance and its climate.
(ii) *Regional climate* (WMO/FAO): promote research on climate and its variability
to provide better evaluations of the effects on critical climatic regions.

SOCIO-ECONOMIC ASPECTS

To provide a useful contribution to policy formulation, the assessment of a potential environmental hazard must include an evaluation of the costs and benefits to

society that would result from a reduction of the hazard. Neither the methodology nor the available data are adequate to properly assess the socio-economic effects of stratospheric pollution or of measures taken to control it. Much work has been done on the costs of alternative courses of action particularly with respect to possible limitations to total CFM emissions. However, further elaboration is needed before a complete evaluation can be made with confidence.

Recommendations

Studies of the socio-economic impact of predicted ozone layer depletions and of alternative courses of actions to limit or control identified ozone-depleting emissions to the atmosphere should be supported at the national and international level. Specifically it is proposed that action be undertaken to:

Production and emmission data (UNEP, ICC, OECD, ICAO). Obtain more detailed data on global production, emission and use of substances that have the potential to affect stratospheric ozone. These data are required to model stratospheric changes as well as to evaluate socio-economic alternatives.

Methodology (UNESA, OECD). Develop improved methodologies for assessing the socio-economic factors related to CFM use, aircraft emissions, nitrogen fertilizers and other potential modifiers of the stratosphere.

INSTITUTIONAL ARRANGEMENTS

1. The Action Plan will be implemented by UN bodies, specialized Agencies, international, national, intergovernmental, and non-governmental organizations and scientific institutions.

2. The UNEP should exercise a broad coordinating and catalytic role aimed at the integration and coordination of research efforts by arranging for:

- collation of information on ongoing and planned research activities;

- presentation and review of the results of research;

- identification of further research needs.

3. In order for UNEP to fulfill that responsibility it should establish a Coordinating Committee on Ozone Matters composed of representatives of the agencies and non-governmental organizations participating in implementing the Action Plan as well as representatives of countries which have major scientific programmes contributing to the Action Plan. The Committee should be provided with adequate secretariat services.

The Committee should make recommendations to the Executive Director relevant to the continuing development and coordination of the Action Plan.

4. While much of the work included in the Action Plan is being and will be undertaken at the national level, and is the financial responsibility of countries, there is a continuing need for coordination of the planning and execution of monitoring and research related to particular segments of the Action Plan. This need can most effectively be met by the specialized Agencies as indicated in the recommendations.

5. Each Agency should arrange for the provision of scientific advice relevant to its needs and those of the Coordinating Committee on Ozone Matters. In addition, the Executive Director of UNEP may from time to time convene a multidisciplinary panel of experts to provide broadly-based scientific advice on the Action Plan.

6. UNEP should consider the need for and feasibility of establishing special

coordinating mechanisms or procedures for certain areas of interdisciplinary work, such as photobiology, which presently lack such coordinating facilities.